**PLACE** IN **RETURN BO** to re ove this ci kout from your record.
TO AVOID FINES etum
DATE DUE

# Process Models
# and
# Theoretical Geomorphology

# British Geomorphological Research Group
# Symposia Series

# Process Models and Theoretical Geomorphology

Edited by
**M. J. Kirkby**
*School of Geography,*
*University of Leeds, UK*

JOHN WILEY & SONS
Chichester · New York · Brisbane · Toronto · Singapore

Copyright © 1994 by John Wiley & Sons Ltd,
Baffins Lane, Chichester,
West Sussex PO19 1UD, England
Telephone (+44) (243) 779777

The copyright of Chapter 17 has been retained by the
Commonwealth of Australia

All rights reserved.

No part of this book may be reproduced by any means,
or transmitted, or translated into a machine language
without the written permission of the publisher.

*Other Wiley Editorial Offices*

John Wiley & Sons, Inc., 605 Third Avenue,
New York, NY 10158-0012, USA

Jacaranda Wiley Ltd, 33 Park Road, Milton,
Queensland 4064, Australia

John Wiley & Sons (Canada) Ltd, 22 Worcester Road,
Rexdale, Ontario M9W 1L1, Canada

John Wiley & Sons (SEA) Pte Ltd, 37 Jalan Pemimpin #05-04,
Block B, Union Industrial Building, Singapore 2057

*Library of Congress Cataloging-in-Publication Data*

Process models and theoretical geomorphology / edited by M. J. Kirkby.
    p.  cm.—(British Geomorphological Research Group symposia
series)
   Includes bibliographical references and index.
   ISBN 0-471-94104-2
   1. Geomorphology.  I. Kirkby, M. J.  II. Series.
GB401.5.P748  1994
551.4′1—dc20                                             93-4581
                                                           CIP

*British Library Cataloguing in Publication Data*

A catalogue record for this book is available from the British Library

ISBN 0-471-94104-2

Typeset in 10/12pt Times by Dobbie Typesetting Limited, Tavistock, Devon
Printed and bound in Great Britain by Bookcraft (Bath) Ltd

# Contents

# List of Contributors

**F. Ahnert**  Lehrstuhl für Physische Geographie, Geographisches Institut der RWTH, Templergraben 55, D-5100 Aachen, Germany

**P. J. Ashworth**  School of Geography, University of Leeds, Leeds LS2 9JT, UK

**J. L. Best**  Department of Earth Sciences, University of Leeds, Leeds LS2 9JT, UK

**A. Billard**  Laboratoire de Géographie Physique, CNRS–URA 141, 1 Place Aristide Briand, Meudon 92190, France

**R. L. Bras**  Department of Civil and Environmental Engineering, Massachusetts Institute of Technology, Boston, MA 02139, USA

**R. W. Brown**  Department of Geological Sciences, University College, Gower Street, London WC1E 6BT, UK

**G. Christaller**  Labor für Sensortechnik, Technische Fachhochschule Berlin, Luxemburgerstr. 10, 1000 Berlin 65, Germany

**C. de Jong**  Institut für Geographische Wissenschaften, Freie Universitat Berlin, Grunewaldstr. 35, 1000 Berlin 41, Germany

**E. Derbyshire**  Department of Geography, Royal Holloway, University of London, Egham, Surrey TW20 0EX, UK

**W. E. Dietrich**  Department of Geology and Geophysics, University of California, Berkeley, CA 94720, USA

**P. E. Ergenzinger**  Institut für Geographische Wissenschaften, Freie Universitat Berlin, Grunewaldstr. 35, 1000 Berlin 41, Germany

**G. W. Geehan**  BP Research, Sunbury-on-Thames, Middlesex TW16 7LN, UK

**A. R. Gilchrist**  Department of Oceanography, Dalhousie University, Halifax, Nova Scotia, Canada B3H 4J1

**A. J. W. Gleadow**  Victorian Institute of Earth and Planetary Sciences, Fission Track Research Group, Department of Geology, La Trobe University, Bundoora 3083, Australia

**A. M. Harvey**  Department of Geography, University of Liverpool, Liverpool L69 3BX, UK

**D. P. Horn**  Cheltenham and Gloucester College of Higher Education, Shaftesbury Hall, St Georges Place, Cheltenham, Gloucestershire GL50 3PP, UK

**K. Kashiwaya**  The Graduate School of Science and Technology, Kobe University, Nada, Kobe 657, Japan

**M. J. Kirkby**   School of Geography, University of Leeds, Leeds LS2 9JT, UK

**J. O. Leddy**   School of Geography, University of Leeds, Leeds LS2 9JT, UK

**J. C. Lin**   Department of Geography, National Taiwan University, Taipei, Taiwan 106

**D. S. Loewenherz-Lawrence**   Department of Geological Sciences, University of Illinois at Chicago, Box 4348 Chicago, IL 60680-4348, USA

**R. Masterman**   Department of Geography, University of Nottingham, Nottingham NG7 2RD, UK

**D. R. Montgomery**   Department of Geological Sciences and Quaternary Research Center, University of Washington, Seattle, WA 98195, USA

**T. Muxart**   Laboratoire de Géographie Physique, CNRS–URA 141, 1 Place Aristide Briand, Meudon 92190, France

**S. Rice**   Department of Geography, University of British Columbia, Vancouver, Canada V6T 1Z2

**S. J. Riley**   Alligator Rivers Region Research Institute, Office of the Supervising Scientist, Jabiru, Northern Territory, Australia

**I. Rodriguez-Iturbe**   Instituto International de Estudios Avanzados, Caracas, Venezuela

**M. A. Summerfield**   Macrogeomorphology Research Group, Department of Geography, University of Edinburgh, Drummond Street, Edinburgh EH8 9XP, UK

**C. R. Thorne**   Department of Geography, University of Nottingham, Nottingham NG7 2RD, UK

**J. Wang**   Geological Hazards Research Institute, Gansu Academy of Sciences, Lanzhou, China

**R. F. Warner**   Department of Geography, University of Sydney, NSW 2006, Australia

**G. Willgoose**   Department of Civil Engineering and Surveying, University of Newcastle, NSW 2308, Australia

# Series Preface

The British Geomorphological Research Group (BGRG) is a national multidisciplinary society whose object is 'the advancement of research and education in geomorphology'. Today, the BGRG enjoys an international reputation and has a strong membership from both Britain and overseas. The Group has been actively involved in stimulating the development of geomorphology and geomorphological societies in several countries. The BGRG was constituted in 1961 but its beginnings lie in a meeting held in Sheffield under the chairmanship of Professor D. L. Linton in 1958. Throughout its development the Group has sustained important links with both the Institute of British Geographers and the Geological Society of London.

Over the past three decades the BGRG has been highly successful and productive. This is reflected not least by BGRG publications. Following its launch in 1976 the Group's journal, *Earth Surface Processes* (since 1981 *Earth Surface Processes and Landforms*) has become acclaimed internationally as a leader in its field, and to a large extent the journal has been responsible for advancing the reputation of the BGRG. In addition to an impressive list of other publications on technical and educational issues, including more than 30 *Technical Bulletins*, the new *Technical and Software Bulletins* and the influential *Geomorphological Techniques*, edited by A. Goudie, BGRG symposia have led to the production of a number of important works. These have included *Nearshore Sediment Dynamics and Sedimentation* edited by J. R. Hails and A. P. Carr; *Geomorphology and Climate*, edited by E. Derbyshire; *Geomorphology, Present Problems and Future Prospects*, edited by C. Embleton, D. Brunsden and D. K. C. Jones; *Mega-geomorphology*, edited by R. Gardner and H. Scoging; *River Channel Changes*, edited by K. J. Gregory; and *Timescales in Geomorphology*, edited by R. Cullingford, D. Davidson and J. Lewin. This sequence of books culminated in 1987 with a publication, in two volumes, of the *Proceedings of the First International Geomorphology Conference*, edited by Vince Gardiner. This international meeting, arguably the most important in the history of geomorphology, provided the foundation for the development of geomorphology into the next century.

This current BGRG Symposia Series has been founded and is now being fostered to help maintain the research momentum generated during the past three decades, as well as to further the widening of knowledge in component fields of geomorphological endeavour. The series consists of authoritative volumes based on the themes of BGRG meetings, incorporating, where appropriate, invited contributions to complement chapters selected from presentations at these meetings under the guidance and editorship of one or more suitable specialists. Although maintaining a strong emphasis on pure geomorphological research, BGRG meetings are diversifying, in a very positive way, to consider links between geomorphology *per se* and other disciplines such as ecology, agriculture, engineering and planning.

The first volume in the series was published in 1988. *Geomorphology in Environmental Planning*, edited by Janet Hooke, reflected the trend towards applied studies. The second volume, edited by Keith Beven and Paul Carling, *Floods – Hydrological, Sedimentological and Geomorphological Implications*, focused on a traditional research theme. *Soil Erosion on Agricultural Land* reflected the international importance of the topic for researchers during the 1980s. This volume, edited by John Boardman, John Dearing and Ian Foster, formed the third in the series. The role of vegetation in geomorphology is a traditional research theme, recently revitalized with the move towards interdisciplinary studies. The fourth in the series, *Vegetation and Erosion – Processes and Environments*, edited by John Thornes, reflected this development in geomorphological endeavour, and raised several research issues for the next decade. The fifth volume, *Lowland Floodplain Rivers – Geomorphological Perspectives*, edited by Paul Carling and Geoff Petts, reflects recent research into river channel adjustments, especially those consequent to engineering works and land-use change. The sixth volume in the series *Geomorphology and Sedimentology of Lakes and Reservoirs*, edited by John McManus and Robert Duck, continued the interdisciplinary theme and the seventh, *Landscape Sensitivity*, edited by David Thomas and Robert Allison, addressed the relationships between geomorphology and environmental change.

The present book, *Process Models and Theoretical Geomorphology*, edited by Mike Kirkby, is the eighth volume in the series and represents a timely review of the development of modern modelling techniques in geomorphology. Drawn from ten countries the volume presents papers from a joint COMTAG/BGRG symposium held in the UK in 1991 and covers tectonics, channel processes, valley head development and local applications. The book provides a stimulating mixture of theoretical and field work and the chapters on numerical methods, in particular, represent important new developments in the subject which should appeal to the widest research community.

Jack Hardisty
BGRG publications

# Preface

In September 1991, a symposium was held in Leeds, jointly organised by COMTAG (IGU Commission on Measurement, Theory and Applications in Geomorphology) and BGRG (British Geomorphological Research Group). The paper sessions were followed by a successful field meeting in Scotland organised through St Andrew's University. The conference was partially supported by grants from the sponsoring organisations, the Royal Society and John Wiley, as publishers of the BGRG journal, *Earth Surface Processes and Landforms*. The main theme of the meeting was Theoretical Geomorphology and its applications, and this book presents 18 of the papers presented at the symposium, selected after undergoing a full, normal refereeing process. It is published in the BGRG Symposium Series. Authors come from ten countries, from China and Australia to Canada, and the papers cover a wide range of themes but with some important convergences of interest on critical topics.

The papers fall into four natural groups, roughly corresponding to sessions within the symposium. There is a group of papers on Tectonic and General Approaches, on Channel Processes, on Valley Heads, and on Applications. In the first group, Ahnert's paper reviews the role of randomness in process models, and concludes that it only has a useful place if constrained within a well defined deterministic framework. The two papers with Summerfield among the authors examine the incorporation of tectonics into geomorphological models for passive continental margins, and the use of apatite fission track analysis as a means of evaluating these models. Kashiwaya provides a novel theoretical analysis of the relationships between tectonic and erosional forces, while Lin adopts a more conventional analysis of neotectonic landforms in the highly active area of Taiwan. Despite some books and papers which have attempted a proper synthesis between geomorphological processes and plate tectonics, it is clear that much more remains to be done before this branch of geomorphology reaches its full potential. At least part of the difficulty in development lies in extending our detailed process understanding to the larger areas over which tectonic events have their greatest long-term significance.

The group of papers on channel processes reflect strong current interest in gravel-bed rivers, related to their widespread occurrence in both mountain areas and in braided systems with potential for preservation as *inter alia*, oil reservoirs. Ashworth *et al.* report flume studies examining loci of fine-grained sedimentation within braided deposits. Ergenzinger *et al.* examine river-bed adjustment during floods, using an electronic detection system to monitor flux of magnetic pebbles. Rice analyses downstream changes in grain size through a drainage basin, paying particular attention to mixing of hillslope and tributary sediment inflows. Warner describes instability in New South Wales coastal rivers to alternating flood and drought-dominated flow regimes, relying on photogrammatic and sedimentary evidence. Masterman and Thorne provide a basis for integrating flow resistance across the gravel bed and vegetated banks found in many rivers. There is a common concern with aspects of grain sorting, both across the channel and floodplain/braidplain, and its cumulative effects on changes in grain size downstream.

The group of papers on valley heads brings together views from several, sometimes conflicting, research approaches. The work of Montgomery and Dietrich is firmly rooted in a body of empirical work which is now coming to fruition with a series of theoretical insights into hollow initiation by landsliding and wash erosion. Harvey's work illustrates historical changes which occur around streamhead areas, emphasising the role of major events in generating a particular spatial realisation. Willgoose *et al.* exemplify the explicit threshold-based M.I.T. group approach to channel initiation, applied here in the context of overland and subsurface flow processes. Kirkby attempts to reconcile threshold approaches with the stability arguments associated with Smith and Bretherton, showing that, for wash processes, each has a domain of dominance. Loewenherz-Lawrence looks at the implications of the Smith and Bretherton stability arguments for the growth and stability of perturbations in the landscape, illustrating an important topic which has implications for the detailed pattern of streamhead incision and refilling. Fractal-based methods, which can also contribute to our understanding of streamhead evolution, are not represented, but there now appears to be some convergence of the other process-based approaches reported here.

The final group of three papers illustrates the range of problems to which theoretical methods in geomorphology can now be applied. Muxart *et al.* describe work on understanding the erosion processes and their variations across the highly erodible Chinese loess area. Riley's paper illustrates a need rather than a solution at this stage, where rehabilitation of mine spoil must not be allowed to threaten the very slow natural evolution of Australia's Northern Territory landscape. Finally, Horn shows how a one-dimensional coastal evolution model can be applied to an understanding of shoreline sediment types.

The papers presented in any symposium provide an instantaneous cross-section of activity in the field. It is encouraging that much of the theoretical work presented is not too far from applicability, although the gaps are revealing. Perhaps the most striking need is for a greater convergence between earth scientists operating from different disciplines and at different scales. Geomorphologists need to work more closely with geologists and geophysicists concerned with the generation and destruction of relief, particularly by plate tectonic processes. Some convergence is also needed to bridge the scales of greatest interest to geomorphologists and geologists, with process geomorphologists needing to re-focus process work to be applicable to larger areas; and geologists needing to incorporate geomorphological processes at continental and regional scales.

M. J. Kirkby
January 1993

# Part 1

## TECTONIC AND GENERAL APPROACHES

# 1 Randomness in Geomorphological Process Response Models

**FRANK AHNERT**
*Geographisches Institut der RWTH, Aachen, Germany*

## ABSTRACT

The effects of random variations of selected system components are examined in several types of geomorphological process response models of landform development, among them slope models, and models of gully or valley systems, of karst landforms and of sorted stone nets. The model landforms and surface patterns generated in this way possess an impressive general similarity to natural landforms and patterns. However, the same degree of similarity may also be reached with models that are purely deterministic. Furthermore, the deterministic models have a greater explanatory value than models with randomized components. It seems, therefore, that assumptions of randomness do not offer any significant advantage in the theoretical analysis and synthesis of landform development.

## INTRODUCTION

The individuality of natural landforms is due to the influence of locally varied components that are not included in general process response models. Often these local factors and conditions are so numerous and varied in their functional interaction, that their combined effect appears to be random within certain limits or constraints. This suggests that they can, therefore, be simulated in otherwise deterministic models by the introduction of random variations of selected model parameters.

The use of randomness in geomorphology is not new. It dates back at least to the random-walk models of drainage systems and longitudinal stream profiles by Leopold & Langbein (1962). Later, other authors (e.g. Shreve, 1966, 1975; Smart, 1968; Werrity, 1972; Abrahams, 1987) investigated the possibility of randomness in the spatial organization of channel networks. A volume edited by Merriam (1976) presented several applications in geology, among them a random-walk model for the simulation of alluvial fan deposition (Price, 1976). Recently Willgoose *et al.* (1991) have simulated valley networks with the help of minor random variations of the initial land surface, and Wright & Webster (1991) have modelled soil erosion in a complex manner with spatially random hydraulic conductivity and temporally random rainfall intensity. Important early critical discussions of the general implications of randomness have been presented by Mann (1970) and Howard (1971, 1972).

*Process Models and Theoretical Geomorphology*. Edited by M. J. Kirkby
©1994 John Wiley & Sons Ltd

In this paper, a series of theoretical model experiments is presented in which selected components of process response models are varied randomly, with the purpose to examine

(a) what specific effect such randomization has on the resulting model development; and
(b) what explanatory or other advantage can be gained from it.

Five types of models serve as examples:

1. Slope profile models with fixed baselevels;
2. slope profile models with fluvial downcutting;
3. three-dimensional models of gully or valley development;
4. three-dimensional models of karst landform development; and
5. models of the development of sorted stone nets.

The first four types are BASIC-modifications of the landform development program SLOP3D (cf. Ahnert, 1976a, 1987, 1988, and for relevant process equations especially Ahnert 1977); the fifth type is a modification of a random-walk model (Ahnert 1981).

## SLOPE PROFILES WITH FIXED BASELEVEL

Slope profile models lend themselves to evaluations of the effects of random variations particularly well because their design is simple and the changes brought about by the random variations are easily identified. A standard SLOP3D profile model of slope development with fixed baselevel is used as a test case. The denudation process is suspended-load wash, which means that the removal of material takes place without redeposition anywhere on the slope. The rate of this process is a function of the local slope gradient and the local runoff discharge. Unless stated otherwise the latter increases progressively downslope. Figure 1.1a shows the successive slope profiles of the fully non-random deterministic model: the initial rectilinear profile becomes smoothly concave, and the relief, the mean slope and the mean denudation rate decrease progressively in the course of development.

In Figure 1.1b, the initial profile has been modified by random variations with the constraint that the slope direction must not be reversed. All other components of the model system are identical with those of the model in Figure 1.1a. It can be seen that the initial profile influences the slope development only during the early phases. Later the profile becomes as smoothly concave as that in Figure 1.1a.

This circumstance indirectly also illustrates the point that, for the interpretation of landform development, prior forms or 'initial' forms need only be considered if there are still remnants (or at least traces of remnants) present.

Spatially and temporally random variations of rainfall intensity cause random variations of runoff. However, their effect is slight, as the runoff still increases downslope and the variations of rainfall on the profile merely vary the amount of the local downslope runoff increment. In Figure 1.1c this increment varies between zero and five times the standard value.

Figure 1.1d illustrates the influence of spatially and temporally random variations of local erodibility of the substrate, also within a range between zero and five times the standard value. The resulting changes in profile shape are quite noticeable. Similar fluctuations of the profile shape occur when independent variations of local runoff and

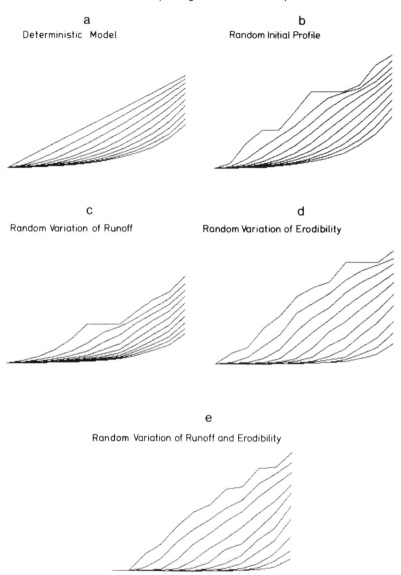

**Figure 1.1** Models of slope profile development by wash denudation with fixed baselevel. (a) Fully deterministic model; (b) with random variation of the initial profile; (c) with random variation of the spatial and temporal distribution of rainfall and, consequently, of runoff; (d) with random variation of the spatial and temporal distribution of erodibility; and (e) with random variation of the initial profile and of the spatial and temporal distribution of both rainfall and erodibility

of local erodibility are combined and interfere with each other (Figure 1.1e). Actual field examples might be found where slope profiles have developed over a short time as, for example, on slopes subject to severe soil erosion. In such cases, temporary variations of runoff and of erodibility would have a greater influence upon the profile shape than in long-term landform development. However, this effect does not require

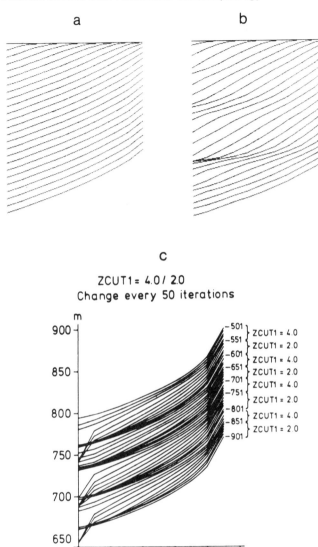

**Figure 1.2** Slope profile development during fluvial downcutting at the slope foot. (a) Constant-rate downcutting; (b) randomly varied rate of downcutting; and (c) periodical variations in the gross rate of downcutting (after Ahnert, 1988, 391. Reproduced by permission). The net rate of downcutting is the difference between the gross rate and the rate of waste supply from the slope. Aggradation occurs if the latter is greater than the former

randomness. Such profile variations can also occur if the variations of runoff and erodibility are, for example, independently periodic ones or in any other way numerically prescribed, as long as their fluctuations do not cancel out each other.

## SLOPE PROFILE DEVELOPMENT DURING FLUVIAL DOWNCUTTING

Another parameter that may be randomly varied is the rate of fluvial incision at the slope foot. Figure 1.2a shows a purely deterministic profile model of slope development with suspended-load wash and a constant rate of fluvial incision. In the early phases of form differentiation the slope is convex, to be replaced progressively from the slope foot upwards by the concave form that is typical of slopes on which wash denudation dominates. Eventually a steady state of constant relief and constant form is reached, which indicates the presence of dynamic equilibrium between weathering, slope denudation and stream incision, so that all points are lowered at the same rate (cf. Ahnert, 1954; Hack, 1960).

In Figure 1.2b, the rate of fluvial incision varies randomly. This results in great variations of form on the lower part of the profile. The upper part, however, is hardly affected at all. The slope crest gets lowered almost uniformly at a rate that is close to the mean rate of incision at the slope foot. The longer the slope profile is, and the more frequent the changes in the rate of downcutting are, the less will these changes affect the denudation rate at the top of the slope (cf. Ahnert, 1988, 389–393). The length of the slope acts in this way as a filter that absorbs these variations during their upslope transmission along the profile in the course of 'regressive', i.e. crestward advancing, denudation (Ahnert, 1954).

Again, however, these phenomena are not limited to random fluctuations. They can also occur if the variations of gross downcutting at the base are periodic (Figure 1.2c) or in some other predetermined manner variable between specific limits. It should be noted that, in contrast to the suspended-load wash in Figures 1.2a and 1.2b, the denudation process in Figure 1.2c is bedload wash, with a point-to-point downslope transfer of transported material so that the net elevation change at the slope foot is equal to the difference between the gross rate of downcutting and the rate at which waste material arrives from the slope. This accounts for the phases of temporary aggradation.

## RANDOM COMPONENTS IN THREE-DIMENSIONAL MODELS OF GULLY OR VALLEY DEVELOPMENT

In Figure 1.3a a plateau-like surface with a planar slope at its margin has been modified by minor spatial random variations of the type (expressed in BASIC):

$$Z(J,K) = Z(J,K) + K*(RND(1) - 0.5) \tag{1}$$

where $Z(J,K)$ is the elevation of surface point $(J,K)$ and K is a small constant. As a consequence, the overland flow will converge in some places and diverge in others. The importance of such minor surface irregularities upon rill initiation has been recognized also in earlier empirical field studies, for example, by Bowyer-Bower and Bryan (1986).

If a threshold value of overland flow discharge is introduced at which areal interrill erosion changes to linear rill erosion, such a model can serve as an example for the development of badlands (Figure 1.3b), here with the development of a pediment-like surface at the slope foot. Figure 1.4 shows a comparable field example from central Tanzania. A similar model approach has been used recently by Willgoose *et al.* (1991).

The badlands model becomes a model of valley development if an initial surface of low relief is being uplifted and progressively dissected. The uplift is conveniently modelled as a lowering of baselevels at constant rates. Figure 1.5 is an example with two baselevels of which one, located at the left forward corner of the block diagram, is lowered three

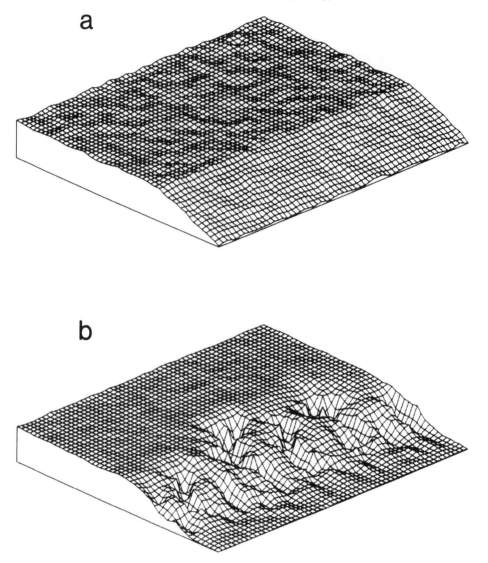

**Figure 1.3**   Three-dimensional model of gully development by suspended-load wash. Sheet wash changes to linear erosion when the runoff discharge exceeds a predetermined threshold. (a) Initial surface; and (b) after 20 time units

times as fast as the other one, located at the left rear corner. As a result, the area which is tributary to the first baselevel is dissected more deeply, with vigorous headward erosion, and becomes progressively enlarged at the expense of the area tributary to the second baselevel. Eventually even the upper reach of the main stream of the second drainage basin is captured.

The evolution of the drainage pattern that accompanies this form of development is shown in Figure 1.6. The gradients of the two trunk rivers at the bottom and at the top of these diagrams are maintained artificially, dependent upon the elevation of their

**Figure 1.4** Badlands at the margin of horizontal deposits, with pedimentation at the foot. Isimila, central Tanzania. Photograph: Ahnert, 1979

respective baselevel. They represent the erosional boundary conditions of the model field. Initially, the drainage is directed merely into the local hollows but later becomes progressively more integrated. With the uplift, steeper gradients and continuous channels develop first at the margins and extend from there progressively by headward erosion and stream capture. The shifting of the main divide is clearly visible, as is the elbow of capture at time $T = 40$.

The random variations of the initial surface in this model lead to the development of dendritic channel patterns which gradually become integrated to form two large drainage basins of the fourth and third order, respectively. Their combined bifurcation ratios vary between 3.0 and 3.5.

The question arises whether such dendritic patterns can only be generated by means of a random model component such as the randomized initial surface. To test this, the model program of Figures 1.5 and 1.6 was altered only in one respect, namely, instead of random variations the initial surface was given a prescribed pattern of shallow depressions spaced at regular intervals.

Figure 1.7 shows the result: a dendritic pattern develops and becomes very similar in general appearance to that in Figure 1.6. In other words, randomness is not required to create such a pattern. The principal difference between the effect of initial randomness and initial non-randomness of the surface is that in the non-random case the resulting drainage pattern will be exactly identical in different model runs made with the same input data, while in the random case the drainage pattern will differ every time the model is run.

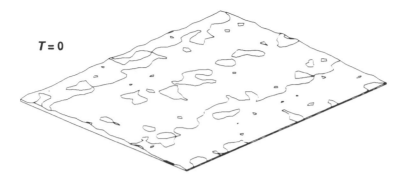

$T = 0$

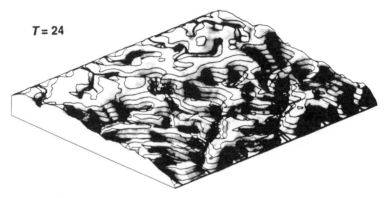

$T = 24$

**Figure 1.5** Model showing the progression development of valley systems during uplift of an initial denudation surface of low relief

## A MODEL OF KARST LANDFORM DEVELOPMENT

The idea of the following model owes much to the discussions and field demonstrations of polygonal karst landforms in New Zealand by Paul W. Williams during the COMTAG Field Symposium in 1988 and to his related papers (cf. Williams, 1985, 1987). This circumstance is gratefully acknowledged.

Dolines on pure, more or less horizontally bedded limestone frequently have no recognizable pattern of distribution which might indicate some structural controls by the direction and spacing of joints. The model (called POLYCON) therefore starts with the assumption that the rock is entirely homogeneous and that there are only local variations of the quantity of available water that acts as a solvent. In the model these variations are produced in the same way as in the cases of Figures 1.3, 1.5 and 1.6 (cf. equation (1)), namely, by random variations of elevation of the initial model surface (Figure 1.8, $T = 0$) which result in convergence of flow into the initial hollows. A similar effect could be obtained by an assumption of spatially random variations in solubility which in turn might be a function of variable density and size of joints.

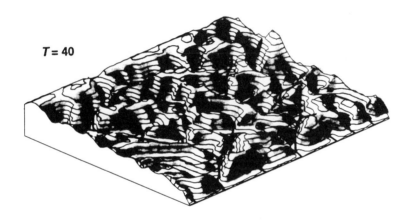

**Figure 1.5**(continued)

Denudation occurs as a removal of rock substance in solution and its rate is a function of the local discharge and the local gradient. According to Williams (1985) the solution removal occurs by subcutaneous water movement in the soil, parallel to the land surface, and continues vertically downward into the rock from the centres of closed depressions.

After four time units of development, several small shallow dolines have formed (Figure 1.8, $T = 4$). In the areas between them, the random variations of the initial surface remain preserved for some time. Doline fields with such a pattern are fairly common.

After $T = 8$ time units the dolines are deeper, and after $T = 16$ time units all vestiges of the initial surface have disappeared and a polygonal or cockpit karst landscape has developed. At an arbitrary elevation a denudational baselevel of karst development is assumed to exist. This level could be the epiphreatic zone (Williams, 1987, 458) or a stratigraphic boundary plane against a less soluble rock below. In either case, the dolines cannot be deepened below this level. Instead, their floors then become widened by basal sapping. Dolines merge to form

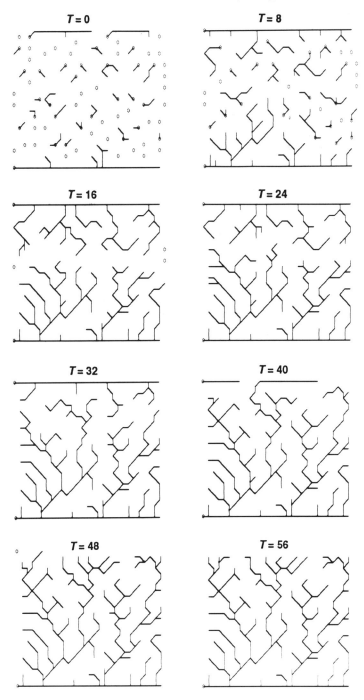

**Figure 1.6**  Evolution of drainage pattern of the model shown in Figure 1.5

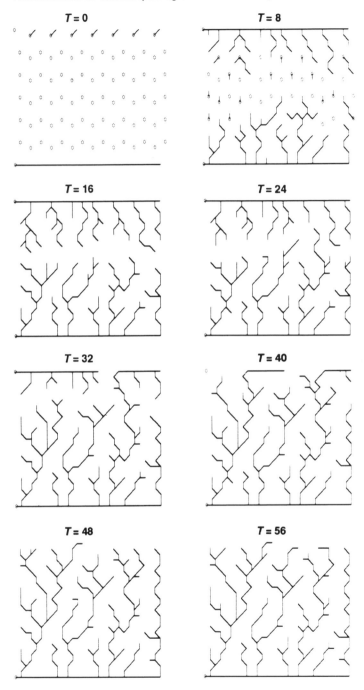

**Figure 1.7** Evolution of drainage pattern of the same model as in Figures 1.5 and 1.6, but without any random variations of components. The initial surface has been varied by deterministic prescription

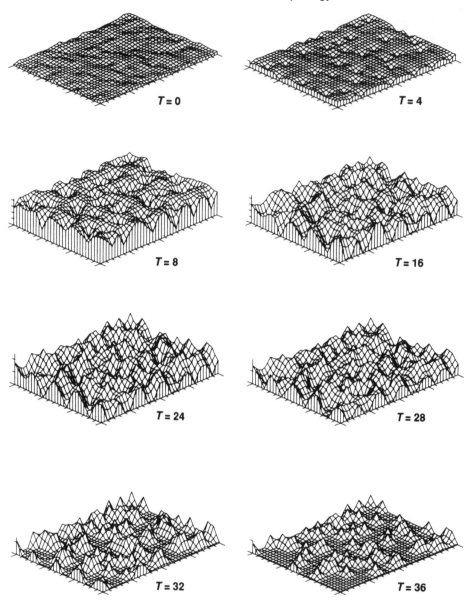

**Figure 1.8**  Development of karst landforms. $T$ = elapsed time units

uvalas, and their floors widen, so that larger karst basins result. These basins are clearly visible in Figure 1.8 after $T = 28$ time units. Their further extension leads to the development of a basal plain with residual karst cones (mogotes, cf. Jennings, 1985, 179) or karst towers, as shown at times $T = 32$ and $T = 36$.

An interesting morphogenetic aspect of this developmental sequence is the fact that it progresses from scattered dolines to cone and tower karst without a particular prescription of climatic conditions or of climatic change. Perhaps cone karst and tower karst are 'typical' in tropical and subtropical regions only because there the development of the present karst landforms has continued much longer than elsewhere, without interruption by process regimes that would tend to destroy these forms. In the middle and higher latitudes, by contrast, times favourable to karst development alternated with times of such interference by glacial or periglacial conditions, including prevalence of mechanical weathering, frost creep and gelifluction, during the Pleistocene.

## MODELS OF STONE NET DEVELOPMENT

Sorted stone nets, polygons or rings (cf. Washburn, 1979, 146–147) are most common in periglacial regions. Similar patterns, but usually with smaller mesh sizes, have also been observed in many non-periglacial areas (cf. Schweinfurth, 1964; Rohdenburg & Walther, 1968; Rolshoven, 1976; Wilson & Clark, 1991).

Most of these occurrences have also been interpreted as the work of freeze–thaw cycles. However, there are also other causes. This writer has observed such patterns on volcanic cinder surfaces in the Auvergne (Central Massif of France), consisting of lapilli and finer-grained volcanic sand. There the lapilli had been moved into ring-like patterns by the footsteps of passing sheep. Another occurrence was observed on a parking lot covered by sand and crushed stone fragments in Aachen, Germany, where the movement of stones is caused by the lateral pressure vector of the tyres of turning cars. Also frequent are sorted stone nets caused by the footsteps of people, for example, on the footpath through the eastern gateway of Portchester Castle near Portsmouth (England), and on a footpath for visitors in the Waiotapu Hot Springs area, North Island of New Zealand (Figure 1.9). Especially in this last case, effects of frost action must be virtually ruled out. The variety of occurrences indicates that the actual cause of stone movement is quite irrelevant for the explanation of the pattern. The only requirements seem to be:

1. An approximately horizontal regolith surface free of vegetation;
2. loose surface material ranging in grain size from small stones to sand or silt, with the stones scattered fairly densely, but not as a complete cover;
3. a cause of stone movement in which impulses from all compass directions have the same chance of occurring;
4. impediment of the mobility of a stone when it lies wedged in between two or more other stones (friction effect).

These requirements have been translated into a model with the following ingredients (cf. Ahnert, 1981):

1. A surface, defined by a square grid of cells.
2. On this surface, a prescribed number of stones is distributed, either randomly or uniformly.
3. No cell can contain more than one stone.
4. Stones are immobile when they lie between two immediately adjacent stones either in a straight line or at an obtuse angle. All other stones are mobile.
5. Mobile stones are moved in individual steps of a random walk, that is, in a randomly chosen direction, to one of the eight adjacent cells if that cell is empty. If that cell is already occupied, the stone remains in place.

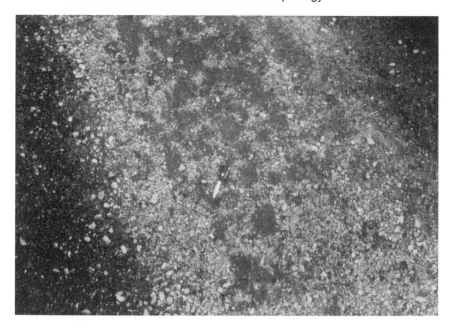

**Figure 1.9**  Sorted stone net on a footpath in the Waiotapu Hot Springs area, North Island, New Zealand. Photograph Ahnert, 1988

6. Stones that move out of the model field return into it at the opposite margin. This way, the total number of stones is preserved. The left marginal area is therefore considered a spatial continuation of the model field beyond the right margin; the same is true for the lower and upper margins. In order to show this spatial continuity, the model field is shown fourfold on each of the plots in Figure 1.10. The fourfold repetition of the individual patterns is merely a consequence of this and must not be mistaken for a spatial periodicity in the sorting process.
7. A time unit of development has passed when each mobile stone has been given a chance of movement as described above.

The results of this model experiment are shown in Figure 1.10. An initial random distribution of 300 stones in a field of 900 cells progressively changes to a sorted net pattern. Stones marked in black are those that have just moved in the time unit indicated.

The pattern developed by the model is very similar to stone net patterns found in the field. However, the question remains whether such a development requires randomness. To test this, a second model was designed, also of 300 stones in a 900-cell field, in which the initial distribution of stones was regular and in which the directions of stone movement were strictly prescribed, with a clockwise succession of all eight possible bearings. Each direction therefore had still an equal chance of occurring, but was chosen in an orderly sequence.

This non-random model still produced a net pattern that was reasonably similar to those found in nature (Figure 1.11). Therefore, it seems that randomness is not needed to generate such seemingly irregular patterns of sorted stone nets.

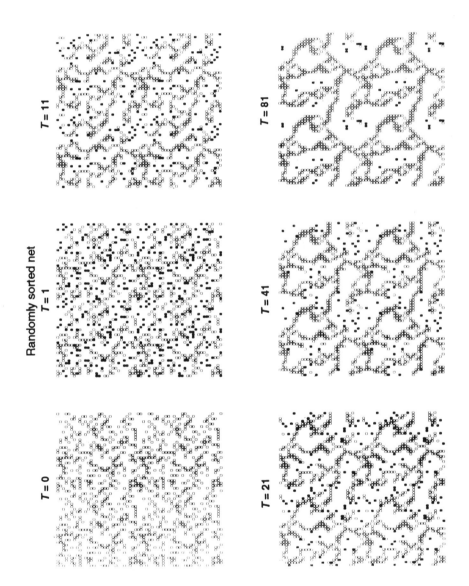

**Figure 1.10** Simulation of stone net development by means of the random-walk model STONQ6

18

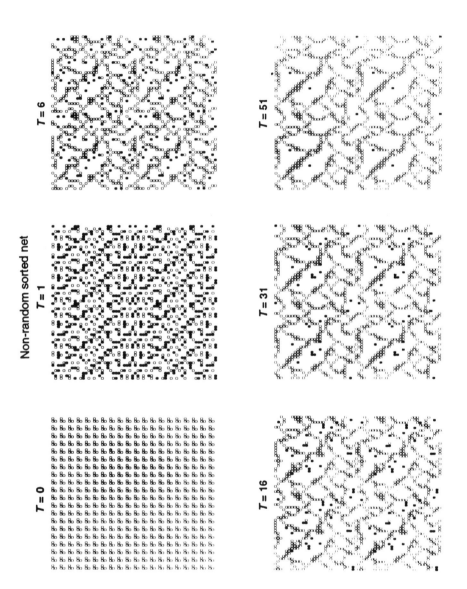

**Figure 1.11**  Deterministic model of stone net development (STREGQ6)

## CONCLUSIONS

In evaluating these model experiments, one can distinguish several types of random variations, applied to several types of model components (Table 1.1). The effects vary accordingly.

Initial spatial variations of the slope profile (Figure 1.1b) disappeared rapidly and proved irrelevant for the further development of the profile form. Essentially the same is true for the initial distribution of stones in Figures 1.10 and 1.11.

In three-dimensional models, on the other hand, the initial variations of the land surface remained effective for a long time in influencing the location of gullies, valleys and dolines, because they lead here to self-perpetuating spatial process differentiations (Figures 1.3–1.8). A permanent spatial variation of resistance parameters would have a similar effect.

Permanent spatial and temporal variations in the rate of baselevel lowering always affect the landform development in the vicinity of the baselevel but are progressively filtered out during their upslope transmission (Figure 1.2). This happens also if the variations are non-random.

Permanent spatial and temporal variations of runoff (Figure 1.1c) have little effect because they occur within an overriding downslope cascade of flow in which the variation of local increments is relatively unimportant.

Variation of erodibility (Figure 1.1d–1.1e) or of stone movements (Figure 1.10–1.11) are not tied into a cascade and therefore have a more decisive influence upon the development of form or pattern. However, they need not be random.

The other question asked at the beginning was whether such partially randomized models offer an explanatory advantage over deterministic models. Undoubtedly, many of the forms and patterns obtained in this way look realistic, perhaps also aesthetically pleasing, to the eyes of the geomorphologist. It may also seem 'like nature' that repeated

**Table 1.1**  Types of variations

| | Spatial (S) | Temporal (T) | Spatial and temporal (ST) |
|---|---|---|---|
| Initial (i) | iS | — | — |
| Permanent (p) | pS | pT | pST |
| Temporary (t) | tS | tT | tST |

| Type of variation | Example(s) |
|---|---|
| Initial spatial variations | Initial slope profile<br>Initial land surface<br>Initial distribution of stones |
| Permanent spatial variations | Rock resistance (e.g. as a function of joint spacing) |
| Permanent temporal variations | Lowering of baselevel |
| Permanent spatial and temporal variations | Runoff<br>Erodibility<br>Stone movements |

runs of the same partially randomized model program produce forms and patterns that, while being in a general way similar to one another, actually differ in many details.

Such general similarity means that the pattern or landform of the model can represent a known type of pattern or form, but not a particular field example. Moreover, this type can also be modelled, as the examples of figures 1.7 and 1.11 show, without resort to randomization.

Specific similarity of a model with a particular field example can only be achieved by introducing additional specific modifications and adaptations of the process, material or other model components that affect the form or pattern. Such an approach has been used already, for example, in models of the Grand Canyon in Arizona (Pollack, 1969; Cunningham & Griba, 1973; Aaronsson & Linde, 1982), of the Kall valley slopes in Germany (Ahnert, 1987, 5–8) and, most recently, in three-dimensional evolution models of specific local landform assemblages on acidic and ultrabasic rocks near Jacupiranga, São Paulo, Brazil (Römer 1993).

A satisfactory theoretical model should account for the past landform development and/or make the future development predictable. It requires that the components and factors are fully identified so that identical premises produce identical results. This requirement favours deterministic modelling. The inclusion of random variables permits at best a statistical predictability within a range of possible results. However, in many cases this range can also be estimated with specific prescribed variations of parameters.

Stochastic methods based on initial assumptions of randomness are of great importance in empirical field studies. These provide, by statistical evaluation, the functional equations which then are brought as deterministic components into theoretical process response models. The reintroduction of some randomness into a model equation which previously has been freed from random influences by the preceding statistical analysis appears as an illogical step backwards in the methodological research sequence. And furthermore, as has been shown, the use of randomness in such theoretical models does not seem to offer any significant explanatory or predictive advantage.

## REFERENCES

Aaronsson, G. and Linde, K. (1982). Grand Canyon – a quantitative approach to the erosion and weathering of stratified bedrock. *Earth Surface Processes and Landforms*, 7, 589–599.
Abrahams, A. D. (1987). Channel network topology: regular or random? In Gardiner, V. (ed.), *International Geomorphology 1986*, Part II, John Wiley & Sons, Chichester, pp. 145–159.
Ahnert, F. (1954). Zur Frage der rückschreitenden Denudation und des dynamischen Gleichgewichts bei morphologischen Vorgängen. *Erdkunde*, **VIII**, 61–64.
Ahnert, F. (1976a). Darstellung des Struktureinflusses auf die Oberflächenformen im theoretischen Modell. *Zeitschrift für Geomorphologie N.F. Suppl.*, 24, 11–22.
Ahnert, F. (1976b). Brief description of a comprehensive three-dimensional model of landform development. *Zeitschrift für Geomorphologie, N.F. Suppl.*, 25, 29–49.
Ahnert, F. (1977). Some comments on the quantitative formulation of geomorphological processes in a theoretical model. *Earth Surface Processes*, 2, 1191–1201.
Ahnert, F. (1981). Stone rings from random walks. *Transactions, Japanese Geomorphological Union*, 2, 301–312.
Ahnert, F. (1987). Approaches to dynamic equilibrium in theoretical simulations of slope development. *Earth Surface Processes and Landforms*, **12**, 3–15.
Ahnert, F. (1988). Modelling landform change. In Anderson, M. G. (ed.), *Modelling Geomorphological Systems*, John Wiley & Sons, Chichester, pp. 375–400.

Bowyer-Bower, T. A. S. and Bryan, R. B. (1986). Rill initiation: Concepts and experimental evaluation on badland slopes. *Zeitschrift für Geomorphologie, N.F. Suppl.*, **60**, 161–175.

Cunningham, F. F. and Griba, W. (1973). A model of slope development and its application to the Grand Canyon, Arizona, USA. *Zeitschrift für Geomorphologie, N.F. Suppl.*, **17**, 43–77.

Hack, J. T. (1960). Interpretation of erosional topography in humid temperature regions. *American Journal of Science*, **258A**, 80–97.

Howard, A. D. (1971). Simulation of stream networks by headward growth and branching. *Geographical Analysis*, **3**, 29–50.

Howard, A. D. (1972). Problems of interpretation of simulation models of geologic processes. In Morisawa, M. (ed.), *Quantitative Geomorphology: Some Aspects And Applications*, Binghamton, NY, pp. 61–81.

Jennings, J. N. (1985). *Karst geomorphology*. Basil Blackwell, Oxford, 293 pp.

Leopold, L. B. and Langbein, W. B. (1962). The concept of entropy in landscape evolution. *US Geological Survey Professional Paper*, **500A**.

Mann, C. J. (1970). Randomness in nature. *Geological Society of America Bulletin*, **81**, 95–104.

Merriam, D. F. (ed.) (1976). *Random Processes in Geology*, Springer-Verlag, Berlin–Heidelberg–New York, 168 pp.

Pollack, H. N. (1969). A numerical model of Grand Canyon. *Four Corners Geological Society Guidebook*, pp. 61–62.

Price, W. E. Jr (1976). A random-walk simulation model of alluvial fan deposition. In Merriam, D. F. (ed.), *Random Processes in Geology*, Springer-Verlag, Berlin–Heidelberg–New York, pp. 55–62.

Rohdenburg, H. and Walther, D. (1968). Rezente Strukturböden in Giessen. *Eiszeitalter und Gegenwart*, **19**, 279–282.

Rolshoven, M. (1976). Aktive Frostmusterung in Augsburg. *Eiszeitalter und Gegenwart*, **27**, 189–192.

Römer, W. (1993). Die Morphologie des Alkalikomplexes von Jacupiranga und seiner Umgebung, *Aachener Geographische Arbeiten*, Heft 26, 300 pp.

Schweinfurth, U. (1964). Ein Polygonboden auf Mt Allen, Stewart Island (Neuseeland). *Zeitschrift für Geomorphologie, N.F.*, **8**, 1–6.

Shreve, R. L. (1966). Statistical law of stream numbers. *Journal of Geology*, **74**, 17–37.

Shreve, R. L. (1975). The probabilistic-topological approach to drainage-basin geomorphology. *Geology*, **3**, 527–529.

Smart, J. S. (1968). Statistical properties of stream lengths. *Water Resources Research*, **4**, 1001–1014.

Smart, J. S. (1978). The analysis of drainage network composition. *Earth Surface Processes*, **3**, 129–170.

Washburn, A. L. (1979). Geocryology. Edward Arnold Ltd, London, 406 pp.

Werrity, A. (1972). The topology of stream networks. In Chorley, R. J. (ed.), *Spatial analysis in Geomorphology*, Harper & Row, New York–Evanston–San Francisco, pp. 167–196.

Willgoose, G., Bras, R. L. and Rodriguez-Iturbe, I. (1991). Results from a new model of river basin evolution. *Earth Surface Processes and Landforms*, **16**, 237–254.

Williams, P. W. (1985). Subcutaneous hydrology and the development of doline and cockpit karst. *Zeitschrift für Geomorphologie N.F.*, **29**, 463–482.

Williams, P. W. (1987). Geomorphic inheritance and the development of tower karst. *Earth Surface Processes and Landforms*, **12**, 453–465.

Wilson, P. and Clark, R. (1991). Development of miniature sorted patterned ground following soil erosion in East Falkland, South Atlantic. *Earth Surface Processes and Landforms*, **16**, 369–376.

Wright, A. C. and Webster, R. (1991). A stochastic distributed model of soil erosion by overland flow. *Earth Surface Processes and Landforms*, **16**, 207–226.

# 2 Apatite Fission Track Analysis: Its Potential for the Estimation of Denudation Rates and Implications for Models of Long-term Landscape Development

**RODERICK W. BROWN**
*Department of Geological Sciences, University College, London, UK*

**MICHAEL A. SUMMERFIELD**
*Department of Geography, University of Edinburgh, UK*

**and**

**ANDREW J. W. GLEADOW**
*Department of Geology, La Trobe University, Australia*

## ABSTRACT

Although a large body of data is now available on present rates of denudation, there are major problems involved in extrapolating such rates to geological time-scales and relating them to models of long-term landscape development. Offshore sediment volume calculations can provide a valuable indication of long-term denudation rates, but the resolution obtainable is limited by the dificulty of accurately defining changes in sediment source areas over time.

An alternative means of documenting long-term denudation rates is provided by thermochronologic techniques which have the advantage of providing specific information on spatial variations in rates within drainage basins. These techniques include radiometric dating and fission-track analysis and are based on the cooling ages of minerals which indicate when the daughter products or fission tracks produced by the decay of radioactive isotopes began to be retained. Depths and rates of denudation can be calculated if a palaeogeothermal gradient is assumed, if a vertical sequence of samples is available, or if ages yielded by systems with different closure temperatures are compared in a single sample. Apatite fission track analysis is a particularly valuable technique because the low temperature at which fission tracks begin to be preserved in apatite ($\sim 110°C$) enable relatively shallow depths of denudation to be identified. Moreover, track-length distribution data can yield additional information about the cooling (denudational) history of a sample.

Comparison of thermochronologic data from orogenic, passive margin and cratonic terrains reveals the large differences in long-term rates of of landscape modification in these contrasting morphological environments. Assessment of these data also makes it possible to discriminate between the applicability of different models of landscape development.

*Process Models and Theoretical Geomorphology.* Edited by M. J. Kirkby
©1994 John Wiley & Sons Ltd

## INTRODUCTION

Although rates of change are necessarily a fundamental component of models of long-term landscape development, the classical models of landscape evolution, which still provide the framework within which long-term changes are discussed, were very poorly constrained by empirical data on denudation and uplift rates. One of the primary reasons for the rejection by many geomorphologists of the Davisian programme of reconstructing landscape history was this very lack of adequate data on rates of geomorphic processes. Nevertheless, the enormous growth in surface process studies since the 1950s in itself failed to yield the kinds of information directly relevant to issues of long-term landscape change, focused, as they almost invariably were, on very limited temporal and spatial scales. When comparisons were made between present-day fluvial denudation rates based on solid and solute load data, and initial estimates of long-term denudation rates derived from geological evidence, it was claimed that contemporary rates are seven to eight times those typical of the geological past (Menard, 1961). Such an apparent discrepancy was readily interpreted as a result of a number of likely factors, including the impact of human activity, isostatic deformation of the crust in response to changes in ice-sheet loading and the rapid fluctuation of sea level during the Quaternary, and changes in the extent of particular morphoclimatic zones and the area of the continents through geological time (Brunsden & Thornes, 1977, p. 109). Indeed it was even argued that, owing to the pervasive effects of human activity, 'present rates of erosion may provide us with no information directly relevant to the past' (Douglas, 1967).

It is critical to know whether present denudation rates differ significantly from those characterizing the geological past for two reasons. First, and most obviously, we need to know whether the present is, in fact, an entirely atypical representation of geological time in terms of rates of landscape modification. Second, we need to know whether we can legitimately extrapolate rates for particular surface processes and geomorphic environments derived from very short periods of measurement to much longer time spans. Clearly this will only be appropriate if the present is broadly representative of the geological record. There is, of course, no disputing the marked enhancement of denudation rates as a result of human activity in particular localities, and of specific climatic and tectonic events during the Quaternary. Such effects, however, may not be quite as widespread or significant as some workers have previously suggested.

Since it is only at the regional and sub-continental scale that we can reasonably consider problems of long-term landscape development, it is at these scales that we need to tackle the question of discrepancies between present and past denudation rates. Two technical developments over the past couple of decades have provided the opportunity of addressing this issue. One has been the rapid growth of information on offshore sediment volumes derived from borehole and seismic data. While little of this has yet been specifically applied to the unravelling of the denudational history of adjacent landmasses (see Matthews, 1975; Poag & Sevon 1989; Rust & Summerfield, 1990, for examples), it has enormous potential as a means of establishing generalized denudation chronologies. The major shortcoming of offshore sedimentary sequences as a basis for reconstructing landscape histories is the problem of constraining the source area of the recorded sediment. The other major development, which helps to mitigate this problem by providing point estimates of depths of rock removal, is the application of thermochronologic techniques to the construction of denudational chronologies (Summerfield, 1991a). All such techniques are based on the principle that radiometric dating 'clocks' are set at rather specific temperatures. If the geothermal gradient can be estimated or calculated then this closure temperature can be related to a specific depth below the landsurface. The radiometric age of a surface sample will then indicate the time elapsed since the sample cooled through the closure temperature, and thereby yield a long-term denudation rate.

Unfortunately, the great majority of radiometric systems have closure temperatures corresponding to depths in excess of 10 km and are therefore of limited use in documenting even very long-term denudation rates outside rapidly eroding orogenic mountain belts. But one technique, known as fission track analysis, can provide information on denudational histories involving much shallower depths of rock removal when applied to the mineral apatite. The sensitivity and broad applicability of this technique are such that it has the potential to revolutionize our knowledge of long-term denudation rates, and to provide critical tests of models of landscape evolution. Here we explain how fission track analysis can be used to derive denudational histories, and illustrate the way in which the technique can be applied to the testing of landscape evolution models in a range of denudational environments.

## APATITE FISSION TRACK ANALYSIS

Apatite fission track analysis (AFTA) is a technique for determining the low-temperature ($<125°C$) thermal history of rocks (Wagner, 1968; Wagner & Reimer, 1972; Naeser, 1979; Gleadow et al., 1986; Green et al., 1989a). Like other radiometric techniques of thermochronology the method relies on the effects of the radioactive decay of a particular element which, in the case of AFTA, is $^{238}U$. However, unlike conventional isotopic dating techniques, AFTA exploits the effects of the spontaneous fission decay (as opposed to alpha decay) of $^{238}U$ atoms. When an atom of $^{238}U$ decays by spontaneous fission it splits into two highly charged fission fragments of approximately equal mass which are propelled in opposite directions. The passage of these highly charged fission fragments through the host crystal causes a linear zone of damage in the crystal lattice. The linear damage trails are known as fission tracks (Fleischer et al., 1975) and appear as randomly oriented tubes intersecting the polished surface of the host crystal after a standardized chemical etching treatment (Green, 1986) (Figure 2.1).

Each fission track is the product of a single spontaneous fission event. Fission tracks are therefore produced within the host mineral at a rate determined by the rate of fission decay, which is statistically constant over the time interval of interest, and the uranium content of the crystal. The probability of a single fission track intersecting a randomly oriented surface through the host crystal is a function of the length of the etchable portion of the fission track (Laslett et al., 1982; Green, 1988). The areal density of fission tracks intersecting a polished and etched surface through the host crystal is therefore a function of time, uranium content and the distribution of etchable track lengths.

Fission track ages are typically calculated for ~ 20 individual apatite grains from each sample by measuring the areal density of spontaneous fission tracks that intersect the polished and etched surface of the grains. In the external detector method (Gleadow, 1981) the induced track density in each apatite grain is measured by securing a thin sheet

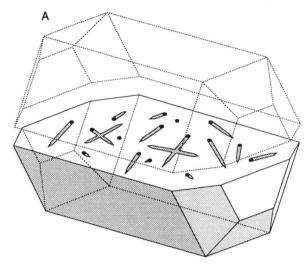

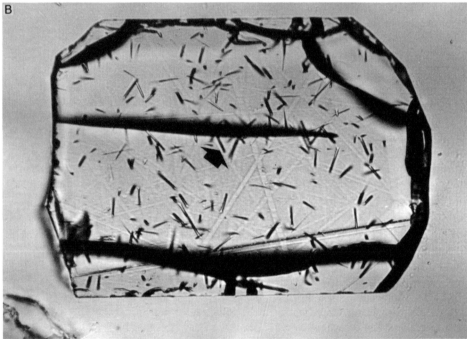

**Figure 2.1**  A schematic diagram (A) of a polished and etched internal surface through an apatite crystal showing the etched fission tracks intersecting the surface as well as two confined tracks intersected by surface tracks. The photomicrograph (B) shows a polished and etched prismatic section through an apatite crystal. The etched fission tracks intersecting the surface are clearly visible, as is a confined track (arrowed). The long axis of the grain is ~150 $\mu$m

of muscovite (the external detector) against the polished mineral surface during thermal neutron irradiation. For each grain the induced tracks recorded by the external detector originate from induced fission decay of $^{235}U$ atoms from the same region of the polished surface on which the areal density of spontaneous tracks is measured. The uranium concentrations are then determined by calibrating the induced track density, measured on the etched muscovite detector, against the induced track density measured on a similar external detector in contact with a standard glass of known uranium concentration exposed to the same irradiation. The system is empirically calibrated against age standards using a technique known as the zeta calibration method (Fleischer & Hart, 1972; Hurford & Green, 1983). Fission track ages of samples are usually reported as a weighted mean (or pooled age) of the individual grain ages along with some statistical measure of the dispersion of individual grain ages (Galbraith, 1990; Hurford, 1991).

## Thermal Annealing of Fission Tracks

All fission tracks have the same etchable track length when they form, and for tracks in apatite this initial etchable length is $\sim 16 \pm 1$ $\mu$m (Gleadow et al., 1986). However, fission tracks undergo a process of thermal annealing whereby the damage that constitutes the latent or unetched fission tracks is gradually repaired. The effect of this thermal annealing is to shorten the etchable length of the tracks. The extent of track shortening is primarily dependent on the maximum temperature experienced by the track, and to a lesser extent on the duration of heating (Fleischer et al., 1965; Green et al., 1989b). The kinetics of the thermal annealing process (track shortening) of fission tracks in apatite have been established by laboratory experiments (Wagner and Reimer, 1972; Laslett et al., 1987; Crowley et al., 1991) and by the analysis of samples from deep hydrocarbon exploration wells where the thermal history of the well is independently constrained (Naeser and Forbes, 1976; Gleadow and Duddy, 1981; Naeser, 1981). These experiments have enabled mathematical models to be developed which are capable of predicting fission track ages and track length distributions for hypothetical thermal histories (Laslett et al., 1987; Carlson, 1990; Crowley et al., 1991). Such fission track modelling procedures have been discussed in detail by Duddy et al. (1988), Green et al. (1989b), Jones & Dokka (1990) and Corrigan (1991). While there are some differences in the thermal calibration of the various annealing models, they are, nevertheless, extremely useful for developing an understanding of AFTA data. This can be achieved by examining the progressive changes in apparent age and track length for a range of hypothetical thermal histories. The modelled AFTA data presented in this paper were generated using the annealing model of Laslett et al. (1987), but the qualitative trends in such data do not depend on the chosen annealing model.

Over geological time scales ($10^6$–$10^7$ a) fission tracks in apatite will be progressively shortened from their initial length of $\sim 16 \pm 1$ $\mu$m with increasing temperature until they are completely annealed at temperatures $> 110 \pm 10°$C. The measured apparent fission track age of a sample decreases systematically with progressive thermal annealing, and at sufficiently elevated temperatures all the tracks are annealed and the apparent age reduced to zero. This occurs because a decrease in the mean track length causes a proportionate reduction in the measured spontaneous track density (Laslett et al., 1982; Green, 1988).

The increase in the rate of thermal annealing with respect to temperature is not linear, and at temperatures greater than $\sim 60°$C the rate rises significantly. The range between $\sim 60°$C and the temperature at which all tracks are completely annealed has been termed the partial annealing zone (PAZ) (Gleadow & Fitzgerald, 1987), or partial stability field (Wagner & Reimer, 1972). It is important to note, however, that fission tracks experience thermal annealing, and hence track shortening, at temperatures below $\sim 60°C$, albeit at a very slow rate. The rate of thermal annealing is also dependent on the chemical composition of the host apatite crystal, specifically on the ratio of $Cl/Cl + F$ (Green et al., 1989b), with higher rates of annealing for fluorine-rich apatites relative to chlorine-rich apatites at a given temperature. This compositional effect implies that individual apatite grains from the same sample can display different degrees of thermal annealing even though they have experienced the same thermal history. This effect is commonly observed for samples of sedimentary rocks that contain apatites from several sources (and thus display a range in apatite composition) which have experienced temperatures near the limit of track stability ($\sim 95-105°C$). This variation in annealing behaviour between single apatite grains thus provides important information about the maximum temperatures experienced by a sample.

## Confined Track Length Distributions

An estimate of the distribution of track lengths in an apatite sample is made by measuring the lengths of $\sim 100$ horizontal, confined tracks (Bhandari et al., 1971; Gleadow et al., 1986; Green et al., 1989a). Confined tracks are those that are completely contained within the host mineral grain and have been etched through the etchant penetrating along contiguous cracks or other tracks that intersect the polished surface. Although biased against sampling short tracks (Laslett et al., 1982, 1984; Green, 1988), this method provides the most direct and reproducible estimate of the true distribution of track lengths within a sample (Gleadow et al., 1986; Green et al., 1989b). As new fission tracks are formed continually in the host crystal by radioactive decay each track experiences a different portion of the total thermal history of the sample. The final distribution of track lengths and the measured apparent fission track age of a sample therefore represent an integrated record of its total thermal history over the temperature range within which fission tracks are preserved. For geological heating times ($10^6-10^7$ a) the upper limit of this temperature range is $\sim 110 \pm 10°C$. It is important to note that an apatite fission track age will only record the time that a sample was last at a temperature of $\sim 110 \pm 10°C$ if, by rapid movement through the PAZ, it has subsequently cooled quickly to a temperature at which the rate of thermal annealing is greatly reduced ($< 60-70°C$), and if it has not thereafter been reheated. It is therefore essential to have information about the distribution of track lengths within a sample in order meaningfully to interpret measured apparent fission track ages (Gleadow et al., 1986; Kohn et al., 1991).

Typical confined track length distributions for three different styles of thermal history are shown in Figure 2.2. The model length distributions were produced from the three thermal histories shown (a, b and c). Samples that have experienced simple thermal histories, such as volcanic rocks that have cooled rapidly and have never been reheated to temperatures greater than $30-40°C$ (model a), will have apatite fission track ages that closely approximate the time of eruption. Their track length distributions will reflect this simple thermal history, with all the tracks being near their initial lengths and giving a mean track length of $\sim 14-15 \mu m$ and a standard deviation of $\sim 1.0 \mu m$. More complex thermal histories will produce similarly more complicated track length distributions. For example, if a sample cools below $\sim 110°C$ but remains at elevated temperatures (say $\sim 90-100°C$) for a substantial period of time before cooling rapidly to typical surface temperatures (model b), then all the fission tracks that formed prior to cooling will

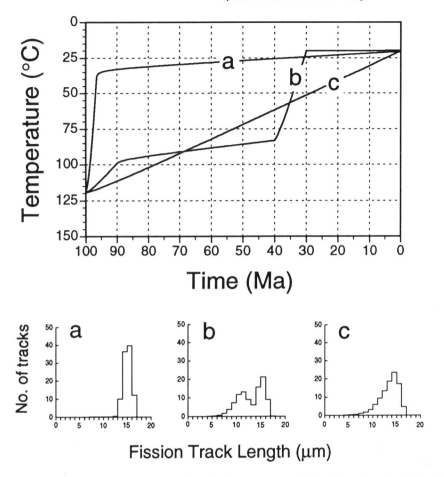

**Figure 2.2** Apatite fission track length distributions characteristic of three hypothetical styles of thermal history (see text for discussion)

have been thermally annealed and thus shortened. Any fission tracks that accumulated after cooling will preserve their maximum length, thus producing a distinctly bimodal track length distribution. The earlier formed annealed tracks give rise to the shorter and the younger unannealed tracks to the longer of the two modal peaks. The apparent fission track age for this type of thermal history will be significantly younger than the time the sample first cooled below ~110°C. This is because a substantial proportion of the tracks preserved since that time will be short and thus only a small fraction of these will contribute to the measured age. So the measured apparent fission track age for a sample of this type will fall somewhere between the time that tracks first began to accumulate and the time the sample finally cooled to the surface temperature (see Green, 1986).

Another situation is represented by a sample that cools monotonically and relatively slowly from above ~110°C to surface temperatures in response to low to moderate rates of denudation (model c). The resulting track length distribution for a sample experiencing this type of thermal history is negatively skewed, reflecting the wider range of temperatures experienced by the preserved tracks, but being biased towards the longer track lengths. These simple examples illustrate how contrasting types of thermal and denudational histories can be inferred from characteristic track length distributions for individual apatite samples.

## Fission Track Crustal Profiles

The limited stability of fission tracks in apatite when exposed to temperatures in the range characteristic of the upper few kilometres of the crust ($\sim 0$–$125°C$) results in systematic variations in track length distribution and apparent fission track age with depth. The form of such a fission track profile reflects the distribution of temperature within the shallow crust and its variation through time, and is therefore characteristic of the style of thermal history experienced by the crustal section. Analysis of samples representing a range of depths within the crust, such as a suite of samples from a deep drill hole (Hammerschmidt *et al.*, 1984) or from outcrops over a range of surface elevations, can therefore be employed to document the vertical pattern of fission track parameters within the crust. Such a fission track profile can then be used to reconstruct the thermal history for the sampled section and thus place quantitative constraints on the chronology and magnitude of the denudational history.

In tectonically stable regions where rates of denudation have been low ($< \sim 30 \, \mathrm{m \, M^{-1}}$) for a prolonged period ($> 10^6$ a), such as cratonic interiors, or where progressive burial has been occurring, such as in sedimentary basins, the typical form of the fission track crustal profile is controlled primarily by the progressive increase of temperature with depth. The pattern produced by this situation of continuous residence at constant or gradually increasing temperature is one of progressive reduction in mean track length and hence fission track age with depth. An example of this type of profile is shown in Figure 2.3 and represents a composite section from several hydrocarbon exploration wells from the Otway Basin in southeastern Australia (Gleadow & Duddy, 1981; Green *et al.*, 1989a). The apatite fission track age decreases systematically with depth from the initial provenance age of $\sim 125$ Ma BP at the surface to zero at a depth of $\sim 3.5$ km forming a characteristic concave-up profile of apparent apatite age (Figure 2.3A). The four representative track length distributions in Figure 2.3B show how the mean track length decreases and how the shape of the track length distribution becomes broader (higher standard deviation) with depth. The increase in the standard deviation of the track length distribution with decreasing mean track length is a consequence of the anisotropy of track shortening (tracks perpendicular to the c-crystallographic axis anneal more quickly than tracks parallel to the c-axis) as well as the variation in apatite composition between grains (Green *et al.*, 1986; Green, 1988).

This continuous and progressive decrease in apparent apatite age and mean track length with depth defines a fission track 'stratigraphy' within the crust (Gleadow, 1990; Brown, 1991) related to depth below the mean landsurface. The ultimate form of the fission track profile will depend on the duration of landsurface stability and on the original age of the apatites. This dependence of form on the period of landsurface stability is illustrated in Figure 2.4A. The five theoretical profiles were constructed using the kinetic annealing model of Laslett *et al.* (1987) and show how the form of the apparent age profile develops with time. The model profiles represent residence at constant temperature for periods of 10 Ma to 50 Ma and assume that all the samples have zero initial age.

The shape of the apparent age profile is quite sensitive to the rate and duration of denudation. Gleadow (1990) demonstrated that the initial concave-up apparent age profile is gradually smoothed into a linear gradient of apparent age with depth for denudation rates $> \sim 30 \, \text{m Ma}^{-1}$ sustained over $\sim 100$ Ma. The gradient of the resulting linear apparent age profile closely approximates the rate of denudation, with the ages of the individual apatite samples recording the approximate time at which they cooled below $110°C$. The effect of denudation rates on the form of the apparent age profile is shown in Figure 2.4B. The modelled profiles represent continuous denudation at various rates ranging from $0 \, \text{m Ma}^{-1}$ to $100 \, \text{m Ma}^{-1}$ over a period of 50 Ma, assuming an initial geothermal gradient of $25°C \, \text{km}^{-1}$. It is clear from these modelled profiles that, given sufficient time, even very slow rates of denudation will ultimately produce linear arrays of apparent age with depth. The critical parameter which governs the curvature of the apparent age profile is the amount of crustal section removed during the period of denudation. As long as approximately three or more kilometres of section are progressively removed (at a constant rate) during the denudational episode then the resulting apparent age profile will be linear and the gradient of apparent age with depth will approximate the rate of denudation. This form of apparent age profile was first documented in the European Alps by Wagner and Reimer (1972) and Wagner *et al.* (1977), and subsequently in a profile from the Eielson drill-hole in Alaska (Naeser, 1979).

More complicated denudational histories will obviously produce correspondingly more complex apatite age profiles. If there is a rapid increase in the rate of denudation after a prolonged period of geomorphic stability, in response to a sudden fall in base level, for example, then the base of the existing concave-up apparent age profile will be shifted upwards towards the new mean topographic surface as denudation proceeds. Apatite samples that were at temperatures $> \sim 110°C$ prior to the acceleration in denudation will have accumulated no fission tracks and hence have zero apparent ages up to that point. On the initiation of cooling produced by the accelerated denudation these samples will begin to retain fission tracks once they cool below $\sim 110°C$, and a new apparent age profile will begin to develop below the earlier profile. If the amount of section removed during this episode is of the order of a few kilometres then part of the earlier apparent age profile may be preserved within the upper sections of the new topographic relief. The transition from the earlier upper profile to the new lower profile will be marked by a pronounced inflection. The location of this inflection marks the depth at which the apparent apatite age was reduced to zero (that is, the depth of the $110°C$ palaeoisotherm) prior to the increase in denudation rate, and the corresponding age approximates the time at which the acceleration in denudation took place.

The form of the new apparent age profile that develops below this inflection will depend on both the rate and duration of the accelerated period of denudation and could be either a linear change in apparent age with depth if sufficient section were removed, or possibly a new concave-up PAZ profile. Clearly, it is possible to produce a composite profile with an inflection comprising two linear components with differing age–depth gradients representing a change from a moderate rate of denudation to a significantly higher rate. Interpreting an observed gradient in apparent apatite age therefore requires discrimination between prolonged annealing at constant temperature (negligible or no denudation) and continuous denudation at some constant rate. Resolving these two

A

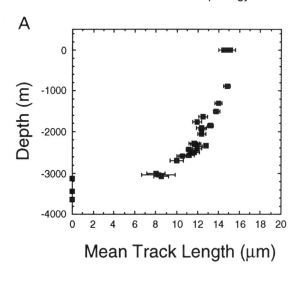

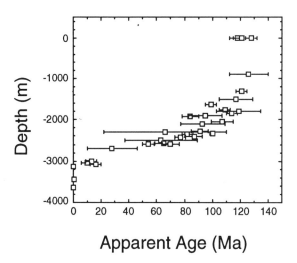

**Figure 2.3** (A) Composite apatite fission track crustal profiles of mean fission track length (solid squares) and apparent apatite fission track age (open squares) plotted against depth for samples from several wells from the Otway Basin in southeastern Australia. These clearly illustrate the progressive decrease in mean track length and apparent apatite fission track age with depth, and the characteristic concave-up form of both profiles

interpretations of an age gradient is obviously a crucial step in arriving at a meaningful interpretation of any apatite fission track crustal profile. Fortunately the relationship between apparent apatite age and mean confined track length for each sample provide a diagnostic pattern for both the linear and preserved PAZ profiles. To illustrate this relationship a simple single-stage ('square pulse') denudational history has been used to generate thermal histories for a series of points spaced regularly at 300 m intervals

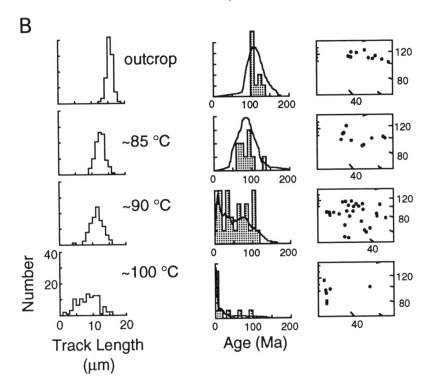

**Figure 2.3**(continued)   (B) Fission track length distributions, single crystal age histograms, and single crystal age radial plots (Galbraith, 1990) for four samples from the Otway Basin wells. These are representative of successive degrees of thermal annealing and illustrate the progressive change in the shape of the track length distribution and the dispersion in apparent single crystal fission track ages with increasing depth. The mean apparent age and the mean track length of the sample decreases with progressive thermal annealing. In addition, the distribution of track lengths and the distribution of single crystal apparent ages becomes broader as the degree of thermal annealing increases. This occurs because fission tracks in the individual apatite crystals anneal at different rates due to the effect of variable chemical composition and annealing anisotropy

over a depth of 7 km (Figure 2.5). The thermal histories were calculated using a numerical solution for the partial differential equation describing the conduction of heat in a moving medium (Carlslaw & Jaeger, 1959, p. 385) and used to generate the theoretical fission track parameters shown in Figures 2.5C and 2.6. The thermal histories for 12 selected points (labelled a–l in Figures 2.5 and 2.6) were used to generate confined track length distributions representative of the two sections of the model profile, points a–f from the upper PAZ section (Figure 2.6A) and g–l from the lower linear apparent age section (Figure 2.6B). The modelled cooling paths are essentially linear, but the effect of including the advection term in the numerical solution produces a small delay in the initial rate of cooling which gives rise to the slight curvature of the cooling paths of the deeper samples.

Several features of fission track crustal profiles are well demonstrated by the simple example shown in Figure 2.5C. The prominent inflection in the apparent apatite age profile occurs at ~ 50 Ma which is the time that the modelled denudational episode

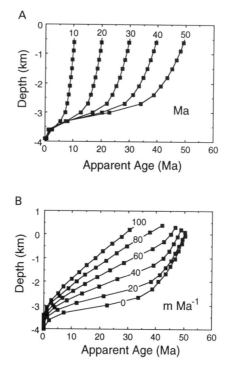

**Figure 2.4** Model profiles (generated using the annealing model of Laslett *et al.* (1987)) showing the effect of time (A) and denudation rate (B) on the form of the apparent apatite fission track age profile

began, and the gradient in apparent age with elevation below the inflection is $\sim 200$ m Ma$^{-1}$ which approximates the rate of denudation (Figure 2.5A). The distinct concave-up form of the upper section of the apparent age profile clearly reflects the form of the preserved PAZ profile formed in the first 50 Ma period of the model (compare with Figure 2.3A). The modelled profile also illustrates the distinctly different patterns in the variation of mean confined track length with depth for the upper and lower sections. The most striking feature of the mean track length profile for the upper section is the pronounced increase in mean track length that occurs at about 4.5 km elevation ($\sim 3$ km palaeodepth) following a gradual decrease in mean length with decreasing elevation (Figure 2.5C). This form is in marked contrast to the monotonic profile of mean track length with depth for the lower section where the mean length remains almost constant at $\sim 14.5\,\mu$m over the whole 3 km section. This difference is clearly reflected by the shape of the plots of mean track length against apparent age shown in Figures 2.6A and 2.6B. These two characteristic relationships result from the contrasting nature of the track length distributions for the two sections of the profile.

The track length distributions from the lower, linear portion of the modelled profile are all remarkably similar to each other (Figure 2.6B). They are unimodal with mean lengths of $\sim 14.5\,\mu$m and standard deviations of a little less than $\sim 2\,\mu$m, and are slightly negatively skewed because of small 'tails' of shorter tracks being present. The simple unimodal track length distribution for these samples (g–l) reflects the simple linear cooling history experienced, from temperatures greater than the limit for track stability ($\sim 110 \pm 10^\circ$C). Figure 2.5B (samples g–l) shows that the apparent fission track age recorded by each sample corresponds aproximately to the time that the sample cooled below $\sim 110^\circ$C. In addition, the proportion of time spent by each sample within the PAZ is relatively short and hence most of the accumulated fission tracks experienced little thermal annealing, as is reflected by the long mean track lengths for these samples. The small 'tail' of shorter tracks represents the small proportion of tracks that formed at elevated temperatures as the samples passed through the PAZ ($\sim 110$–$60^\circ$C).

In contrast to the unimodal track length distributions of the lower sections of the profiles, the track length distributions from the upper, preserved annealing zone sections are more complex and variable (Figure 2.6A). These comprise two track length components, one representing tracks formed during the first 50 Ma at constant temperature, and the other tracks formed during the cooling stage once denudation had begun. The mean length of the early component of tracks in each sample reflects the temperature at which they accumulated during the first 50 Ma interval. The later component of tracks will have a similar length distribution and mean to samples from the lower section of the profile described above. For shallow samples, such as sample a, little thermal annealing will have occurred and so both the early and later component of tracks will have long mean track lengths close to the maximum of $\sim 15\,\mu$m. The resulting composite track length distribution will have a small standard deviation ($\sim 1\,\mu$m) and a mean length of $\sim 15\,\mu$m. With increasing palaeodepth the mean length of the early track length component will decrease, reflecting the higher temperatures at which these earlier tracks accumulate. The resulting composite track length distributions for deeper samples become broader and eventually distinctly bimodal as the difference in the mean lengths of the two components of the track length distribution becomes larger with increasing palaeodepth (samples b to d).

For maximum palaeotemperatures $> \sim 95^\circ$C the lengths of the earlier formed tracks are such that the proportion of these tracks that is sampled is dramatically reduced. This is due to the track length measuring procedure being strongly biased against sampling severely shortened tracks ($< \sim 10\,\mu$m) (Laslett *et al.*, 1982; Green, 1988). The effect of this sampling bias is that the track length distributions for the lowest samples from the upper section of the modelled profile (samples d to f) become increasingly dominated by the later, long track length component, and so the mean lengths of the resulting composite track length distributions begin to increase. However, the apparent fission track age will continue to decrease because the proportion of earlier formed tracks that intersect the etched surface, and thus contribute to the age measurement, continues to decrease with increasing maximum palaeotemperature. Once the maximum palaeotemperature experienced by a sample reaches the limit of track stability ($\sim 110 \pm 10^\circ$C) then none of the earlier formed tracks will be preserved.

The relationship between apatite age and mean track length is clearly an important aspect of interpreting AFTA data. It can be effectively analysed by plotting the mean and standard deviation of track length against apparent fission track age, as shown in Figure 2.6. The resulting diagrams are commonly known as 'boomerang plots' due to the distinctive trajectory of apatite age relative to mean track length that characterizes the PAZ section of the profile (Figure 2.6A). Boomerang plots clearly illustrate the progressive reduction in both the relative proportion and mean length of the older

A

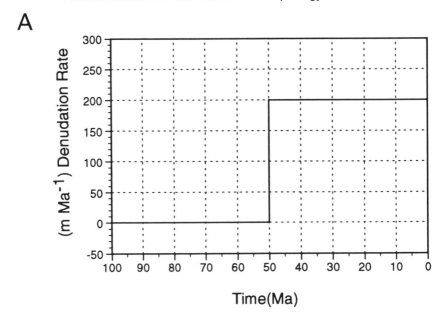

B

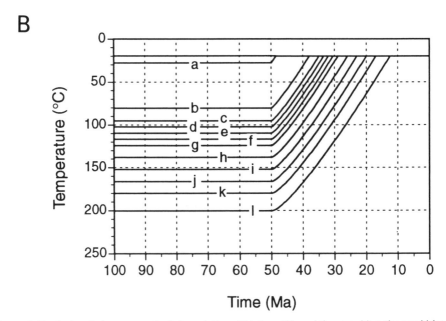

**Figure 2.5** A simple 'square pulse' denudational history (A) and the resulting thermal histories for a series of points at increasing depths (B). The thermal histories in (B) (labelled a–l) were used to predict the apparent apatite ages and mean track lengths forming the model profiles illustrated in (C). The apparent ages and mean track lengths are plotted relative to the initial depth prior to the denudational episode (palaeodepth) and relative to a hypothetical elevation scale (elevation). The horizontal dashed line at approximately 4 km palaeodepth is the depth at which the apparent apatite age would have been zero prior to the initiation of the denudation episode and represents the approximate palaeodepth of the 110°C isotherm

C

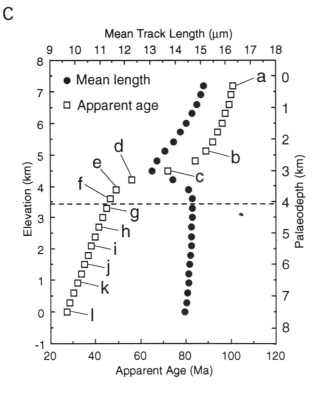

**Figure 2.5**(continued)

component of tracks as well as apatite age. In contrast, when plotted against mean track length the linear section of the profile has a monotonic apatite age trajectory reflecting the uniform shape of the track length distributions (Figure 2.6B). These plots can therefore effectively discriminate between the two types of profile.

## APPLICATION TO PROBLEMS OF LONG-TERM LANDSCAPE DEVELOPMENT

In this section we discuss several examples of the application of AFTA to problems of long-term landscape development, with particular attention being given to some of the apparent inconsistencies that exist between various interpretations of AFTA results. Examples of AFTA studies from three contrasting denudational environments are considered representing passive continental margin, orogenic mountain belt, and cratonic intraplate settings.

### Passive Continental Margins

#### Southeastern Australia

The morphology of the southeastern Australian continental margin is typical of many passive margins in that it is characterized by an interior highland region (the southern

38

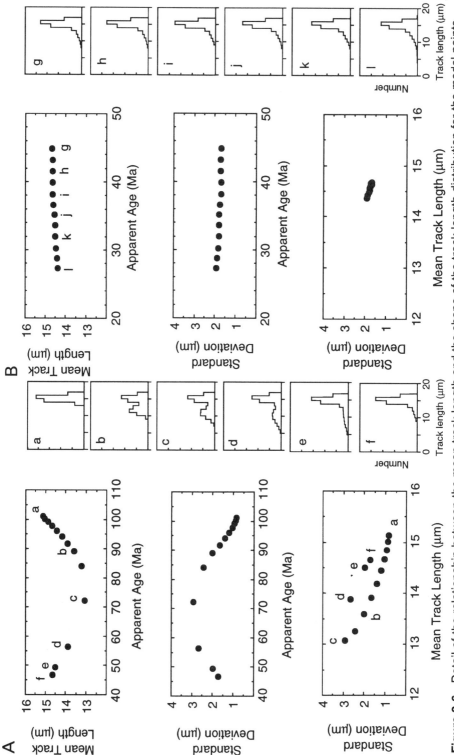

**Figure 2.6**  Detail of the relationship between the mean track length and the shape of the track length distribution for the model points from the upper, partial annealing zone section (points a–f) (A), and from the lower, linear section (points g–l) (B) of the modelled profile shown in Figure 5C. These diagrams are commonly referred to as 'boomerang plots' and are discussed in detail in the text

part of the East Australian Highlands) separated from a dissected coastal zone by a prominent erosional escarpment (Bishop, 1988). Its denudational history and morphological development have been the subject of numerous studies which have led to widely differing models of long-term landscape development (see Bishop '1988' for review). An extensive AFTA data set now exists for the southeastern Australian continental margin (Moore *et al.*, 1986; Dumitru *et al.*, 1991). Although most of the landscape development models focus on the timing, rate and amount of surface uplift of the margin, the thermochronologic data can provide a valuable means of evaluating these models through the resulting temporal and spatial patterns of denudation that they predict.

In the model proposed by Stephenson & Lambeck (1985) and Lambeck & Stephenson (1986), for instance, the morphological evolution of southeastern Australia is controlled by the pattern of denudation, where the rate of denudation is assumed to be proportional to elevation, and the resulting Airy isostatic rebound of a pre-existing orogenic mountain belt. The variation in the mount of predicted denudation is consequently strongly correlated with elevation, with the maximum depth (1.5–2.0 km) occurring in the interior of the Highlands, and almost zero denudation near the present coastline (Figure 2.7A). They concluded that the dominant process involved in the morphological evolution of the region since the Early Mesozoic has been Airy isostatic rebound in response to the relatively slow denudation (rates of $\sim 5\,\mathrm{m\,Ma^{-1}}$) of an existing highland terrain. Lambeck & Stephenson (1986) argued specifically against significant rejuvenation of the East Australian Highlands in the Late Mesozoic or Tertiary.

The AFTA data from southeastern Australia presented by Moore *et al.* (1986) and Dumitru *et al.* (1991) document a strong correlation between apparent fission track age and present-day elevation, with the oldest apparent ages generally occurring at the highest elevations. The apparent ages range from $\sim 350\,\mathrm{Ma\,BP}$ in the Highlands to $\sim 100\,\mathrm{Ma\,BP}$ at the lowest elevations near the coast (Figure 2.7B). The confined track length distributions from the youngest samples are narrow with mean track lengths of $\sim 14\,\mu\mathrm{m}$ indicating that the samples cooled rather rapidly from temperatures greater than $\sim 110°\mathrm{C}$ to less than $\sim 70°\mathrm{C}$ at $\sim 100\,\mathrm{Ma\,BP}$. The track length distributions for the samples with intermediate apparent ages are broader and have shorter mean track lengths ($\sim 11$–$12\,\mu\mathrm{m}$). They are also distinctly bimodal in some cases, in contrast to the track length distributions for the oldest samples which have slightly negatively skewed, unimodal distributions with mean track lengths of $\sim 13.5\,\mu\mathrm{m}$. The data suggest that the oldest samples have not been significantly heated since cooling to near surface temperatures at about 350 Ma BP, and that the intermediate ages represent samples that have remained at elevated temperatures until cooling at $\sim 100\,\mathrm{Ma\,BP}$.

This pattern in the variation of mean track length with apparent age is remarkably similar to that of the upper section of the model profile in Figures 2.5 and 2.6, and this strongly supports the interpretation of Moore *et al.* (1986) that the maximum temperatures experienced by the samples since $\sim 350\,\mathrm{Ma\,BP}$ range from $<50°\mathrm{C}$ for the oldest samples, to $50$–$110°\mathrm{C}$ for the intermediate age samples, and $>110\,°\mathrm{C}$ for the youngest samples, and that the youngest samples cooled relatively rapidly (within 20 Ma) from their maximum temperatures to $\sim 70°\mathrm{C}$ beginning $\sim 100\,\mathrm{Ma\,BP}$.

This thermal history interpretation of the AFTA data implies that the maximum amount of denudation occurred along the eastern margin of the southeastern Highlands during the mid-Cretaceous, with the lowest amounts predicted for the interior regions.

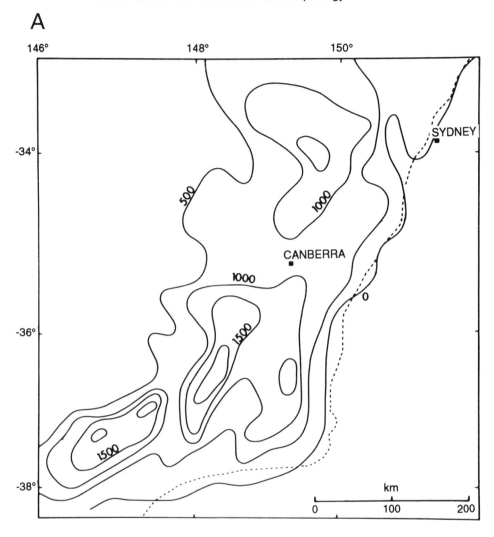

**Figure 2.7** Contour diagrams for southeastern Australia of estimated denudation since 250 Ma BP according to Lambeck & Stephenson (1986) (Reproduced by permission of Blackwell Scientific Publications (Aust)) (A), and apatite fission track age presented by Dumitru *et al.* (1991) and Moore *et al.* (1986) (Reproduced by permission of Elsevier Science Publishers BV) (B). Note the positive correlation between regions of oldest apatite fission track age and areas with the highest predicted denudation which demonstrates that the hypothesized denudation pattern of Lambeck & Stephenson (1986) is invalid *B, Benalla syenite, *M, Myalla Road syenite

Notwithstanding the problem of constraining the palaeogeothermal gradient, the maximum palaeotemperatures estimated from the AFTA data indicate that several kilometres of denudation occurred along the southeastern continental margin during the mid-Cretaceous, and that very little (< ~1 km) denudation appears to have occurred in the interior regions where the oldest apparent apatite ages are found. This regional pattern in the amounts of predicted denudation based on the AFTA data, which broadly corresponds with that expected from an escarpment retreat mode of landscape

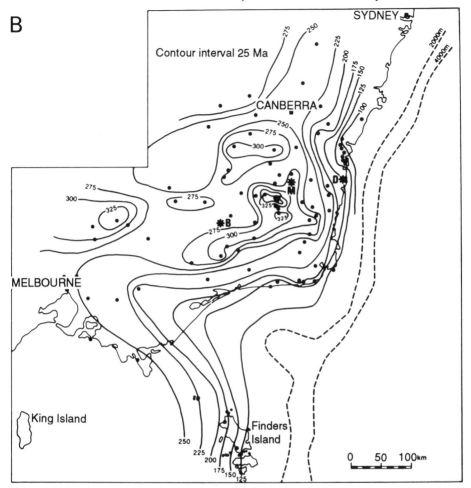

**Figure 2.7**(continued)

development (Summerfield, 1991b), contrasts markedly with that proposed by Lambeck & Stephenson (1986). Furthermore, the timing of the episode of denudation indicated by the AFTA data suggests that maximum rates of denudation were substantially higher than $\sim 5\,\mathrm{m\,Ma^{-1}}$, and in fact probably reached $100\,\mathrm{m\,Ma^{-1}}$. That this phase of accelerated denudation began at 100 Ma BP indicates that the landscape of the Highlands was rejuvenated in the Late Mesozoic. However, this period of accelerated denudation did not necessarily occur in response to surface uplift since it could have simply been promoted by new, lower base levels established at about this time in response to rifting during the early stages of the opening of the Tasman Sea (Gilchrist and Summerfield, this volume, Chapter 3).

### The Transantarctic Mountains

The Transantarctic Mountains consist of major escarpments flanking a significant crustal upwarp along the western margin of the East Antarctic Shield adjacent to the Ross Sea

Basin. They are formed mainly of Precambrian to Early Palaeozoic metamorphic and igneous rocks overlain by several kilometres of Devonian to Triassic sedimentry rocks of the Beacon Group. Elevations attain a maximum of 4.5–5 km near the Ross Sea coast, but decrease rapidly oceanward of the escarpment. There is a less dramatic decrease in elevation inland where the landsurface becomes progressively buried beneath the ice cover of the Wilkes Glacial Basin.

In an AFTA study of the Transantarctic Mountains carried out in South Victoria Land, Fitzgerald *et al.* (1986/87) and Gleadow & Fitzgerald (1987) have documented a fission track crustal profile (Figure 2.8) with a prominent inflection in the apparent age profile at ~50 Ma BP, which is similar in form to the model profile illustrated in Figures 2.5 and 2.6. The inflection occurs at 800 m above sea level at Mt Doorly near the coast, but it appears at successively lower elevations further inland. It was inferred from these AFTA results that the Transantarctic Mountains were formed at 50 Ma BP in response to the rifting that led to the formation of the Ross Sea Basin. The axis of maximum rock uplift was located along the rift margin, ~20 km inland from the present coastline, with decreasing amounts of rock uplift further inland (Gleadow and Fitzgerald, 1987).

The fission track crustal profiles presented by Fitzgerald *et al.* (1986/87) and Gleadow and Fitzgerald (1987) provide strong evidence of substantial denudation along the Ross Sea margin on the scale of several kilometres beginning ~50 Ma BP, and also indicate that the depth of denudation decreases inland from a maximum near the coast of ~4–5 km. This pattern and chronology of denudation is strong evidence for the formation of significant relief during the Early Cenozoic. Such an interpretation is therefore not easily reconciled with palaeontological data from the Transantarctic Mountains which have been regarded by some researchers as indiating substantial surface uplift (~0.5–2 km) of the region as recently as the Late Pliocene–Pleistocene (Webb *et al.*, 1984; Webb & Harwood, 1987). It must be emphasized, however, that the AFTA data only constrain the chronology and amount of denudation (rock uplift relative to the landsurface) that has occurred and do not in themselves require that most of the topography of the present Transantarctic Mountains was created during the Early Cenozoic. Isostatic rebound of the eroding topography would allow several kilometres (4–5 km) of removal of section to be achieved during the Cenozoic at the modest rates of ~100 m Ma$^{-1}$ implied by the AFTA data in response to relatively minor initial relief (~0.5–1 km) along the rift margin. This could have been generated by modest surface uplift coincident with rifting, by the creation of a new, lower base level due to subsidence of the Ross Sea Basin, or by a combination of both of these effects. This interpretation of the AFTA data is not necessarily inconsistent with the proposal that the mean elevation of the Transantarctic Mountains was relatively low (possibly <1 km) prior to the hypothesized major episode of surface uplift occurring during the Late Pliocene–Pleistocene suggested by Webb *et al.* (1986), Behrendt & Cooper (1991) and McKelvey *et al.* (1991). But neither does it support such a chronology of surface uplift since AFTA data alone provide information on vertical movement of rocks relative to the landsurface, not absolute changes in surface elevation (Summerfield, 1991a).

Equating estimates of rates of surface uplift with rates of denudation calculated from AFTA data will inevitably lead to apparently incompatible interpretations. The current debate concerning the morphological development of the Transantarctic Mountains demonstrates this dilemma, which is concisely summarized by the following statement from McKelvey *et al.* (1991):

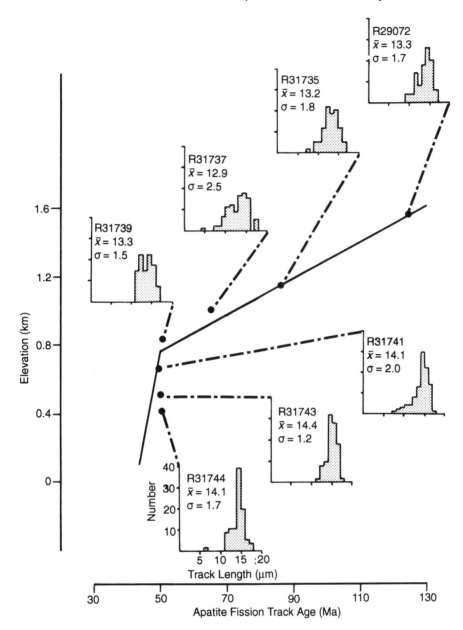

**Figure 2.8** Apparent apatite fission track age profiles and track length distributions from the Transantarctic Mountains in South Victoria Land (Gleadow & Fitzgerald, 1987. Reproduced by permission of Elsevier Science Publishers BV). This clearly illustrates the inflection separating an upper partial annealing zone section of the age profile from a lower linear portion of the age profile. The age at which the inflection occurs (~50 Ma) is approximately the time at which cooling began, possibly in response to accelerated denudation. Compare the track length distributions with those of the modelled profile illustrated in Figures 2.5 and 2.6

... this palaeogeographic scenario necessitates minimum uplift in the order of 1300 m at some time during the last 2.5–3 Ma, a rate much greater than the 100 m/Ma average calculated from fission-track data (Gleadow & Fitzgerald, 1987) for much of the Transantarctic Mountains.

McKelvey *et al.* (1991) have clearly interpretated the mean denudation rate of $\sim 100 \, \text{m Ma}^{-1}$ calculated from AFTA data to be equivalent to a mean rate of surface uplift. This interpretation conflicts with their proposal for much higher surface uplift rate during the Late Pliocene–Pleistocene of $1000 \, \text{m Ma}^{-1}$. As argued above, this discrepancy is apparent rather than real. This kind of confusion concerning the appropriate use of AFTA data within the context of reconstructing palaeotopographies has been encouraged, at least in part, by the common practice of reporting AFTA data in terms of 'uplift' rates. None the less, by considering the geomorphic implications of the denudational information obtainable from AFTA results it may be possible to gain insights into the nature of landscape development, and thus, indirectly, into the history of surface uplift and generation of topography (Summerfield & Brown, 1993). In any case, this example underlines the need to distinguish carefully between empirical results and the interpretations derived from them.

## Modes of Landscape Development

The AFTA data from southeastern Australia and the Transantarctic Mountains provide support for models of long-term landscape development for passive margins that envisage maximum denudation rates occurring early in the development of rifts. In addition, it appears that the greatest amount of denudation, of at least several kilometres, occurs near newly formed passive margins and probably propagates inland via some form of escarpment retreat, in some instances coupled with major drainage restructuring, such as appears to be the case in southeastern Australia (Ollier, 1982, 1985) and southern Africa (Summerfield, 1985; Partridge & Maud, 1987; Rust & Summerfield, 1990). Several other AFTA studies of passive margins have documented similar denudational histories, for instance those on the Appalachians in the eastern USA (Miller & Duddy, 1989), the Red Sea (Bohannan *et al.*, 1989; Omar *et al.*, 1989), and southwestern Africa (Brown *et al.*, 1990).

## Orogenic Mountain Belts

The topography of orogenic mountain belts contrasts markedly with that of passive continental margins. This morphological difference is largely a result of the contrasting tectonic mechanisms responsible for the formation of the topography within these two denudational environments. In convergent plate boundary settings rates of horizontal plate motion can be up to several tens of kilometres per million years and hence rates of crustal thickening within orogenic mountain belts can also be very high. It follows that maximum rates of crustal uplift (uplift of the crust relative to a fixed datum) along convergent plate margins are also likely to be high since they will be largely controlled by rates of crustal thickening. Consequently, rates of denudation in these environments must be similarly high because elevations would soon reach unrealistic values if this were not the case (Brunsden & Lin, 1991; Summerfield 1991a). For example, a crustal uplift rate of 5000 m Ma$^{-1}$ would create a surface uplift of 5 km within 1 Ma in the absence of denudation. The fact that such high rates of plate convergence and crustal shortening have been operating for several million years in some orogenic terrains, such as Taiwan (Suppe, 1987) and the Southern Alps of New Zealand (Adams, 1985; Kamp et al., 1989), implies that maximum rates of denudation must be broadly similar to rates of crustal uplift in these tectonic settings.

Estimates of present-day denudation rates in these environments are indeed high, exceeding 5000 m Ma$^{-1}$ in some cases (Li, 1976; Whitehouse, 1988), while fission track crustal profiles from orogenic mountain belts confirm, at least in some cases, that comparable rates have been sustained over time spans of the order of $10^5$–$10^6$ a. Such profiles are typically similar in style to the lower section of the model profile in Figure 2.5C, exhibiting an approximately linear relationship between apparent apatite age and elevation, and relatively long mean track lengths (Wagner et al., 1977; Hurford, 1986; Benjamin et al., 1987) (Figure 2.9).

Some of the first applications of fission track dating were carried out in the European Alps (Wagner and Reimer, 1972; Schaer et al., 1975; Wagner et al., 1977) and excellent reviews of these early studies together with more recent work have been presented by Hurford et al. (1989) and Hurford (1991). It was in this initial fission track research in the Central Alps (Wagner & Reimer, 1972; Wagner et al., 1977) that the significance of apparent age gradients, where the apparent apatite ages increase with increasing elevation, was first recognized. Such studies also led to the first estimates based on thermochronologic data of denudation rates over geological time scales for the Central Alps (Figure 2.9A). More recent research (Hurford, 1986, 1991; Hurford et al., 1991) has confirmed the generally high rates of denudation (500–1000 m Ma$^{-1}$) across the Alps. However, using the additional information provided by confined track length measurements, these later studies have enabled closer constraints to be placed on the rates and chronology of cooling and thus have led to the refinement of denudation rate estimates.

In other Alpine-type terrains, such as the Nanga Parbat region of the Himalayas (Zeitler et al., 1982; Zeitler, 1985), and the Bolivian (Benjamin et al., 1987; Figure 2.9B) and Venezuelan Andes (Kohn et al., 1984), AFTA results have recorded characteristically high rates of denudation (up to several $10^3$ m Ma$^{-1}$), and generally linear apparent apatite age profiles. In a recent fission track study of the Southern Alps of New Zealand, Kamp et al. (1989) calculated that adjacent to the Alpine Fault in the Mount Cook area denudation has occurred at a mean rate of 2300–2500 m Ma$^{-1}$ from 5.5 to 1 Ma BP, and at 6200 m Ma$^{-1}$ over the past 1 Ma (Figure 2.10). This latter rate is remarkably close to the 5500–7800 m Ma$^{-1}$ rate of surface uplift estimated from apparent remnants of Late Quaternary wave-cut terraces by Bull and Cooper (1986). Given the strongly supported suggestion that the central Southern Alps are in an approximate topographic steady state (Adams, 1985; Whitehouse, 1988; Brunsden & Lin, 1991) the equivalence of these two rates is highly significant. The particular value of fission track analysis here arises both from the independent support it provides for denudation rates derived from geomorphic evidence (Whitehouse, 1988), and from the much longer period of time to which it applies.

Kamp et al. (1989), among others, have highlighted an important aspect of using thermochronologic data, such as AFTA, to derive denudation rates, namely the need to assess the effects of advection of heat and consequent increase of the geothermal

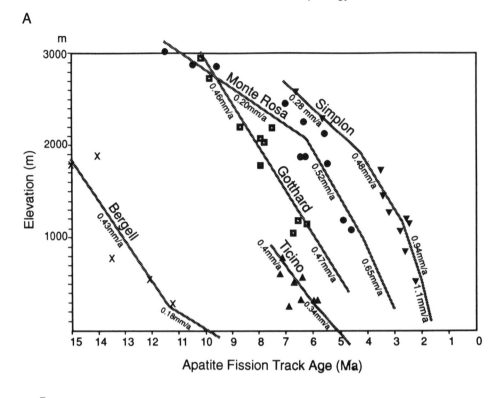

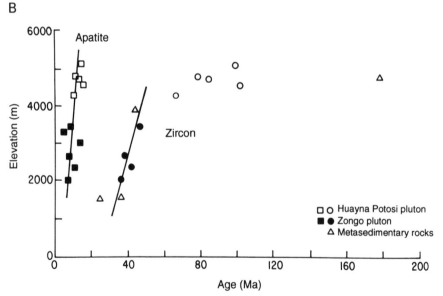

**Figure 2.9**   Apparent apatite age profiles from Alpine-type terrains which demonstrate the essential linear apparent age profiles typical of these regions. Profiles are from the European Alps (Wagner *et al.*, 1977. Reproduced by permission of Memoire di Scienze Geologiche) (A) and the Andes (Benjamin *et al.*, 1987. Reproduced by permission) (B)

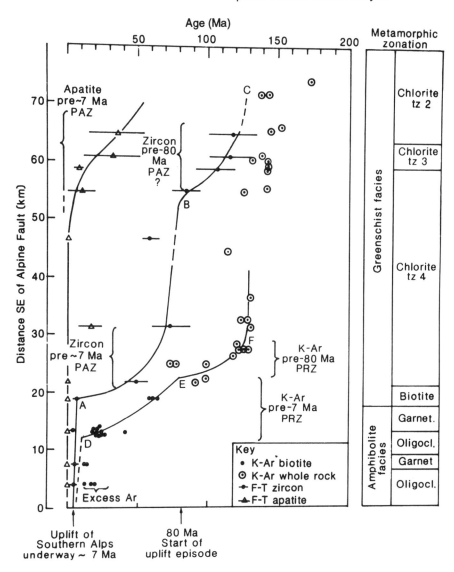

**Figure 2.10**   Thermochronologic data including apatite and zircon fission track ages and K–Ar ages on whole rocks and on biotites from the Southern Alps, New Zealand (Kamp *et al.*, 1989. Copyright by the American Geophysical Union). Note that when age is plotted against distance from the Alpine Fault (surrogate for depth) the apatite fission track ages form what appears to be a composite age profile, similar to that for the Transantarctic Mountains shown in Figure 2.8 but with the inflection at only 1 Ma BP

gradient caused by the high rates of denudation associated with Alpine terrains. They used the two-dimensional thermal model developed for the Southern Alps by Koons (1987) to quantify the effect that denudation would have on the geothermal gradient, and calibrated their denudation estimates accordingly. Developing this point Summerfield & Brown (1993) have recently provided a detailed quantitative assessment of the thermal effects of advection and the implications for transforming thermochronologic data into information about rates of denudation.

## Cratonic Intraplate Settings

Landscapes typical of tectonically stable, cratonic interiors are generally characterized by extensive landsurfaces of minimal local relief, such as the Yilgarn Block in Western Australia, the Canadian Shield, and parts of the interior region of the Kaapvaal Craton in South Africa. These landsurfaces have been interpreted by some workers as extremely ancient landforms, with sub-aerial exposure purportedly dating back, in one case at least, to the Early Palaeozoic (Stewart *et al.*, 1986). However, AFTA data from the Kaapvaal Craton in South Africa (R. W. Brown, unpublished data), and from the southern part of the Yilgarn Block (Fergusson, 1981), place some constraints on long-term denudation rates from such intracratonic environments. In both areas the oldest apparent apatite ages are recorded from the interior of the cratons, but these are surprisingly young ($\sim 500$ Ma BP) considering the Archean age of the host rocks. The apparent ages decrease as the margins of the craton are approached and reach 250–300 Ma BP at the southern margin of the Yilgarn Block, and 90 Ma BP along the eastern margin of the Kaapvaal Craton. The younger apparent ages along cratonic margins are most reasonably interpreted as being the result of denudation and cooling, possibly in association with tectonic events such as the Pan-African age metamorphic episode and subsequent rifting related to the break-up of Gondwana.

These apparent apatite fission track ages of $\sim 500$ Ma BP for some cratonic interiors are not easily reconciled with some prevailing interpretations of long-term landscape development for such environments which envisage repeated phases of exhumation and shallow re-burial, with extremely low mean rates of net denudation ($0.1$–$0.2$ m Ma$^{-1}$) (Fairbridge & Finkl, 1980). Nor do they accord with the very great ages attributed to some intracratonic erosional features. If these apparent AFTA ages of 500 Ma do indeed represent slow cooling ages rather than a distinct thermal overprint at about this time, then they suggest substantially higher long-term denudation rates for cratonic interior regions than have generally been postulated.

Geothermal gradients from cratonic interiors are typically of the order of $10$–$15^\circ$C km$^{-1}$. These imply a minimum depth of denudation of $7.5$–$10$ km since $\sim 500$ Ma BP, giving mean denudation rates of $15$–$20$ m Ma$^{-1}$. These estimates would be increased by any intervening episodes of burial by sediments which were sufficiently shallow not to leave a thermal imprint detectable by AFTA. The alternative explanation, which requires much less denudation, involves appealing to higher palaeogeothermal gradients, possibly associated with the Pan-African age ($\sim 500$ Ma BP) tectonism which created the metamorphic mobile belts surrounding the cratons. For some cratons, such as the Kaapvaal Craton in South Africa, this alternative interpretation would be difficult to reconcile with the abundant evidence indicating that sub-cratonic lithosphere is typically old, cold and thick (Jordan, 1975, 1978; Boyd *et al.*, 1985; Boyd & Gurney, 1986; Jones, 1988). Although the limited amount of AFTA data from these terrains prohibits any conclusive interpretations, it certainly suggests that the idea of very long-term landsurface stability in cratonic interiors should be re-examined, and that further AFTA studies need to be undertaken.

# SUMMARY

The unique low-temperature ($< \sim 125°C$) sensitivity of apatite fission track analysis provides extremely valuable, quantitative estimates of the amount, rate and chronology of regional, long-term denudation. Such information on denudational histories over geological time scales is generally unobtainable by other means, and so AFTA can provide crucial data relevant to assessing theoretical models of long-term landscape development. It is unfortunate that the common practice of reporting the results of AFTA studies in the form of variably, and sometimes incorrectly defined 'uplift' rates has led to substantial confusion among many users concerning their meaning (Brown, 1991; Summerfield, 1991a; Summerfield & Brown, 1993). Nevertheless, the rapidly growing global AFTA data set provides the best opportunity yet to address the question of rates of long-term landscape development on a quantitative basis.

In this chapter we have highlighted the important aspects of AFTA data interpretation, and demonstrated its usefulness in constraining rates and amounts of denudation. Its successful application within a wide range of geomorphic environments emphasizes the valuable and practical nature of the technique within the context of reconstructing denudation chronologies. In addition, AFTA has raised some new questions about traditional models of landscape development, such as the long-term geomorphic stability of cratonic interiors, and of the role of escarpment retreat in the development of passive margin topography.

## ACKNOWLEDGEMENTS

Fission track research at the Victorian Institute of Earth and Planetary Sciences is supported by the Australian Research Council and the Australian Institute for Nuclear Science and Engineering. R.W.B. was supported by an Australian Postgraduate Research Award and acknowledges the British Council for a Visiting Bursary. This research was partly supported by the Natural Environment Research Council through grant number GR3/6693 to M.A.S. Additional support to M.A.S. was provided by the Royal Society through an Anglo-Australian Visiting Fellowship, and the Royal Society of Edinburgh through a Support Research Fellowship. M.A.S. is also grateful to the Rockefeller Foundation at whose Bellagio Study Center parts of this paper were drafted.

## REFERENCES

Adams, J. (1985). Large-scale tectonic geomorphology of the Southern Alps, New Zealand. In Morisawa, M. and Hack, J. T. (eds), *Tectonic Geomorphology*, Allen and Unwin, Boston, pp. 105–128.

Ballard, S. and Pollack, H. N. (1987). Diversion of heat by Archean cratons: a model for southern Africa. *Earth and Planetary Science Letters*, **85**, 253–264.

Behrendt, J. C. and Cooper, A. (1991). Evidence of rapid Cenozoic uplift of the shoulder escarpment of the Cenozoic West Antarctic rift system and a speculation on possible climate forcing. *Geology*, **19**, 315–319.

Benjamin, M. T., Johnson, N. M. and Naeser, C. W. (1987). Recent rapid uplift in the Bolivian Andes: evidence from fission track dating. *Geology*, **15**, 680–683.

Bhandari, N., Bhat, S. G., Lal, D., Rajagopalan, G., Tamhane, A. S. and Venkatavaradan, V.S. (1971). Fission fragment tracks in apatite: recordable track lengths. *Earth and Planetary Science Letters*, **13**, 191–199.

Bishop, P. (1988). The eastern highlands of Australia: the evolution of an intraplate highland belt. *Progress in Physical Geography*, **12**, 159–181.

Bohannan, R. G., Naeser, C. W., Schmidt, D. L. and Zimmermann, R. A. (1989). The timing of uplift, volcanism and rifting peripheral to the Red Sea: A case for passive rifting? *Journal of Geophysical Research*, **94**, 1683–1701.

Boyd, F. R. and Gurney, J. J. (1986). Diamonds and the African lithosphere. *Science*, **232**, 472–477.

Boyd, F. R., Gurney, J. J. and Richardson, S. H. (1985). Evidence for a 150–200 km thick Archean lithosphere from diamond inclusion thermobarometry. *Nature*, **315**, 387–389.

Brown, R. W. (1991). Backstacking apatite fission-track 'stratigraphy': A method for resolving the erosional and isostatic rebound components of tectonic uplift histories. *Geology*, **19**, 74–77.

Brown, R. W., Rust, D. J., Summerfield, M. A., Gleadow, A. J. W. and De Wit, M. C. J. (1990). An Early Cretaceous phase of accelerated erosion on the southwestern margin of Africa: Evidence from apatite fission track analysis and the offshore sedimentary record. *Nuclear Tracks and Radiation Measurements*, **17**, 339–351.

Brunsden, D. and Lin, J-C. (1991). The concept of topographic equilibrium in neotectonic terrains. In Cosgrove, J. and Jones, M. (eds), *Neotectonics and Resources*, Belhaven Press, London, pp. 120–143.

Bull, W. B. and Cooper, A. F. (1986). Uplifted marine terraces along the Alpine Fault, New Zealand. *Science*, **234**, 1225–1228.

Carlson, W. D. (1990). Mechanisms and kinetics of apatite fission-track annealing. *American Mineralogist*, **75**, 1120–1139.

Carslaw, H. S. and Jaeger, J. C. (1959). *Conduction of Heat in Solids*, 2nd edition, Oxford University Press, New York, 510 pp.

Corrigan, J. (1991). Inversion of apatite fission track data for thermal history information. *Journal of Geophysical Research*, **96**, 10347–10360.

Crowley, K. D., Cameron, M. and Schaefer, R. L. (1991). Experimental studies of annealing of etched fission tracks in apatite. *Geochimica et Cosmochimica Acta*, **55**, 1449–1465.

Douglas, I. (1967). Man, vegetation and the sediment yields of rivers. *Nature*, **215**, 925–928

Duddy, I. R., Green, P. F. and Laslett, G. M. (1988). Thermal annealing of fission tracks in apatite. 3. Variable temperature behaviour. *Chemical Geology (Isotope Geoscience)*, **73**, 25–38.

Dumitru, T. A., Hill, K. C., Coyle, D. A., Duddy, I. R., Foster, D. A., Gleadow, A. J. W., Green, P. F., Kohn, B. P., Laslett, G. M. and O'Sullivan, A. J. (1991). Fission track thermochronology: Application to continental rifting of south-eastern Australia. *APEA Journal*, **31**, 131–142.

Fairbridge, R. W. and Finkl, C. W. Jr (1980). Cratonic erosional unconformities and peneplains. *Journal of Geology*, **88**, 69–86.

Fergusson, K. U. (1981). Fission track dating of shield areas, Australia: relationships between tectonic and thermal histories and fission track age distributions. Msc Thesis, University of Melbourne, 194 pp.

Fitzgerald, P. G., Sandiford, M., Barrett, P. J. and Gleadow, A. J. W. (1986/87). Asymmetric extension associated with uplift and subsidence of the Transantarctic Mountains and Ross Embayment. *Earth and Planetary Science Letters*, **81**, 67–78.

Fleischer, R. L. and Hart, H. R. Jr (1972). Fission track dating: techniques and problems. In Bishop, W. W., Miller, D. A. and Cole, S. (eds), *Calibration of Hominoid Evolution*, Scottish Academic Press, Edinburgh, pp. 135–170.

Fleischer, R.L., Price, P. B. and Walker, R. M. (1965). Effects of temperature, pressure and ionisation of the formation and stability of fission tracks in minerals and glasses. *Journal of Geophysical research*, **70**, 1497–1502.

Fleischer, R. L., Price, P. B. and Walker, R. M. (1975). *Nuclear Tracks in Solids: Principles and Applications*, University of California Press, Berkeley, 605 pp.

Galbraith, R. F. (1990). The radial plot: graphical assessment of spread in ages. *Nuclear Tracks and Radiation Measurements*, **17**, 207–214.

Gleadow, A. J. W. (1981). Fission track dating methods: What are the real alternatives?. *Nuclear Tracks and Radiation Measurements*, **5**, 3–14.

Gleadow, A. J. W. (1990). Fission track thermochronology – reconstructing the thermal and tectonic evolution of the crust. In *Proceedings of the Pacific Rim '90 Congress*, Australian Institute of Mining and Metallurgy. pp. 15–21.

Gleadow, A. J. W. and Duddy, I. R. (1981). A natural long-term annealing experiment for apatite. *Nuclear Tracks and Radiation Measurements*, **5**, 169–174.

Gleadow, A. J. W. and Fitzgerald, P. G. (1987). Uplift history and structure of the Transantarctic Mountains: New evidence from fission track dating of basement apatites in the Dry Valleys area, Southern Victoria Land. *Earth and Planetary Science Letters*, **82**, 1–14.

Gleadow, A. J. W., Duddy, I. R., Green, P. F. and Lovering, J. F. (1986). Confined fission track lengths in apatite: A diagnostic tool for thermal history analysis. *Contributions to Mineralogy and Petrology*, **94**, 405–415.

Green P. F. (1986). On the thermotectonic evolution of Northern England: evidence from fission track analysis. *Geological Magazine*, **123**, 493–506.

Green, P. F. (1988). The relationship between track shortening and fission track age reduction in apatite: Combined influences of inherent instability, annealing anisotropy, length bias and system calibration. *Earth and Planetary Science Letters*, **89**, 335–352.

Green, P. F., Duddy, I. R., Gleadow, A. J. W., Tingate, P. R. and Laslett, G. M. (1986). Thermal annealing of fission tracks in apatite. 1. A qualitative description. *Chemical Geology (Isotope Geoscience)*, **59**, 237–253.

Green, P. F., Duddy, I. R., Gleadow, A. J. W. and Lovering, J. F. (1989a). Apatite fission track analysis as a palaeotemperature indicator for hydrocarbon exploration. In Naeser, N. D. and McCulloh, T. H. (eds), *Thermal History of Sedimentary Basins: Methods and Case Histories*, Springer-Verlag, New York, pp. 181–195.

Green, P. F., Duddy, I. R., Laslett, G. M., Hegarty, K. A., Gleadow, A. J. W. and Lovering, J. F. (1989b). Thermal annealing of fission tracks in apatite. 4. Quantitative modelling techniques and extension to geological timescales. *Chemical Geology (Isotope Geoscience)*, **79**, 155–182.

Hammerschmidt, K., Wagner, G. A. and Wagner, M. (1984). Radiometric dating on research drill core Urach III: a contribution to its geothermal history. *Journal of Geophysics*, **54**, 97–105.

Hurford, A. J. (1986). Cooling and uplift patterns in the Lepontine Alps South Central Switzerland and an age of vertical movement of the Insbruc fault line. *Contributions to Mineralogy and Petrology*, **93**, 413–427.

Hurford, A. J. (1991). Uplift and cooling pathways derived from fission track analysis and mica dating: a review. *Geologische Rundschau*, **80**, 349–368.

Hurford, A. J. and Green, P. F. (1983). The zeta age calibration of fission track dating. *Chemical Geology (Isotope Geoscience)*, **1**, 285–317.

Hurford, A. J., Flisch, M. and Jaeger, E. (1989). Unravelling the thermo-tectonic evolution of the Alps: a contribution from fission track analysis and mica dating. In Coward, M. P., Dietrich, D. and Park, R. G. (eds), *Alpine Tectonics*, Geological Society Special Publication No. 45, pp 369–398.

Hurford, A. J., Hunziker, J. C. and Stockhert, B. (1991). Constraints on the late thermotectonic evolution of the Western Alps: evidence for episodic rapid uplift. *Tectonics*, **10**, 758–769.

Jones, M. Q. W. (1988). Heat flow in the Witwatersrand basin and environs and its significance for the South African shield geotherm and lithospheric thickness. *Journal of Geophysical Research*, **93**, 3243–3260.

Jones, S. M. and Dokka, R. K. (1990). Modelling fission track annealing in apatite: an assessment of uncertainties. *Nuclear Tracks and Radiation Measurements*, **17**, 255–260.

Jordan, T. H. (1975). The continental lithosphere. *Reviews of Geophysics and Space Physics*, **13**, 1–12.

Jordan, T. H. (1978). Composition and development of the continental lithosphere. *Nature*, **274**, 544–548.

Kamp, P. J., Green , P. F. and White, S. H. (1989). Fission track analysis reveals character of collisional tectonics in New Zealand. *Tectonics*, **8**, 169–195.

Kohn, B. P. and Eyal, M. (1981). History of uplift of the crystalline basement of Sinai and its relation to opening of the Red Sea as revealed by fission track dating of apatites. *Earth and Planetary Science Letters*, **52**, 129–141.

Kohn, B. P., Shagam, R., Banks, P. O. and Burkley, L. A. (1984). Mesozoic–Pleistocene fission track ages on rocks of the Venezuelan Andes and their tectonic implications. *Geological Society of America Memoir*, **162**, 365–384.

Kohn, B. P., Brown, R. W. and Gleadow, A. J. W. (1991). Comment on 'Northern Monashee Mountains, Omineca crystalline belt, British Columbia: Timing of metamorphism, anatexis and tectonic denudation'. *Geology*, **19**, 763–764.

Koons, P. O. (1987). Some thermal and mechanical consequences of rapid uplift: an example from the Southern Alps, New Zealand. *Earth and Planetary Science Letters*, **86**, 307–319.

Lambeck, K. and Stephenson, R. (1986). The post-Palaeozoic uplift history of south-eastern Australia. *Australian Journal of Earth Sciences*, **33**, 253–270.

Laslett, G. M., Kendall, W. S., Gleadow, A. J. W. and Duddy, I. R. (1982). Bias in measurement of fission track length distributions. *Nuclear Tracks and Radiation Measurements*, **6**, 79–85.

Laslett, G. M., Gleadow, A. J. W. and Duddy, I. R. (1984). The relationship between fission track length and density in apatite. *Nuclear Tracks and Radiation Measurements*, **9**, 29–38.

Laslett, G. M., Green P. F., Duddy, I. R. and Gleadow, A. J. W. (1987). Thermal annealing of fission tracks in apatite. 2. A quantitative analysis. *Chemical Geology (Isotope Geoscience)*, **65**, 1–13.

Li, Y. H. (1976). Denudation of Taiwan Island since the Pliocene Epoch. *Geology*, **4**, 105–107.

Matthews, W. H. (1975). Cenozoic erosion and erosion surfaces of eastern North America. *American Journal of Science*, **275**, 818–824.

McKelvey, B. C., Webb, P. N., Harwood, D. M. and Marin, M. C. G. (1991). The Dominion Range Sirius Group – A record of the late Pliocene – early Pleistocene Beardmore glacier. In Thomson, M. R. A. *et al.* (eds), *Geological Evolution of Antarctica*, Cambridge University Press, Cambridge, pp. 675–682.

Menard, H. W. (1961). Some rates of regional erosion. *Journal of Geology*, **69**, 154–161.

Miller, D. and Duddy, I. R. (1989). Early Cretaceous uplift and erosion of the northern Appalachian basin, New York, based on apatite fission track analysis. *Earth and Planetary Science Letters*, **93**, 35–49.

Moore, M. E., Gleadow, A. J. W. and Lovering, J. F. (1986). Thermal evolution of rifted continental margins: New evidence from fission tracks in basement apatites from southeastern Australia. *Earth and Planetary Science Letters*, **78**, 255–270.

Naeser, C. W. (1979). Fission track dating and geologic annealing of fission tracks. In Jaeger, E. and Hunziker, J. C. (eds), *Lectures in Isotope Geology*, Springer-Verlag, Heidelberg, pp. 154–169.

Naeser, C. W. (1981). The fading of fission tracks in the geologic environment – data from deep drill holes. *Nuclear Tracks and Radiation Measurements*, **5**, 248–250.

Naeser, C. W. and Forbes, R. B. (1976). Variation of fission track ages with depth in two deep drill holes. *Transactions of the American Geophysical Union (Abstract)*, **57**, 353.

Ollier, C. D. (1982). The Great Escarpment of eastern Australia: tectonic and geomorphic significance. *Journal of the Geological Society of Australia*, **29**, 13–23.

Ollier, C. D. (1985). Morphotectonics of continental margins with great escarpments. In Morisawa, M. and Hack, J. T. (eds), *Tectonic Geomorphology*, Allen and Unwin, Boston, pp. 3–25.

Omar, G. I., Steckler, M. S., Buck, W. R. and Kohn, B. P. (1989). Fission-track analysis of basement apatites at the western margin of the Gulf of Suez rift, Egypt: Evidence for synchroneity of uplift and subsidence. *Earth and Planetary Science Letters*, **94**, 316–328.

Partridge, T. C. and Maud, R. R. (1987). Geomorphic evolution of southern Africa since the Mesozoic. *South African Journal of Geology*, **90**, 174–208.

Poag, C. W. and Sevon, W. D. (1989). A record of Appalachian denudation in postrift Mesozoic and Cenozoic sedimentary deposits of the U.S. middle Atlantic continental margin. *Geomorphology*, **2**, 119–157.

Rust, D. J. and Summerfield, M. A. (1990). Isopach and borehole data as indicators of rifted margin evolution in southwestern Africa. *Marine and Petroleum Geology*, **7**, 277–287.

Schaer, J. P., Reimer, G. M. and Wagner, G. A. (1975). Actual and ancient uplift rate in the Gotthard region, Swizz Alps: A comparison between precise leveling and fission track apatite age. *Tectonophysics*, **29**, 293–300.

Stephenson, R. and Lambeck, K. (1985). Erosion-isostatic rebound models for uplift: an application to south-eastern Australia. *Geophysical Journal of the Royal Astronomical Society*, **82**, 31–55.

Stewart, A. J., Blake, D. H. and Ollier, C. D. (1986). Cambrian river terraces and ridgetops in central Australia: Oldest persisting landforms? *Science*, **233**, 758–761.

Summerfield, M. A. (1985). Plate tectonics and landscape development on the African continent. In Morisawa, M. and Hack, J. T. (eds), *Tectonic Geomorphology*, Allen and Unwin, Boston, pp. 27–51.

Summerfield, M. A. (1991a). *Global Geomorphology*, Longman, London, and Wiley, New York, 537 pp.

Summerfield, M. A. (1991b). Sub-aerial denudation of passive margins: regional elevation versus local relief models. *Earth and Planetary Science Letters*, **102**, 460–469.

Summerfield, M. A. and Brown, R. W. (1993). Uplift, denudation and the interpretation of thermochronologic data. *Earth and Planetary Science Letters* (submitted).

Suppe, J. (1987). The active Taiwan mountain belt. In Schaer, J-P. and Rodgers, J. (eds), *The Anatomy of Mountain Ranges*, Princeton University Press, Princeton, pp. 277–293.

Thornes, J. B. and Brunsden, D. (1977). *Geomorphology and Time*, Methuen, London, 208 pp.

Wagner, G. A. (1968). Fission track dating of apatites. *Earth and Planetary Science Letters*, **4**, 411–415.

Wagner, G. A. and Reimer, G. M. (1972). Fission track tectonics: the tectonic interpretation of fission track ages. *Earth and Planetary Science Letters*, **14**, 263–268.

Wagner, G. A., Reimer, G. M. and Jaeger, E. (1977). Cooling ages derived by apatite fission track, mica Rb/Sr and K/Ar dating: the uplift and cooling history of the Central Alps. *Memorie Degli Instituti di Geologia e Mineralogia dell' Universita di Padova*, **30**, 1–27.

Webb, P. N. and Harwood, D. M. (1987). Terrestrial flora of the Sirius Formation: Its significance for late Cenozoic glacial history. *Antarctic Journal of the United States*, **22**, 7–11.

Webb, P. N., Harwood, D. M., McKelvey, B. C., Mercer, J. H. and Stott, L. D. (1984). Cenozoic marine sedimentation and ice-volume variation on the East Antarctic craton. *Geology*, **12**, 287–291.

Webb, P. N., Harwood, D. M., McKelvey, B. C., Mabin, M. C. G. and Mercer, J. H. (1986). Late Cenozoic tectonic and glacial history of the Transantarctic Mountains. *Antarctic Journal of the United States*, **21**, 99–100.

Whitehouse, I. E. (1988). Geomorphology of the central Southern Alps, New Zealand: the interaction of plate collision and atmospheric circulation. *Zeitschrift für Geomorphologie Supplement*, **69**, 105–116.

Zeitler, P. K. (1985). Cooling history of the NW Himalaya, Pakistan. *Tectonics*, **4**, 127–151.

Zeitler, P. K. Johnson, N. M., Naeser, C. W. and Tahirkheli, R. A. K. (1982). Fission track evidence for Quaternary uplift of the Namga Parbat region, Pakistan. *Nature*, **298**, 255–257.

# 3 Tectonic Models of Passive Margin Evolution and their Implications for Theories of Long-term Landscape Development

**ALAN R. GILCHRIST**
*Department of Oceanography, Dalhousie University, Halifax, Canada*

and

**MICHAEL A. SUMMERFIELD**
*Department of Geography, University of Edinburgh, UK*

## ABSTRACT

A wide range of numerical models exist which attempt to account for the primary structural, stratigraphic and morphological characteristics of passive margins. A number of these models predict significant surface uplift along the margins of newly developing rifts as a consequence of thermal and mechanical effects. Such rift flank uplift is confined to the zone immediately adjacent to the rift and its amplitude decreases with time due to sediment loading within the rift and the decay of thermal anomalies associated with lithospheric extension.

Marginal upwarps are, however, also found on some mature rifted margins (>60 Ma) where their axes are typically located >100 km inland of the hinge zone. These features cannot be accounted for by the thermal or dynamic effects of the rifting event itself but are predicted as a consequence of the flexural isostatic response of the lithosphere to the differential unloading promoted by escarpment retreat.

These flexural effects in combination with long-term uplift, generated by the underplating associated with mantle plumes, play a key role in influencing the evolution of continental landscapes through the control they exert on regional base levels and the development of major drainage systems. Moreover, the understanding of the tectonic background of landscape evolution afforded by recent tectonic models provides, in combination with thermochronologic and offshore sediment volume data, clear constraints on long-term rates of uplift and denudation in catchment areas draining to passive margins.

## INTRODUCTION

Geomorphology experienced a marked shift during the 1950s and 1960s from an overriding concern with regional-scale, long-term landscape change to a strong emphasis on small-scale, short-term studies of surface processes. Many reasons have been suggested for the eclipse of the Davisian programme of constructing landscape histories from the

*Process Models and Theoretical Geomorphology.* Edited by M. J. Kirkby
©1994 John Wiley & Sons Ltd

identification of surviving remnants of denudational episodes set in the context of a simple, but flexible, model of long-term landscape evolution. Perhaps the most frequently cited has been the neglect of the detailed operation of surface processes by Davis, King and other practitioners of the evolutionary approach. But an equally problematic aspect of classical landscape evolution models was their lack of any coherent tectonic framework. Although made largely for pedagogic reasons, Davis's generally applied simplifying assumption that a cycle of erosion would be initiated by a rapid episode of surface uplift subsequently acquired almost the status of an axiom of landscape evolution.

Although both the tectonic models and quantitative data on rates of crustal deformation necessary for an adequate understanding of landscape change are now becoming available, remarkably little interest has been shown by most geomorphologists in their incorporation into models of long-term landscape development. Similarly, geomorphologists have been slow to appreciate the implications of the thermochronologic techniques developed and refined over the past decade which, in conjunction with the wealth of data becoming available on offshore sediment sequences, are at last providing critical information on regional-scale rates of denudation over the time spans of millions of years relevant to an understanding of long-term landscape evolution (Brown *et al.*, this volume, Chapter 2). There is now the opportunity of combining such data with morphological observations and geomorphological modelling, and of incorporating these components into tectonic models. Moreover, landscape analysis can itself potentially yield valuable constraints on the range of tectonic models currently being proposed by providing critical data on the location and timing of episodes of surface elevation change and denudation.

Although the most obvious topographic expression of tectonic processes is found at plate boundaries, the proportion of the continents directly affected by interplate tectonics is relatively small. A much larger area is located in intraplate settings where the role of tectonics may be less immediately apparent, but is certainly no less significant. Here our primary aim is to assess the wide range of models currently being proposed to describe the tectonic evolution of passive continental margins in relation to the mode of morphological development that they predict. We also briefly assess some implications of such models for theories of long-term landscape development. More specifically, we focus on the problem of explaining the morphology of high-elevation rifted margins, and on the significance of the generation and persistence of high terrain along passive margins for landscape evolution over the adjacent continental interior. In considering this problem the key questions that need to be addressed include the timing and magnitude of surface uplift, the location of the axis of maximum surface uplift relative to the rift hinge, the lateral extent of the uplifted zone, the degree and timing of the denudational modification of tectonically created topography, and the nature of the isostatic response of the lithosphere to denudation.

## PASSIVE MARGIN MORPHOLOGY

Passive margins exhibit a wide range of morphologies (Figure 3.1). Nevertheless, a distinction can be drawn between high-elevation and low-elevation margins (Gilchrist & Summerfield, 1990). High-elevation passive margins, such as those of eastern Brazil,

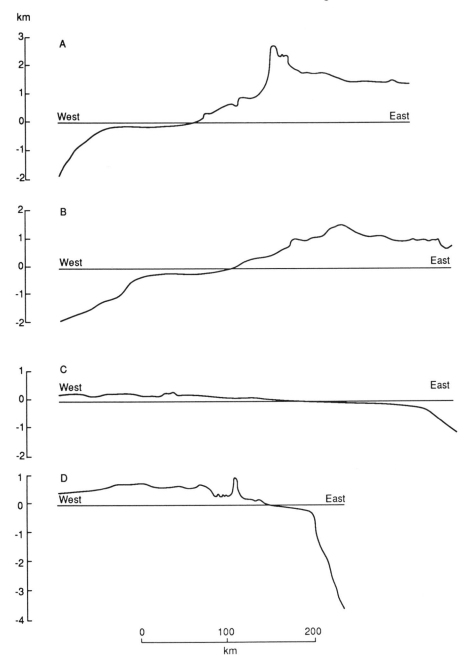

**Figure 3.1**  Typical cross-sectional forms of passive continental margins: (A) Arabian margin of the Red Sea at latitude 18°N; (B) high-elevation southwest African (Namibia) margin at latitude 27°S; (C) conjugate low-elevation eastern South American (Uruguay) margin at latitude 33°S; and (D) eastern Australian margin at latitude 30°S. Profile irregularities are due primarily to major margin-parallel drainage incision. Profiles constructed from 1:1M Operational Navigation Charts and 1:10M General Bathymetric Chart of the Oceans

southwestern Africa and the flanks of the Red Sea, are characterized by a generally narrow continental shelf and a dissected coastal zone of significant relief which is separated from an interior plateau by an escarpment (Great Escarpment), or series of escarpments. The height of the interior plateau forming the hinterland of such margins often exceeds 500 m, but there is commonly a marked increase in elevation in the vicinity of the main escarpment which creates a distinct marginal upwarp. Where this elevated zone is immediately adjacent to the rift it is usually termed a rift-flank uplift. In low-elevation passive margins, such as those of southern Australia and Argentina, a generally broad continental shelf leads to a coastal plain which rises gradually to a relatively low-lying interior surface. Between these two end members there is a wide range of passive margin morphology. In particular, pre-rifting structures may play a dominant role in the morphological development of a margin, as exemplified by the influence of Appalachian structures on the landscape of the eastern seaboard of North America. Another feature of considerable interest is the topographic asymmetry between previously conjugate margins (Etheridge *et al.*, 1989; Lister *et al.*, 1991). A clear example is provided by the South Atlantic margins of Africa and South America. Here high topography does not lie symmetrically astride the site of continental break-up but switches from one side to the other (Figure 3.1).

There already exists a substantial geomorphological literature proposing a range of chronological schemes for, and interpretations of, the morphological development of particular passive margins (see, for example, De Swardt & Bennet, 1974; Partridge & Maud, 1987, 1988; Summerfield, 1985; Rust & Summerfield, 1990, on southern Africa; Gardner & Sevon, 1989, on the eastern margin of the USA; and the review by Bishop, 1988, on eastern Australia). Indeed, contrasts in the chronologies that have been presented indicate that identification of the denudational history of passive margins largely on the basis of conventional geomorphological evidence is problematic. Although it is not possible to review this literature here, it is apparent that previous work by geomorphologists has generally been deficient in incorporating geophysical data and concepts. Where attempts have been made to explore the role of geophysical factors, such as the isostatic response to denudation considered by King (1955) and Pugh (1955), these have, in some cases at least, been beset by misunderstanding and confusion (Gilchrist & Summerfield, 1991). However, it is also appropriate to note that much previous modelling of the tectonic evolution of passive margins by geophysicists has ignored or misinterpreted morphological evidence of potential relevance to the evaluation of their models.

## TECTONIC MODELS OF RIFTED MARGIN EVOLUTION

The past decade or so has seen a rapid growth in the quantitative modelling of the tectonic evolution of rifted margins (see Allen & Allen (1990) and Keen & Beaumont (1990) for reviews). Both because of the information on margin subsidence provided by offshore sedimentary sequences, and because of the hydrocarbon potential of some passive margins, such modelling initially focused on the prediction of spatial and temporal patterns of post-rift sediment accumulation. Only more recently has the form of the onshore topography been seriously taken into account and regarded as an important component that needs to be explained by rifting models.

The range of mechanisms that are thought to play a role in the tectonic evolution of passive margins is now sufficiently great that a simple typology of processes is not possible. Although a distinction can be made between active rifting processes (in which continental break-up is initiated through the effects of sub-lithospheric thermal anomalies, or mantle plumes), and passive rifting mechanisms (in which rifting is regarded as a response to stretching of the lithosphere (Şengör & Burke, 1978)), it is probably more realistic to consider that both processes may be relevant in a single rifting event. The differences between these two types of rifting are, nevertheless, critical in their morphological effects. Active rifting models predict surface uplift occurring prior to rifting, whereas in passive rifting models any surface uplift will occur after rifting has begun (Keen, 1985). Although the relative timing of rifting and surface uplift is not always apparent from studies of particular continental margins, in specific cases, such as the northern Red Sea region, there does seem to be convincing evidence from the presence of elevated marine sediments deposited just prior to the time of rifting for rift-flank uplift occurring immediately following the rifting event (Steckler, 1985; Garfunkel, 1988).

In active rifting the lithosphere can be thinned from below as a result of increased heat flow at the base of the lithosphere due to the presence of a mantle plume, and by the physical removal of material from the base of the lithosphere by convective flow. In addition, deformation of the lithosphere can result from the dynamic effects of convective flow in the asthenosphere. Although the theoretical magnitude of length scales of elevation changes to be expected as a result of active rifting are not well constrained, surface uplifts of several kilometres are probably possible (Keen & Beaumont, 1990), and plume-related domes may be up to 2000 km across (White & McKenzie, 1989).

Since conductive re-heating of the lithosphere is too slow to explain the growth of mid-plate swells on fast-moving oceanic lithosphere, penetrative magmatism has been favoured as a means of rapid heat transfer (Crough, 1983; Fleitout *et al.*, 1986). Although the association of volcanism with some domal uplifts on continental crust, such as those in East Africa, provides direct evidence of penetrative magmatism, other domal uplifts lack evidence of volcanic activity coincident with surface uplift. Both penetrative magmatism and the dynamic effects of convective flow could generate a geologically very rapid increase in surface elevation. However, in such cases the ultimate constraint on the rate of surface uplift will be the velocity of the continental lithosphere with respect to the underlying mantle plume since this will determine the rate at which a particular part of the lithosphere is progressively exposed to the thermal and dynamic effects of convective upwelling. The rapid movement of the Pacific Plate over the vigorous Hawaiian hot spot provides a reasonable upper estimate of the rate of surface uplift to be expected and this has been calculated to be about $200 \, m \, Ma^{-1}$ (Crough, 1983).

In the initial attempt to model quantitatively passive rifting, McKenzie (1978) assumed instantaneous, uniform stretching of the lithosphere. Surface elevation changes during extension in this model result simply from the isostatic response to lithospheric thinning (expressed as an extension factor $\beta$), and the replacement, within the stretched region, of cool, lower lithosphere by hot asthenosphere, and of crust by denser lithospheric mantle (Figure 3.2). For probable initial ratios of crustal to lithospheric thickness, however, subsidence prevails from the beginning of extension and so this model cannot account for the development of marginal upwarps (Figure 3.3A). Modelling extension more realistically as occurring over a period of time comparable to the thermal time

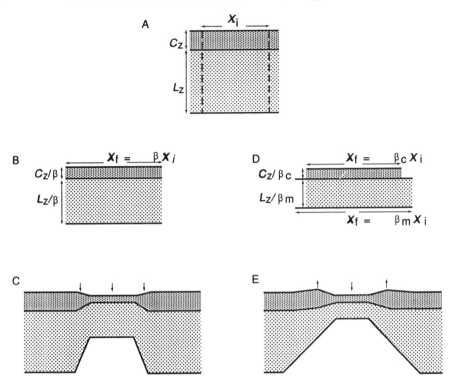

**Figure 3.2** Two-dimensional extension models: (A) initial state of the lithosphere, where $C_z$ is the thickness of the crust, $L_z$ is the thickness of the sub-crustal lithosphere, and $X_i$ is the initial horizontal length; (B) uniform extension of initial state (A) by $\beta = X_f/X_i$, where $X_f$ is the final horizontal length; (C) geometry immediately after instantaneous uniform stretching; (D) discontinuous depth-dependent extension where the stretching factors for the crust ($\beta_c$) and the sub-crustal lithosphere ($\beta_m$) are independent, but uniform; (E) geometry immediately after discontinuous depth-dependent extension. Arrows indicate vertical displacement vectors for the crust in (C) and (E). (Based on Rowley & Sahagian, 1986, Fig. 1, p. 32. Reproduced by permission)

constant of the lithosphere simply increases the amount of initial subsidence because of the loss of heat through conductive cooling (Jarvis & McKenzie, 1980).

Rift-flank surface uplift can, however, be generated if it is assumed that stretching involves a decoupling within the lithosphere so that the amount of extension varies with depth. Usually this decoupling is assumed to be located at the Moho, with the crust thinning proportionally less than the sub-crustal lithosphere (Figure 3.2). If the sub-crustal lithosphere can be stretched without extending the overlying crust then the replacement of asthenosphere for lithosphere at depth can generate surface uplift of up to 1.5 km (Royden & Keen, 1980) (Figure 3.3B). Although such depth-dependent extension results in a space problem, solutions to this difficulty have been proposed (Rowley & Sahagian, 1986; White & McKenzie, 1988). The relative amounts of extension assumed in two-layer models are usually selected in order to generate reasonable amounts of rift-flank uplift, although such models also provide a first order approximation of the variations with depth in the rheological properties of the lithosphere. Standard stretching models assume local (Airy) isostasy, although there is good evidence that the

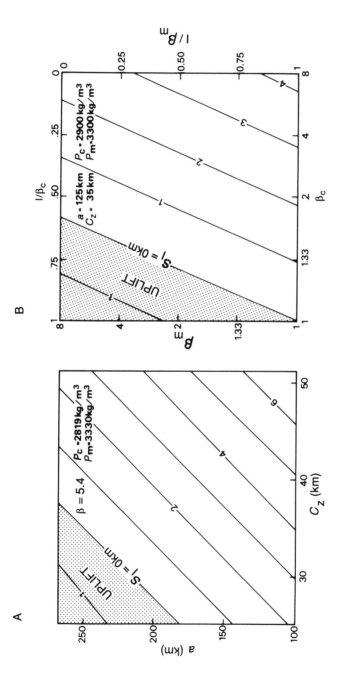

**Figure 3.3** Sensitivity of uplift and subsidence to variations in stretching factors, and initial crustal and lithospheric thickness, for the uniform extension and depth-dependent extension models. (A) shows the sensitivity of the initial subsidence/uplift ($S_i$) to changes in lithospheric ($a$) and crustal ($C_z$) thickness for instantenous uniform extension. (B) illustrates the initial subsidence/uplift for the depth-dependent extension model as a function of the degree of crustal ($\beta_c$) and sub-crustal lithosphere ($\beta_m$) stretching. (Based on Keen & Beaumont, 1990, Fig. 9, p. 422. Reproduced by permission of Supply and Services, Canada)

lithosphere, at least inland of the immediate vicinity of rifting, has significant strength and exhibits a flexural response to loads (Ebinger *et al.* 1989; Zuber *et al.*, 1989; Bechtel *et al.*, 1990). Application of flexural isostasy improves the prediction of the distribution of surface elevation changes across passive margins in response to changes in mass distribution, but the actual variations in flexural rigidity, both spatially and temporally, are poorly constrained for most margins. A secondary effect which could potentially generate rift-flank uplift is lateral flow of heat from the warm, extended lithosphere of the rift zone to the cool, unstretched lithosphere of the adjacent continent. Such lateral heat flow and the associated surface uplift would be at a maximum for narrow rifts across which the extension factor $\beta$ changes abruptly. If flexural isostatic effects are included, however, the predicted magnitude of uplift is significantly reduced.

In these passive rifting models the thermal and isostatic consequences of lithospheric extension, and the resulting topographic response, are approximately symmetrical across conjugate margins. By contrast, in asymmetric rifting models an upper-plate margin consisting of crust above a hypothesized deep detachment lies adjacent to a lower-plate margin comprising faulted upper-plate remnants overlying the footwall of the detachment (Lister *et al.*, 1986; Etheridge *et al.*, 1989). On upper-plate margins one of the factors contributing to rift-flank surface uplift is the rise of the mantle geotherm during rifting which induces positive buoyancy through thermal expansion and a reduction in density (Lister & Etheridge, 1989; Lister *et al.*, 1991). The apparent surface uplift evident on some lower-plate margins is also thought to be partly due to thermal effects.

The kinematic models described above make assumptions about the way in which deformation is distributed throughout the lithosphere during extension. Dynamical models of passive rifting, on the other hand, assume that the lithosphere exhibits certain rheological properties and determine how it deforms in response to extensional far-field forces (Braun & Beaumont, 1989a,b). Changes over time in the thermal state and surface elevation of the lithosphere are therefore determined without making arbitrary assumptions about the way the lithosphere responds to stretching. It is thought that the lithosphere has a layered rheology with strong layers resistant to deformation. Necking of the lithosphere during extension is focused by these layers and, depending upon their depth, they control the flexural state of the lithosphere (Braun & Beaumont, 1989a). The depth of strong layers is crucial in determining the amplitude of rift-flank uplifts. Although there are disadvantages in these approaches, most notably our poor knowledge of the rheology of the lithosphere, dynamical models have revealed important mechanisms, capable of generating significant surface uplift, which are not included in the kinematic models (Keen & Beaumont, 1990).

In simple terms the rheological properties of the lithosphere can be represented in terms of an upper 20 to 40 km thick brittle layer located within the crust and a hotter, underlying region in the mantle which deforms by ductile creep. Where the viscosity of this ductile zone is relatively low, the lateral temperature gradients created by rifting and lithospheric extension are apparently capable of generating significant flow in the mantle (Keen, 1985; Steckler, 1985; Buck, 1986). This secondary convection, which advects heat into rift flanks in addition to that arising from lithospheric stretching alone, is not predicted by kinematic models. The resulting thermal uplift is capable of increasing rift-flank elevation by up to around 1000 m and this mechanism has been suggested to explain the post-rift surface uplift of the flanks of the Gulf of Suez (Figure 3.4) (Steckler, 1985). An important consequence of secondary convection is a delay to post-rift

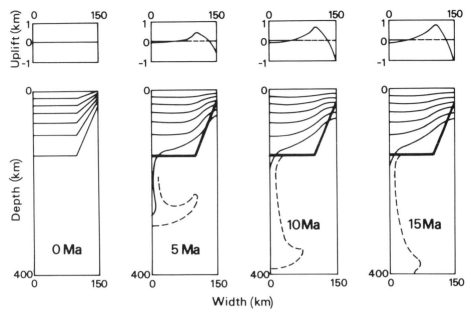

Width (km)

**Figure 3.4** Modelling of the development of secondary convection beneath a rift. The upper boxes illustrate the change in topography resulting from conduction and convection (with no allowance for sub-aerial denudation). The lower boxes show half of a rift which is assumed to be symmetrical at 0, 5, 10 and 15 Ma after rifting. The fluid in the boxes has a temperature- and pressure-dependent olivine rheology and the isotherms are plotted at 200 °C intervals. The 1250 °C isotherm (dashed line) indicates the convective flow and its initial position is shown as the thicker line in the 5–15 Ma plots. (From Buck, 1986, Fig. 5, p. 368. Reproduced by permission)

subsidence arising from cooling of the rift flanks. Near the edge of the rift zone surface uplift will continue after rifting for at least 15 Ma. As with active rifting, the dynamic effects of flow in the mantle caused by secondary convection can also contribute to surface uplift along rift flanks. It is difficult to provide good estimates of the amount of uplift that might be generated but such dynamic uplift may override the initial thermal subsidence of the rift margins and promote net uplift (Keen & Beaumont, 1990).

Whereas surface uplift generated as a result of thermal effects or by the dynamic effects of convective flow will decay with time, those arising from isostatic adjustments to deformation during rifting can create so-called permanent uplift. For instance, it has been suggested that the increase in flexural rigidity with time associated with the cooling of oceanic lithosphere after a thermal event can contribute to the locking in of an originally thermally induced surface uplift (Karner, 1985). However, even if such thermo-mechanical hysteresis is also applicable to the cooling of rifted continental lithosphere, it appears to be capable of accounting for only modest amounts of surface uplift at passive margins. The extreme example is represented by sheared margins where thermal effects and their morphological expression are maximized in comparison with rifted margins with their lower $\beta$ factors. Here it can be shown that, assuming equivalent behaviour of continental and oceanic lithosphere, surface uplift generated by thermo-mechanical hysteresis is limited, after thermal equilibration, to a maximum of less

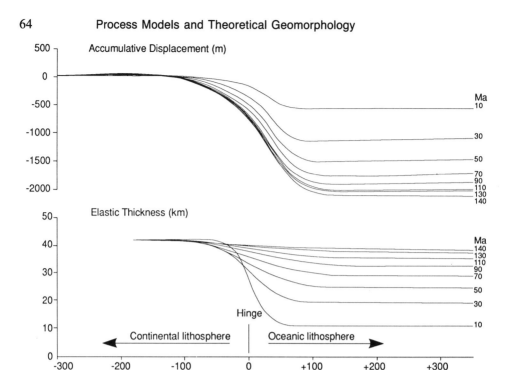

**Figure 3.5** Passive margin morphology predicted from the thermal component of a sheared margin model. The lithosphere is assumed to be 120 km thick and heat transfer has been modelled conductively in two dimensions. The initial temperature structure has been calculated on the basis of a constant geothermal gradient and it is assumed that no thinning of the continental lithosphere has occurred ($\beta = 1$) and that new oceanic lithosphere has been formed adjacent to it ($\beta = \infty$). The isostatic response of the lithosphere is modelled flexurally, the $T_e$ being defined by the 450 °C isotherm. Crustal thickness is 36 km and crustal and mantle densities are, respectively, 2800 kg m$^{-3}$ and 3300 kg m$^{-3}$. The upper boundary of the model is fixed at 0 °C and the lower boundary at 1333 °C. Lateral boundaries are reflective

than 100 m located approximately 200 km inland of the continent–ocean boundary (Figure 3.5).

Permanent surface uplift of rift flanks can also be generated through the isostatic compensation of unloading events for a lithosphere of finite strength. For example, displacement along major normal faults during rifting causes a redistribution of mass which unloads the footwall due to extension (Weissel & Karner, 1989). However, the wavelength of footwall uplift is low (< 100 km) and the axis of surface uplift is located adjacent to the rift hinge marking the boundary between rifted and unrifted continental lithosphere. Moreover, although this kind of model can account for significant surface uplifts of the order of kilometres, in each case the amplitude of residual surface uplift is significantly reduced by sediment loading within the rift.

More generally, marginal upwarps along mature, as opposed to young, passive margins cannot be adequately explained by the dynamic effects of rifting because the axis of maximum surface uplift along most mature margins is now located 100 km or more inland of the rift hinge. In contrast to the dynamic effects of rifting, magmatic underplating associated with lithospheric extension can account for substantial permanent surface uplift through crustal thickening which extends hundreds of kilometres inland of the rift hinge (McKenzie, 1984; Lister *et al.*, 1986; Lister & Etheridge, 1989; White & McKenzie, 1989). Although, according to Latin & White (1990), little melt can be produced in association with the detachment rifting mechanism proposed by Lister *et al.* (1986), White & McKenzie (1989) have argued that significant underplating can occur in association with symmetrical rifting at locations where lithospheric extension is coincident with a mantle plume. In these circumstances decompression melting promoted by lithospheric extension combined with anomalously hot mantle promotes surface uplift. This occurs both through the emplacement of magmatic material, which is of lower density than the underlying asthenosphere, at the base of, and into, the lower crust, and by its extrusion on to the surface to form extensive continental flood basalt provinces.

Underplating associated with mantle plumes would be expected to give rise to domal uplifts up to about 2000 km in diameter (White & McKenzie, 1989) which would be converted into rifted half-domes after continental break-up (Cox, 1989). Although providing an attractive explanation for the extensive elevated areas characteristic of the hinterlands of some high-elevation margins, this mechanism alone is unable to account for the upwarps occurring on many rifted margins which are parallel to the rift and which have a shorter wavelength (50–300 km) than that predicted for the half-dome features associated with magmatic underplating.

Rifted basins formed above extended lithosphere serve as receptacles for sediment shed from the sub-aerial part of the continental margin. Infilling of these structural depressions loads the lithosphere and causes subsidence, while denudation of the adjacent landmass causes unloading and a compensating isostatic uplift of the crust. It has been suggested that the flexural rotation due to such a denudation/sedimentation coupled system, pivoting around the rift hinge, may be a potentially important factor in causing surface uplift on passive margins (Watts, 1982; Summerfield, 1985; Thomas & Summerfield, 1987). However, given the likely flexural rigidity of lithosphere at rifted margins and the distribution of sediments and hence loading on the oceanward flank of the rift hinge, significant surface uplift (> 100 m) on the landward margin would not be anticipated on mature passive margins as a result of sediment loading alone (Gilchrist, 1991). The question remains, then, whether a flexural isostatic response to denudational unloading, as opposed to sediment loading, is capable of generating the magnitude of surface uplift observed on some mature passive margins.

The preceding discussion has demonstrated that although tectonic models of passive margin evolution provide a wide range of possible mechanisms for rift-flank uplift along young passive margins, they fail to account for marginal upwarps along mature passive margins which can be located more than 100 km inland. Thermal models fail both because the thermal effects promoting surface uplift decay with a time constant of around 60 Ma, and because significant heating of the margin is limited to the immediate vicinity of the rift and so cannot account for many upwarps which may be located more than 100 km from the rift hinge. Similarly, conventional mechanical models fail because, with the exception of crustal thickening resulting from underplating, their effects do not extend significantly inland of the rift hinge. However, underplating itself cannot account for the short wavelength topography characteristic of marginal upwarps, although it does provide a possible means of generating broad areas of high elevation along passive margins.

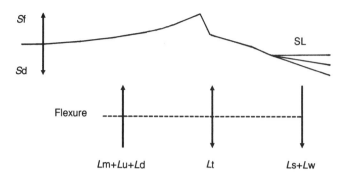

**Figure 3.6** Conceptual morphotectonic model of rifted margin evolution. $L$ represents a load comprising: m, mechanical unloading due to lithospheric extension; u, underplating; d, denudation; t, thermal effects due to lithospheric thinning and thermal anomalies; s, sediment infill of offshore basins; and w, water infill to sea level (SL). These loads are isostatically compensated regionally by lithospheric flexure. The dashed line represents a reference datum, and $S$ represents surface elevation change due to: d, denudation; and f, extrusion of flood basalts in association with underplating. The arrows indicate the direction in which each mechanism generally acts

Virtually all previous tectonic models of passive margin evolution have assumed that sub-aerial denudation has played no significant role in generating gross patterns of surface uplift. But it is becoming apparent that, on some mature high-elevation passive margins at least, the depth and spatial distribution of post-rift denudation has been such as to result in significant isostatic rebound. On the western margin of southern Africa, for instance, both offshore sediment volume estimates (Rust & Summerfield, 1990) and apatite fission track data (Brown *et al.*, 1990) indicate up to several kilometres of post-rifting denudation. Similarly, fission track data from the southeast Australian margin indicate significant depths of denudation since the formation of the Tasman Sea (Dumitru *et al.*, 1991). In situations where isostatic compensation occurs regionally through flexure of the lithosphere, rather than locally (Airy isostasy), then spatial discontinuities in denudation rates, such as would be anticipated across major escarpments, are a potential mechanism for generating surface uplift. The isostatic response to denudation is therefore a potentially important factor in explaining the generation of marginal upwarps on mature high-elevation passive margins and must be incorporated into any comprehensive tectonic model (Figure 3.6).

## MODELLING THE FLEXURAL ISOSTATIC RESPONSE TO DENUDATION ON THE SOUTHWEST AFRICAN MARGIN

Most quantitative models of landscape evolution presented over the past two decades or so have been developed primarily with respect to change at scales ranging from small drainage basins to individual slopes. Although there have been some efforts to model denudation and landscape development at a larger scale (Ahnert, 1970, 1984, 1987), there has been a lack of quantitative modelling of the kinds of coupled denudational–isostatic systems applicable to passive margin settings (Moretti & Turcotte, 1985; Lambeck & Stephenson, 1986). Such models need to address generalized patterns of denudation over areas of regional extent ($>10^5 \, \text{km}^2$) and time scales of $10^6$–$10^8$ a. Although qualitative models have been proposed predicting the episodic upwarping of passive continental margins in response to denudational unloading driven by escarpment retreat (King, 1955, Pugh, 1955), these are based on a misunderstanding of the way in which the lithosphere responds flexurally to changes in load as originally summarized by Gunn (1949) (see Gilchrist & Summerfield, 1991).

In an attempt to assess the likely mode of landscape evolution across passive margins at the large scale following supercontinent break-up we have devised a simple denudational–flexural model linking denudation and isostasy which can account for the gross form of marginal upwarps located more than 100 km inland of the rift hinge on high-elevation mature rifted margins (Gilchrist & Summerfield, 1990). The model assumes that the lithosphere has a constant finite rigidity and therefore responds flexurally to the differential unloading arising from the contrast in denudation rates between the coastal flanks of rifted margins and their continental hinterlands. The model thus has two components; a dual terrain denudational model, and a flexural isostatic model.

## Rationale for the Model

Modelling the gross geomorphic response to continental break-up and the creation of new continental margins necessitates assumptions as to the pre-rift topography and the mode of subsequent landscape modification by denudation. The topography of the supercontinents of Gondwana and Laurasia prior to rifting is at present very poorly constrained. Rifting models rarely assume any significant continental elevation prior to the initiation of lithospheric extension, but there are two reasons for believing that the central areas of these supercontinents may have lain at a significant elevation prior to break-up even in the absence of the various mechanisms potentially capable of inducing surface uplift associated with the rifting event itself.

One reason is that mean river gradients within such supercontinents would probably have been low simply because distances from source to mouth would have been significantly greater than for most cratonic drainage systems in present-day continents. Such low-gradient fluvial systems would have had a limited capacity for incision and sediment transport, so regions of relatively high elevation generated by previous crustal thickening events may well have persisted, at least at locations remote from the coast.

The resulting low rates of denudation and landscape modification would have been further retarded by a second factor – that of aridity. Prevailing aridity immediately prior to the eruption of the Karoo basalts is indicated by the widespread occurrence of aeolian deposits, such as those in the Clarens Formation (Late Triassic) in southern Africa. In addition, climatic modelling has predicted extreme continentality during the period immediately prior to break-up in the Late Permian and the Triassic for both the Laurasian and Gondwanan components of the Pangaean megacontinent (Kutzbach & Gallimore, 1989). Hot summers, cold winters and perennial, or seasonal, aridity are indicated for all areas except the eastern and tropical western coasts, and regions poleward of around 40° latitude. Although denudation rates may have been high locally, overall the Pangaean river systems with their low discharges and minimal channel gradients would have been capable of only very slow rates of landscape modification, perhaps equivalent to those apparently prevailing over much of the interior of Australia and southern Africa today. Indeed, internal drainage systems may have been widespread, as they are in parts of present-day Africa and Australia, and may have therefore simply promoted localized deposition with minimal sediment transport to the oceans.

The mean elevation of any residual regions of relatively high terrain located at sites of supercontinent break-up could have been further increased by thermal and mechanical processes related to the rifting event, such as plume-related underplating, or thermally or dynamically driven uplift. The crucial point, however, is that such mechanisms would not have been necessary to initiate a phase of rapid denudation after rifting if the pre-rifting elevations were already relatively high. The existence of a period of accelerated denudation associated with rifting has now been documented for a number of passive margins by apatite fission track data (Moore *et al.*, 1986; Gleadow & Fitzgerald, 1987; Bohannon *et al.*, 1989; Brown *et al.*, 1990, this volume, Chapter 2; Dumitru *et al.*, 1991). However, careful interpretation of these data, involving a detailed understanding of the geomorphic response to continental break-up, is necessary in order to distinguish between rifting-related surface uplift and pre-existing residual high terrain as the factor initiating such pulses of rapid denudation.

What is clear is that supercontinent break-up results in the establishment of new base levels for rivers draining smaller continental landmasses. It is likely, therefore, that two distinct denudational systems will develop. Lowered base levels in fault-bounded rift basins will lead to rapid incision of the relatively steep rift flanks and this zone will be isolated from the largely unaffected continental interior by a retreating erosional escarpment, or series of escarpments (De Swardt & Bennet, 1974; Summerfield, 1985). This simple dual terrain model is represented in Figure 3.7 and applies in situations where major pre-rifting drainage systems are unable to maintain themselves during supercontinent break-up. The prevalence of drainage orientated away from many rifted margins and the persistence of large areas of internal drainage, such as the Kalahari Basin in southern Africa, suggest that this may generally be the case (Summerfield, 1991a).

## Application of the Model to Southwestern Africa

We have applied our denuational–flexural model to the evolution of the Early Cretaceous western margin of southern Africa (Gilchrist & Summerfield, 1990). It has been calibrated from estimated variations in denudation rates derived from local relief, post-rifting offshore sediment volumes and morphological constraints provided by the present topography of the margin. Additional constraints on spatial variations in long-term denudation have been derived from apatite fission track data for the region. The use of local relief as a surrogate for prevailing denudation rates is supported by the strong empirical relationship between the two variables established by Ahnert (1970).

### Denudational Model

Process response models have generally been utilized to examine the role of transport-limited processes on single hillslopes or small catchments under constant climatic and simple topographic conditions. For the regional-scale case considered here the initial 'syn-rift' topography is not well constrained and, although poorly known, climatic variations appear to have been significant (Ward *et al.*, 1983). In addition, weathering-limited systems, which are clearly important in the present-day landscape of southern

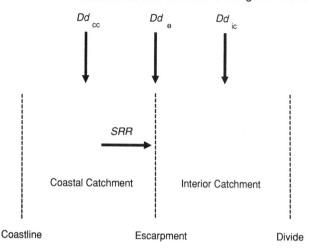

**Figure 3.7** Conceptual denudational model for high-elevation passive margins. Individual components are: $Dd_{cc}$, coastal catchment denudation rate; $Dd_e$, escarpment denudation rate; $Dd_{ic}$, interior catchment denudation rate; *SRR*, escarpment retreat rate

Africa, have not been effectively treated in process response models. Such systems are less well constrained because of poorly quantified variations in weathering rates in contrasting lithologies and under different climates. Surface processes generally operate at a scale of $10^{-1}$–$10^1$ m, and it has been suggested that for a surface process model realistically to simulate denudational processes the model resolution should be no more than $1 \times 1$ m (Parsons, 1988).

Given the difficulties of applying process response models at the temporal and spatial scales applicable to the denudational history of passive margins, empirical relationships between denudation rate and local relief provide an alternative basis for modelling the former, especially where they can be constrained by independent estimates of long-term denudation rates. The validity of such local relief estimates, however, depends on our knowledge of the way spatial variations in local relief may have varied over time. On the western margin of southern Africa there is a clear discontinuity between low local relief values in the elevated continental interior, and relatively high values oceanward of the main escarpment zone, with the main escarpment itself having very high local relief (Figure 3.8A) (Gilchrist & Summerfield, 1990; Summerfield, 1991b). It is interesting to note that a similar pattern appears to apply to other high-elevation passive margins, such as the ~95 Ma old eastern Australia margin and the ~110 Ma old eastern Brazil margin. Significantly, it also occurs along young rift-flank uplifts, such as the ~20 Ma old southern margin of the Gulf of Aden (Figure 3.8B), suggesting that a differential in local relief, and therefore denudation rates, is a feature which persists from the nascent to mature stages of passive margin evolution.

In our dual terrain model differential rates of denudation across the southwest African margin are represented as regions of high and low local relief separated by a moving boundary (a main escarpment). The differential in present mean local relief calculated from the NGDC 10-minute topographic database between the coastal zone and the interior plateau landward of the Great Escarpment is about 2.4 times. This ratio is taken to indicate an equivalent differential in mean denudation rates. It is important to note,

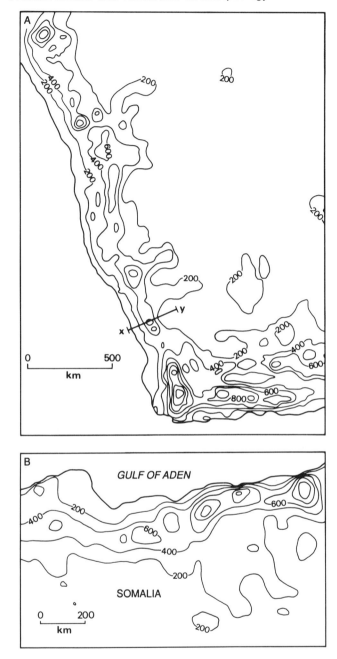

**Figure 3.8**   Local relief derived from NGDC 10-minute digital data (approximately 18 × 18 km areas) for southwestern Africa (A) and the southern periphery of the Gulf of Aden (B). The contour interval is 200 m. The modelled profile in Figure 3.9 is marked by the line X–Y in (A)

therefore, that we are using mean local relief as an estimator of *relative differences* in denudation rates, rather than as a basis for their direct estimation.

That such a differential in denudation rates has probably been essentially sustained since rifting is supported by evidence indicating significantly greater depths of post-rifting denudation near the coast. Offshore sediment volume data for the Cape Basin show that the mean depth of post-rifting denudation for all Atlantic-draining catchments in southern Africa is around 1800 m (Rust & Summerfield, 1990). This estimate may be an underestimate as it takes no account of sediment lost from the Cape Basin, or of material exported from within the present area of Atlantic-draining catchments to the eastern margin of southern Africa. Nevertheless, this figure is well below the upwards of 2.5 km of post-rift denudation suggested by apatite fission track data for the western coastal zone (Brown *et al.*, 1990). In addition, geological evidence further confirms greater depths of post-rifting denudation in the coastal zone through the progressive exposure of older Karoo sediments and basement rocks towards the Atlantic coastline.

The volume of offshore sediment accumulated since the formation of the Cape Basin around 150 Ma BP, combined with the differential in denudation rates between coastal and interior catchments, provide estimates of mean long-term denudation rates of 16.5 and $6.9 \, \text{m} \, \text{Ma}^{-1}$, respectively. Escarpment retreat is modelled as a single moving boundary retreating from the hinge line to its present position at a constant rate of $667 \, \text{m} \, \text{Ma}^{-1}$. Given the current profile form the present escarpment can be maintained by a denudation rate of $900 \, \text{m} \, \text{Ma}^{-1}$. The area of the coastal catchments has been assumed to have increased linearly over time with the retreat of the Great Escarpment from zero at the initiation of rifting to the present value of $6.86 \times 10^5 \, \text{km}^2$, whereas the total area of Atlantic-draining catchments has been assumed to have remained constant at $1.6 \times 10^6 \, \text{km}^2$. These are conservative scenarios as far as our model is concerned since there is some evidence of a significant delay to the full integration of the interior drainage of southern Africa with the Atlantic drainage (Summerfield, 1985; Rust & Summerfield, 1990). Such a delay would have enhanced the differential in denudation rates across the main escarpment, but given the uncertainties in the drainage history of southern Africa we retain the simpler assumptions.

As noted above, the precise pre-rifting topography of southern Africa is unknown and, therefore, has to be inferred. However, a crude estimate can be made by reloading the average amount of material removed from the Atlantic-draining catchments and making an appropriate allowance for isostatic compensation. Here a mean depth of 1800 m is used based on the offshore isopach data (Rust & Summerfield, 1990). Because of the extensive nature of this assumed loading event an Airy calculation suffices. The mean elevation of the present-day Atlantic-draining catchments, as determined from 10-minute digital data, is approximately 930 m. The pre-rifting elevation will therefore have been higher by $h$, where $h = d(\rho_m - \rho_c)/\rho_m$. Employing a crustal density $\rho_c$ of $2800 \, \text{kg} \, \text{m}^{-3}$ and a mantle density $\rho_m$ of $3300 \, \text{kg} \, \text{m}^{-3}$, the mean pre-rifting elevation of the Atlantic-draining area of southern Africa is estimated to have been about 1200 m. This estimate, of course, assumes no crustal deformation in addition to isostatic adjustments due to denudation or reloading.

The figure of 1200 m is used in the model as the reference elevation of the pre-rifting landsurface. A constant elevation is assumed with the initial topography represented as a block terminating at the rift hinge. It is appreciated that this approximation is highly simplistic, but in the absence of data on the detailed pre-rift morphology of the region it is employed as a first order approximation. However, absolute elevation does not modify the denudational model unless base level is attained, and in the case study presented here this occurs only in a small region for the best modelled fit. In general, the results of the model quantifying the isostatic response to denudation, and its effect on landscape development, are more important than the detailed matching of modelled and observed topographic profiles.

### Flexural Model

Inclusion of the isostatic response to denudation is crucial to the analysis of macroscale landscape evolution over cyclic time scales (Gilchrist & Summerfield, 1991). Two basic variants of the flexural isostatic model exhibiting different responses to loading events have commonly been employed. In the elastic model deformation is maintained, whereas in the viscoelastic case the initial elastic response is modified by viscous relaxation. In addition, complex layered rheological models have been implemented to explore the depth dependency of deformation. However, in this study, where we wish to explore the basic regional isostatic response to differential denudation, the elastic plate model suffices.

The isostatic response to denudational unloading has been modelled (Gilchrist & Summerfield, 1990) assuming a thin elastic plate of constant thickness overlying an inviscid fluid according to the equation

$$\frac{\partial^2}{\partial x^2}\left(D\frac{\partial^2 w}{\partial x^2}\right) + (\rho_m - \rho_i)gw = l(x)$$

where $w$ is the vertical flexure, $D$ is the flexural rigidity, $\rho_m$ is the mantle density, $\rho_i$ is the density of the flexure infill (in this case air), $g$ is the acceleration due to gravity, and $l(x)$ is the denudational load as a function of $x$ (Nadai, 1963). The flexural rigidity is given by

$$D = \frac{ET_e^3}{12(1 - \nu^2)}$$

where $E$ is Young's modulus, $T_e$ is the effective elastic thickness of the lithosphere, and $\nu$ is Poisson's ratio. Vertical flexure has been calculated in the wavenumber domain from

$$W(k) = R(k) \cdot L(k)$$

where $R(k)$ is the isostatic response function so that

$$R(k) = \frac{1}{(\rho_m - \rho_i)g + Dk^4}$$

where $W(k)$ and $L(k)$ are the Fourier transforms of $w(x)$ and $l(x)$ and $k$ is wavenumber given by $k = 2\pi/\lambda$ where $\lambda$ is the wavelength. This isostatic model acts as a low-pass filter to an applied load, so the spatial resolution of the denudational model is not critical in determining the isostatic response to denudation provided that deviations from it, which have no doubt occurred, have been relatively minor.

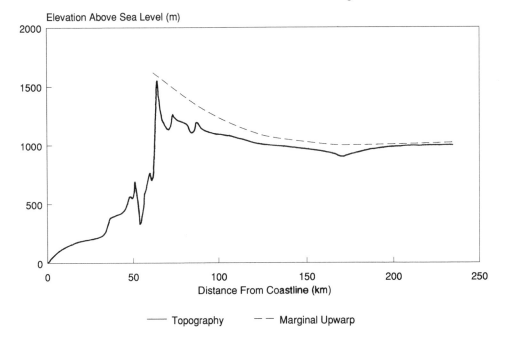

**Figure 3.9** Typical profile for the western margin of southern Africa, a high-elevation rifted margin. The location of the profile is shown in Figure 3.8A

## Denudational-flexural model results

The model has been applied to a representative topographic profile across the western margin of southern Africa (Figure 3.9). In Figure 3.10 this profile is compared with predicted topographic profiles arising from the flexural isostatic response to the modelled pattern of denudation through time employing different values for the effective elastic thickness of the lithosphere ($T_e$) which is treated as a free parameter. Varying the $T_e$ does not appreciably affect the amplitude of the marginal upwarp generated, but does change its wavelength. In addition, lower values of $T_e$ tend to promote surface uplift in the coastal catchment region. The best fit across the topographic profile is achieved with a $T_e$ of 16.5 km.

The accumulative denudation across the profile is illustrated in Figure 3.11. This shows that the greatest denudation occurs in the coastal region and declines sharply inland of the retreating escarpment to a constant value. The accumulative flexural uplift (Figure 3.12) indicates that the axis of maximum uplift develops progressively and the marginal upwarp migrates inland as the escarpment retreats (Figure 3.13). This is a significant result as the progressive uplift of the margin does not accord with a recent synthesis of the landscape history of southern Africa which envisages episodic surface uplift, possibly in lagged response to denudation unloading (Partridge & Maud, 1987). Our quantitative model is apparently able to account for the presence of marginal upwarps on mature passive margins. None the less, the effect of variations in both the isostatic and denudational parameters need to be evaluated in order to assess the sensitivity of the model.

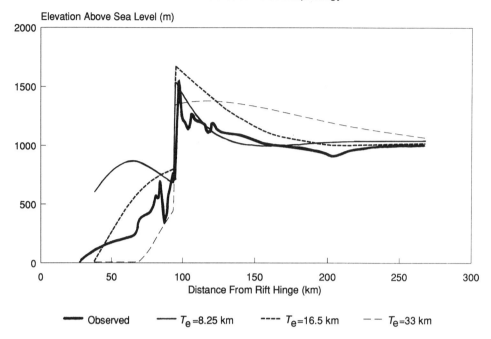

**Figure 3.10**   Calculated topographic profiles for various $T_e$ values compared with observed topography (illustrated in Figure 3.9). Flexural model parameters are $E = 1 \times 10^{11}\,\mathrm{N\,m^{-2}}$, $\nu = 0.25$, $g = 9.81\,\mathrm{m\,s^{-2}}$, $\rho_m = 3300\,\mathrm{kg\,m^{-3}}$, and $\rho_i = 0\,\mathrm{kg\,m^{-3}}$

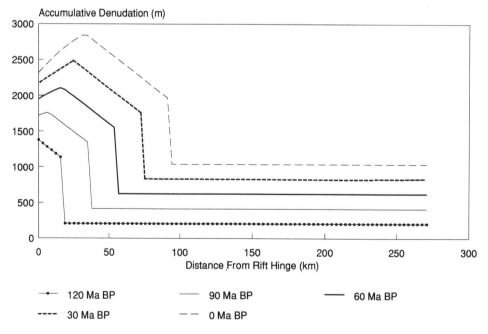

**Figure 3.11**   Accumulative denudation through time for the $T_e = 16.5\,\mathrm{km}$ modelled profile in Figure 3.10

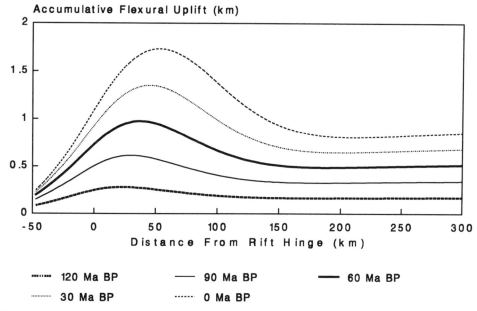

Figure 3.12  Accumulative flexural uplift through time for the $T_e = 16.5$ km modelled profile in Figure 3.10

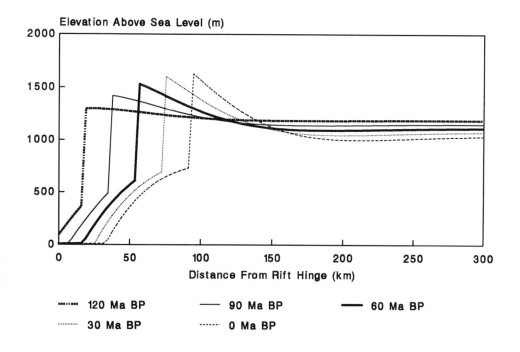

Figure 3.13  Change through time in the $T_e = 16.5$ km modelled profile in Figure 3.10

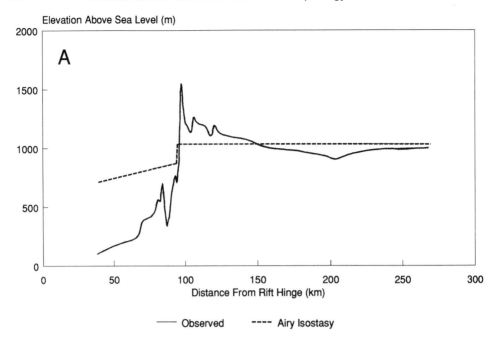

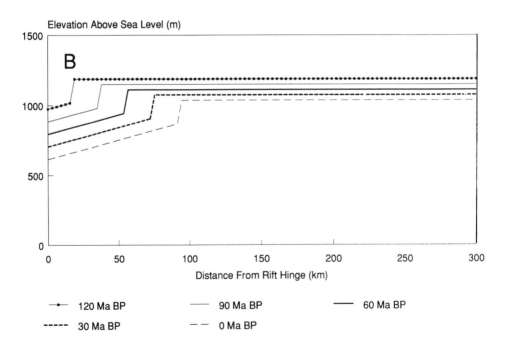

**Figure 3.14** Modelled profile for $T_e = 0$ km (Airy case) compared with observed profile in Figure 3.10 (other parameters the same as given in caption to Figure 3.10) (A); change of profile through time (B)

The pattern of morphological change using a range of flexural rigidities has already been discussed (Figure 3.10). In order to evaluate the end-member states of the isostatic system the Airy and the infinite rigidity cases have been assessed. In the Airy case the lithosphere is assumed to exhibit a lateral rigidity of zero (Figure 3.14), whereas in the infinite rigidity case there is no isostatic response to denudation (Figure 3.15). In both cases denudational unloading does not generate a marginal upwarp. The inclusion of flexural isostasy is therefore a necessary component of the modelling of the isostatic response to denudation across passive margins in order to account for marginal upwarps.

In our dual terrain model it is assumed that the coastal catchment denudation rate is 2.4 times that of the interior catchment and that this differential is maintained throughout the evolution of the margin. In order to test the sensitivity of the model to the differential between denudation rates in these two terrains the alternative assumption that the coastal catchment is graded to sea level throughout the evolution of the model can be made. In this case the amplitude of the marginal upwarp is augmented, although the wavelength is changed only slightly if the same isostatic model parameters are employed (Figure 3.16). However, the increase in amplitude is less than twice that generated in the original, preferred model ($T_e = 16.5$ km) (Figure 3.10) with respect to the elevation of the continental interior.

Modelling of the topography of the western margin of southern Africa demonstrates that the creation of marginal upwarps on mature passive margins can occur as a result of the linkage between differential denudation and flexural isostasy. Moreover, as demonstrated by the above analysis the amplitude and wavelength of such marginal upwarps are relatively insensitive to a range of values for the effective elastic thickness of the lithosphere and the denudation rate differential. Nevertheless, it is important to emphasize that this modelling of the topographic evolution of the western margin of southern Africa represents a drastic simplification of its probable development since continental break-up. For instance, progressive exposure towards the west coast of lower Karoo and basement rocks along this margin strongly suggests the removal of a considerable thickness of section both oceanward, and inland, of the present position of the Great Escarpment. This is supported by apatite fission track data which indicates around 2.5 km of denudation extending some distance inland of this major topographic discontinuity (Brown et al., 1990). These data also indicate that much of this denudation occurred during a relatively brief interlude during the Early Cretaceous, that is roughly coincident with the initiation of rifting, so our model probably misrepresents to some degree the early morphological development of the margin. This may well have involved the rapid retreat of escarpments formed in upper Karoo strata which, since they lacked the high density of resistant dolerite intrusions characteristic of lower Karoo units, would have had a higher erodibility. Development of the margin since the Late Cretaceous is therefore reasonably approximated by a single escarpment model, although the rate of escarpment retreat has probably not been constant, as has been assumed here.

Our primary aim in developing the denudational–flexural model discussed above is not to replicate in detail the morphological evolution of a particular margin, but to demonstrate the inevitable effects of flexural isostasy in landscapes characterized by a sustained differential in denudation across a major topographic discontinuity. The relevance of the mechanism has now also been demonstrated with respect to the southeast Australian highlands where, in an environment characterized by very low rates of denudation, flexural isostatic rebound to denudational unloading has

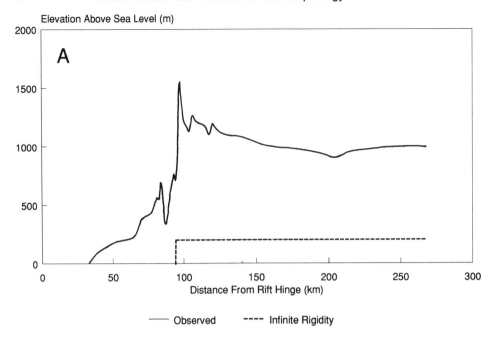

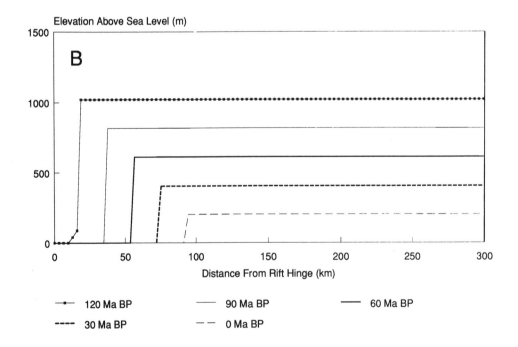

**Figure 3.15** Modelled profile for $T_e = \infty$ km (infinite rigidity case) compared with observed profile in Figure 3.10 (other parameters the same as given in caption to Figure 3.10) (A); change of profile through time (B)

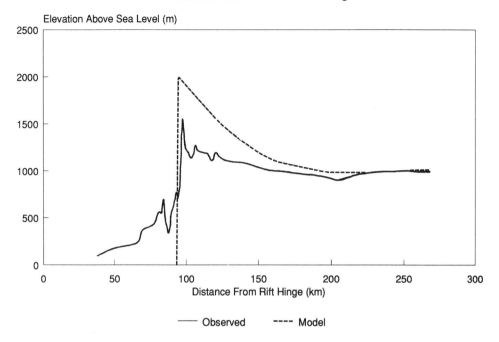

**Figure 3.16**   Modelled profile for $T_e = 16.5$ km when $Dd_{cc} \gg Dd_{ic} = 6.9$ m Ma$^{-1}$, so that the coastal region is graded to sea level. Compare with observed profile in Figure 3.10

been recognized as an important factor in maintaining highland elevation (Bishop & Brown, 1992).

## SOME IMPLICATIONS FOR THEORIES OF LONG-TERM LANDFORM DEVELOPMENT

There is now a growing appreciation that tectonic models need to be fully integrated with surface process theories of landform change if we are to develop a comprehensive understanding of landscape evolution. Brunsden & Lin (1991) have outlined some of the factors relevant to plate boundary settings, but the assessment of passive margin tectonic models presented here indicates some important differences in the relationship between tectonics and landforms in these two contrasting tectonic environments. The tectonic evolution of passive margins can potentially influence the mode of landscape evolution directly, by engendering particular temporal and spatial patterns of surface uplift and subsidence, and indirectly through the effect that such changes in elevation might have on the post-rifting evolution of drainage patterns. It is not possible to examine these issues in depth here, but a number of likely consequences of the tectonic evolution of passive margins can be suggested.

The elevation of the landsurface prior to continental break-up is likely to be critical to the mode of post-rifting landscape evolution. If low-elevation margins are created from the time of rifting then a major episode of landscape rejuvenation would not be anticipated. However, the formation of high-elevation margins will result in elevated landsurfaces along new continental margins lying significantly above base level. As is suggested by the data from apatite fission track analysis for presently high-elevation passive margins (e.g. Brown *et al.*, 1990; Dumitru *et al.*, 1991), this creation of new base levels led to a phase of rapid denudation approximately coincident with rifting. An unresolved problem is whether high-elevation margins are formed as the result of tectonic mechanisms associated with rifting, or whether they simply represent new continental margins formed at locations within a supercontinent with a high residual elevation. In regions for which there is no clear evidence of the presence of a mantle plume at the time of rifting, a residual high elevation may be the most likely explanation.

Irrespective of the mechanism generating high-elevation passive margins, once formed they experience significant denudational modification. For instance, apatite fission track data for southwestern Africa, which reveal a considerable depth of post-rifting denudation both oceanward and inland of the current position of the Great Escarpment, demonstrate that this feature does not simply represent the present manifestation of an escarpment initiated at the time of rifting, as suggested by Ollier & Marker (1985). However, since much of the denudation close to the newly formed Atlantic Ocean, apparently involving the stripping of much of the Karoo sequence, appears to have occurred soon after rifting, the present topographic discontinuity represented by the Great Escarpment in southern Africa is undoubtedly an ancient feature. If this is the case then the flexural upwarp resulting from isostatic compensation of this feature has also probably existed for much of the history of the margin.

Even in the absence of a thermally induced upwarp associated with the rifting event itself, the generation of this marginal bulge, which seems to be characteristic of high-elevation passive margins (Ollier, 1985; Summerfield, 1985, 1991a) is likely to have had a significant impact on drainage evolution. The idea proposed by De Swardt & Bennet (1974) of a dual drainage system in southern Africa as a result of the separation of coastal and inland drainage by a marginal upwarp may also be applicable, at least to some extent, to other passive margins. The development of two distinct denudational systems, one of which is isolated from base-level changes at the continental margin, has profound implications for the way in which continental-scale landscape evolution will occur. It also has critical repercussions for the temporal and spatial pattern of sediment delivery to passive margins, a factor which is highly relevant to the hydrocarbon potential of offshore sedimentary sequences.

In contrast to tectonically highly active plate boundary settings, rates of crustal uplift in passive margin environments are modest. Even with active thermal uplift, rates are unlikely to exceed $200 \, \text{m} \, \text{Ma}^{-1}$; that is, an order of magnitude less than typical values in active orogenic belts. Most other mechanisms capable of generating surface uplift in passive margins would produce rates which are considerably lower. The Davisian notion of rapid surface uplift followed by a prolonged period of denudation is therefore quite inapplicable to passive margin settings; nevertheless, the rapid creation of new base levels which may lie some distance below existing landsurfaces provides conditions for a significant episode of landscape rejuvenation. Again, in contrast to plate boundary settings where an approximate steady state between rates of crustal uplift and denudation may be achieved at a regional scale (Brunsden & Lin, 1991; Summerfield, 1991a), landscape development on passive continental margins is time dependent. In a sense the rupturing of the megacontinent of Pangaea represents a 'singularity' in landscape history, an event which continues to leave its mark in the present landscapes of the

Gondwanan and Laurasian continents. It remains a major challenge to geomorphologists to unravel the complex interrelationships between tectonics and surface geomorphic processes evident in this history.

## SUMMARY

Our primary aim has been to evaluate a range of tectonic models of passive margin evolution in terms of their ability to account for the morphology of high-elevation margins. Thermal uplift associated with active rifting, depth-dependent extension or secondary convection can account for significant increases in elevation of up to 1000 m or more. Although probably important in generating some young rift-flank uplifts, thermal effects decay with a time constant of ~ 60 Ma and therefore such mechanisms cannot explain marginal upwarps on mature passive margins (> 100 Ma old). Isostatic adjustments to deformation, such as thermo-mechanical hysteresis, offshore sediment loading, and those associated with displacements along major normal faults, are capable of generating 'permanent' uplift. However, thermo-mechanical hysteresis and sediment loading can produce only minor amounts of uplift, while isostatic adjustments to fault displacements generate uplift close to the rift hinge which, in any case, is invariably significantly reduced by sediment loading in the adjacent rift. Moreover, since the axis of maximum uplift of marginal upwarps on mature passive margins is located upwards of 100 km inland from the rift hinge, such features cannot be related directly to the dynamic effects of rifting. Magmatic underplating can account for surface uplifts extending hundreds of kilometres inland from the rift hinge but do not provide an adequate explanation of marginal upwarps since the latter have a significantly shorter wavelength.

Given the failure of these models to explain marginal upwarps along mature passive margins, we have explored the isostatic response to denudational unloading associated with escarpment retreat on high-elevation margins. Our denudational–flexural model predicts an evolving marginal upwarp which persists into the mature passive margin state. Applying a dual terrain denudational model to the western margin of southern Africa the denudational–flexural model generates a marginal upwarp which corresponds well to a representative topographic profile across the margin. Further refinement of the model will be possible with increasing empirical data on the denudational history and lithospheric properties of a range of passive margins.

The tectonic evolution of passive margins affects the mode of long-term landscape development directly through the generation of particular spatio-temporal patterns of surface uplift and subsidence, and indirectly through the influence of crustal deformation on the evolution of drainage systems. Although there is considerable variation in both the present morphology and topographic development of different passive margins, the presence of major escarpments and well-defined marginal upwarps on many high-elevation margins suggests that pervasive mechanisms are operating which may yield general models of landscape development for these tectonic environments.

## ACKNOWLEDGEMENTS

This research was supported by the Natural Environment Research Council through grant number GR3/6693 to M.A.S. and a research studentship to A.R.G. M.A.S. is also grateful to the Royal

Society of Edinburgh for a Support Research Fellowship during the tenure of which some of the ideas presented here were developed, and to the Rockefeller Foundation at whose Bellagio Study and Conference Center a part of the paper was drafted. The paper was completed while A.R.G. held a NATO Post-Doctoral Research Fellowship.

## REFERENCES

Ahnert, F. (1970). Functional relationships between denudation, relief and uplift in large mid-latitude drainage basins. *American Journal of Science*, **268**, 243–263.

Ahnert, F. (1984). Local relief and the height limits of mountain ranges. *American Journal of Science*, **284**, 1035–1055.

Ahnert, F. (1987). Process-response models of denudation at different spatial scales. *Catena Supplement*, **10**, 31–50.

Allen, P. A. and Allen, J. R. (1990). *Basin Analysis: Principles and Applications*, Blackwell Scientific Publications, Oxford, 451 pp.

Bechtel, T. D., Forsyth, D. W., Sharpton, V. L. and Grieve, R. A. F. (1990). Variations in effective elastic thickness of the North American lithosphere. *Nature*, **343**, 636–638.

Bishop, P. (1988). The eastern highlands of Australia: the evolution of an intraplate highland belt. *Progress in Physical Geography*, **12**, 159–181.

Bishop, P. and Brown, R. (1992). Denudational isostatic rebound of an intraplate highland belt (the Southeast Australian Highlands): Field data, modelling and implications. *Earth Surface Processes and Landforms*, **17**, 345–360.

Bohannon, R. G., Naeser, C. W., Schmidt, D. L. and Zimmermann, R. A. (1989). The timing of uplift, volcanism, and rifting peripheral to the Red Sea: A case for passive rifting. *Journal of Geophysical Research*, **94**, 1683–1701.

Braun, J. and Beaumont, C. (1989a). A physical explanation of the relation between flank uplifts and the breakup unconformity at rifted continental margins. *Geology*, **17**, 760–764.

Braun, J. and Beaumont, C. (1989b). Dynamical models of the role of crustal shear zones in asymmetric continental extension. *Earth and Planetary Science Letters*, **93**, 405–423.

Brown, R. W., Rust, D. J., Summerfield, M. A., Gleadow, A. J. W. and De Wit, M. C. J. (1990). An Early Cretaceous phase of accelerated erosion on the southwestern margin of Africa: Evidence from apatite fission track analysis and the offshore sedimentary record. *Nuclear Tracks and Radiation Measurement*, **17**, 339–351.

Brunsden, D. and Lin, J-C. (1991). The concept of topographic equilibrium in neotectonic terrains. In Cosgrove, J. and Jones, M. (eds), *Neotectonics and Resources*, Belhaven Press, London, pp. 120–143.

Buck, W. R. (1986). Small-scale convection induced by passive rifting: the cause for uplift of rift shoulders. *Earth and Planetary Science Letters*, **77**, 362–372.

Cox, K. G. (1989). The role of mantle plumes in the development of continental drainage patterns. *Nature*, **274**, 47–49.

Crough, S. T. (1983). Hotspot swells. *Annual Review of Earth and Planetary Sciences*, **11**, 165–193.

De Swardt, A. M. J. and Bennet, G. (1974). Structural and physiographic development of Natal since the late Jurassic. *Transactions of the Geological Society of South Africa*, **77**, 309–322.

Dumitru, T. A., Hill, K. C., Coyle, D. A., Duddy, I. R., Foster, D. A., Gleadow, A. J. W., Green, P. F., Kohn, B. P., Laslett, G. M. and O'Sullivan, A. J. (1991). Fission track thermochronology: Application to continental rifting of south-eastern Australia. *APEA Journal*, **31**, 131–142.

Ebinger, C. J., Bechtel, T. D., Forsyth, D. W. and Bowin, C. O. (1989). Effective elastic plate thickness beneath the East African and Afar Plateaus and dynamic compensation of the uplifts. *Journal of Geophysical Research*, **94**, 2883–2901.

Etheridge, M. A., Symonds, P. A. and Lister, G. S. (1989). Application of the detachment model to reconstruction of conjugate passive margins. In Tankard, A. J. and Balkwill, H. R. (eds), *Extensional Tectonics and Stratigraphy of the North Atlantic Margins*, American Association of Petroleum Geologists Memoir 46, pp 23–40.

Fleitout, L., Froidevaux, C. and Yuen, D. (1986). Active lithospheric thinning. *Tectonophysics*, **132**, 271–278.

Gardner, T. W. and Sevon, W. D. (eds) (1989). *Appalachian Geomorphology*, Elsevier, Amsterdam, 318 pp.

Garfunkel, Z. (1988). Relation between continental rifting and uplifting: evidence from the Suez rift and northern Red Sea. *Tectonophysics*, **150**, 33–49.

Gilchrist, A. R. (1991). Morphotectonics of passive continental margins: application to south-western Africa, Unpublished Ph.D. Thesis, University of Liverpool, 232 pp.

Gilchrist, A. R. and Summerfield, M. A. (1990). Differential denudation and flexural isostasy in formation of rifted-margin upwarps. *Nature*, **346**, 739–742.

Gilchrist, A. R. and Summerfield, M. A. (1991). Denudation, isostasy and landscape evolution. *Earth Surface Processes and Landforms*, **16**, 555–562.

Gleadow, A. J. W. and Fitzgerald, P. G. (1987). Uplift history and structure of the Transantarctic Mountains: new evidence from fission track dating of basement apatites in the Dry Valleys area, southern Victoria Land. *Earth and Planetary Science Letters*, **82**, 1–14.

Gunn, R. (1949). Isostasy – extended. *Journal of Geology*, **57**, 263–279.

Jarvis, G. T. and McKenzie, D. P. (1980). Sedimentary basin formation with finite extension rates. *Earth and Planetary Science Letters*, **48**, 42–52.

Karner, G. D. (1985). Thermally induced residual topography within oceanic lithosphere. *Nature*, **318**, 527–532.

Keen, C. E. (1985). The dynamics of rifting: deformation of the lithosphere by active and passive driving forces. *Geophysical Journal of the Royal Astronomical Society*, **80**, 95–120.

Keen, C. E. and Beaumont, C. (1990). Geodynamics of rifted continental margins. In Keen, M. J. and Williams, G. L. (eds), *Geology of the Continental Margin of Eastern Canada*, Geology of Canada no. 2, Geological Survey of Canada, pp. 391–472.

King, L. C. (1955). Pediplanation and isostasy; an example from South Africa. *Quarterly Journal of the Geological Society*, **111**, 353–359.

Kutzbach, J. E. and Gallimore, R. G. (1989). Pangaean climates: Megamonsoons of the megacontinent. *Journal of Geophysical Research*, **94**, 3341–3357.

Lambeck, K. and Stephenson, R. (1986). The post-Palaeozoic uplift history of south-eastern Australia. *Australian Journal of Earth Sciences*, **33**, 253–270.

Latin, D. and White, N. (1990). Generating melt during lithospheric extension: pure shear vs. simple shear. *Geology*, **18**, 327–331.

Lister, G. S. and Etheridge, M. A. (1989). Detachment models for uplift and volcanism in the Eastern Highlands, and their application to the origin of passive margin mountains. In Johnson, R. W. (ed.), *Intraplate Volcanism in Eastern Australia and New Zealand*, Cambridge University Press, Cambridge, pp. 296–313.

Lister, G. S., Etheridge, M. A. and Symonds, P. A. (1986). Detachment faulting and the evolution of passive continental margins. *Geology*, **14**, 246–250.

Lister, G. S., Etheridge, M. A. and Symonds, P. A. (1991). Detachment models for the formation of passive continental margins. *Tectonics*, **10**, 1038–1064.

McKenzie, D. (1978). Some remarks on the development of sedimentary basins. *Earth and Planetary Science Letters*, **40**, 25–32.

McKenzie, D. (1984). A possible mechanism for epeirogenic uplift. *Nature*, **307**, 616–618.

Moore, M. E., Gleadow, A. J. W. and Lovering, J. F. (1986). Thermal evolution of rifted continental margins: new evidence from fission tracks in basement apatites from southeastern Australia. *Earth and Planetary Science Letters*, **78**, 255–270.

Moretti, I. and Turcotte, D. L. (1985). A model for erosion, sedimentation, and flexure with application to New Caledonia. *Journal of Geodynamics*, **3**, 155–168.

Nadai, A. (1963). *Theory of Flow and Fracture of Solids*, McGraw-Hill, New York.

Ollier, C. D. (1985). Morphotectonics of Passive Continental Margins. *Zeitschrift für Geomorphologie Supplement*, **54**, 1–9.

Ollier, C. D. and Marker, M. E. (1985). The Great Escarpment of southern Africa. *Zeitschrift für Geomorphologie Supplement*, **54**, 37–56.

Parsons, A. J. (1988). *Hillslope Form*, Routledge, London, 212 pp.

Partridge, T. C. and Maud, R. R. (1987). Geomorphic evolution of southern Africa since the Mesozoic. *South African Journal of Geology*, **90**, 179–208.

Partridge, T. C. and Maud, R. R. (1988). The geomorphic evolution of southern Africa: A comparative review. In Dardis, G. F. and Moon, B. P. (eds), *Geomorphological Studies in Southern Africa*, Balkema, Rotterdam, pp. 5–15.

Pugh, J. C. (1955). Isostatic readjustment in the theory of pediplanation. *Quarterly Journal of the Geological Society*, **111**, 361–369.

Rowley, D. B. and Sahagian, D. (1986). Depth-dependent stretching: a different approach. *Geology*, **14**, 32–35.

Royden, L. and Keen, C. E. (1980). Rifting processes and thermal evolution of the continental margin of eastern Canada determined from subsidence curves. *Earth and Planetary Science Letters*, **51**, 343–361.

Rust, D. J. and Summerfield, M. A. (1990). Isopach and borehole data as indicators of rifted margin evolution in southwestern Africa. *Marine and Petroleum Geology*, **7**, 277–287.

Şengör, A. H. C. and Burke, K. (1978). Relative timing of rifting and volcanism on earth and its tectonic implications. *Geophysical Research Letters*, **5**, 419–421.

Steckler, M. S. (1985). Uplift and extension at the Gulf of Suez: indications of induced mantle convection. *Nature*, **317**, 135–139.

Summerfield, M. A. (1985). Plate tectonics and landscape development on the African continent. In Morisawa, M. and Hack, J. T. (eds), *Tectonic Geomorphology*, Allen and Unwin, Boston, pp. 27–51.

Summerfield, M. A. (1991a). *Global Geomorphology*, Longman, London, and Wiley, New York, 537 pp.

Summerfield, M. A. (1991b). Sub-aerial denudation of passive margins: regional elevation versus local relief models. *Earth and Planetary Science Letters*, **102**, 460–469.

Thomas, M. F. and Summerfield, M. A. (1987). Long-term landform development: key themes and research problems. In Gardiner, V. *et al.* (eds), *International Geomorphology 1986: Proceedings of the First International Conference on Geomorphology*, Part II, Wiley, Chichester, pp. 935–956.

Ward, J. D., Seely, M. K. and Lancaster, N. (1983). On the antiquity of the Namib. *South African Journal of Science*, **79**, 175–183.

Watts, A. B. (1982). Tectonic subsidence, flexure and global changes of sea level. *Nature*, **297**, 469–474.

Weissel, J. K. and Karner, G. D. (1989). Flexural uplift of rift flanks due to mechanical unloading of the lithosphere during extension. *Journal of Geophysical Research*, **94**, 13919–13950.

White, N. and McKenzie, D. (1988). Formation of the 'steer's head' geometry of sedimentary basins by differential stretching of the crust and mantle. *Geology*, **16**, 250–253.

White, R. and McKenzie, D. (1989). Magmatism at rift zones: The generation of volcanic continental margins and flood basalts. *Journal of Geophysical Research*, **94**, 7685–7729.

Zuber, M. T., Bechtel, T. D. and Forsyth, D. W. (1989). Effective elastic thickness of the lithosphere and mechanisms of isostatic compensation in Australia. *Journal of Geophysical Research*, **94**, 9353–9367.

# 4  A Quantitative Expression for External Forces

**KENJI KASHIWAYA**
*Department of Earth Sciences, Kanazawa University, Japan*

## ABSTRACT

Mathematical considerations of models of erosional processes of a mountain and a drainage basin reveal that the resulting sediment discharge is related to an external force which erodes or deforms earth surfaces through landslides, rocky-mud flows, etc. The change in sediment discharge is reflected in the grain size distribution. An approximate equation for the external force in central Japan is constructed by using the grain size distribution in lake sediment, and shows that the pattern of fluctuation in the force is very similar to that of global climatic changes: the force was strong during interglacial periods, weak during glacial periods.

## INTRODUCTION

Quantitative expressions for external forces, physical properties of materials which constitute landforms, etc. are critical for modelling the developmental processes of erosional features. The physical properties of landform material are closely related to the nature of rocks and change over time through weathering. The external force or erosional force, which is a general expression for the various complex forces which erode or deform earth surfaces, is usually related to climatic conditions. It is defined mathematically in some theoretical models (e.g. Kashiwaya, 1986, 1987).

Over the short term, for example in laboratory experiments, this force can be divided into some physical factors (e.g. discharge, rainfall intensity, etc.), assuming the physical properties of the materials that constitute the experimental landforms can be controlled (e.g. Kashiwaya, 1979, 1980; Schumm *et al.*, 1987). However, it is very difficult to obtain long-term quantitative data on both the forces and physical properties of materials from field observation because, in most cases, it reveals only the resultant features.

Nevertheless, we have to try to reconstruct a mathematical formulation of past erosional forces if we want to trace the developmental processes of landforms, properly model them and predict landform changes from the models.

In this chapter, we discuss the effect of external force on landform changes and the relationship between the force and sediment discharge, introducing some simple models in which the external forces are explicitly considered. We also attempt to gain an impression of the erosional force which acted on some of the land features and discuss the eroded material and grain size distribution.

*Process Models and Theoretical Geomorphology*. Edited by M. J. Kirkby
©1994 John Wiley & Sons Ltd

## THE MODEL

First, I shall introduce a simple model showing the erosional process of a mountain (Kashiwaya, 1986). The model is based on two assumptions: (1) the rate of change in the volume of the mountain is proportional to the volume of the erodable portion; and (2) the proportionality factor consists of the external force and the physical properties of the material (cf. Figure 4.1). That is

$$dV(t)/dt = -\beta(t)\gamma\{V(t) - V_c(t)\} \quad \text{for } V(t) > V_c(t) \tag{1}$$

where $\beta(t)$ is the external force, $\gamma$ is a coefficient concerning the physical properties of the mountain, $V(t)$ is the volume of the mountain at time $t$, and $V_c(t)$ is the unerodable portion of the mountain. This can be solved by setting the initial condition $(V(0) = V_0)$;

$$V(t) = \exp\{-\int_0^t \beta'(\tau)d\tau\} \cdot [\int_0^t \beta'(\tau)V_c(\tau)\exp\{\int^\tau \beta'(x)dx\}d\tau + V_0] \tag{2}$$

where $\beta'(t)$ equals $\beta(t)\gamma$ (relative external force). Therefore, sediment discharge $Q(t)$ is expressed by

$$Q(t) = -dV(t)/dt = \beta'(t)\{V(t) - V_c(t)\}$$
$$= \beta'(t)\exp\{-\int_0^t \beta'(\tau)d\tau\} \cdot [\int_0^t \beta'(\tau)V_c(\tau)\exp\{\int^\tau \beta'(x)dx\}d\tau + V_0]$$
$$- \exp\{-\int_0^t \beta'(\tau)d\tau\} \cdot [\beta'(t)V_c(t)\exp\{\int_0^t \beta'(\tau)d\tau\}]. \tag{3}$$

This indicates that sediment discharge from the mountain is closely related to the external force:

$$\beta'(t) = Q(t)/\{V(t) - V_c(t)\}. \tag{4}$$

In the simple case where the whole mountain is erodable, i.e., $V_c(t) = 0$

$$\beta'(t) = Q(t)/V(t). \tag{5}$$

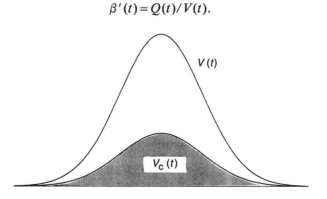

**Figure 4.1**   A model mountain. $V(t)$: volume of the mountain, $V_c(t)$: volume of unerodable portion (after Kashiwaya, 1986. Reproduced by permission of the Japanese Geomorphological Union)

In general, since the erodable part of a mountain $V(t) - V_c(t)$ is much larger than $Q(t)$, and change in $V(t) - V_c(t)$ is much slower than $Q(t)$, $V(t) - V_c(t)$ can be assumed constant compared with $Q(t)$

$$\beta'(t) \propto Q(t). \tag{6}$$

This shows that we can estimate the external force from sediment discharge in this case.

Next, I shall introduce a theoretical model which shows a temporal change for stream nets (Kashiwaya, 1987) on the basis of the two following assumptions: (1) the stream net in a basin extends with time until it reaches a specific upper limit for the basin; and (2) the probability of stream growth in a basin is proportional to the number and total length of streams in the basin (cf. Figure 4.2).

$$dD(t)/dt = \beta(t)\gamma\{1 - D(t)/\delta(t)\}\, D(t) \qquad \text{for } 0 < D(t) < \delta(t) \tag{7}$$

where $D(t)$ is the drainage density at time $t$, $\beta(t)$ is the external force which causes the development of rivers of the basin or surface at time $t$, $\gamma$ is a coefficient concerning the physical properties of the materials that constitute the basin, and $\delta(t)$ is the maximum drainage density. By setting the initial condition $D(0) = D_0$, we obtain:

$$D(t) = \exp\{\int_0^t \beta'(\tau)d\tau\}/[\int_0^t \beta'(\tau)/\delta(\tau)\exp\{\int^\tau \beta'(x)dx\}d\tau + 1/D_0] \tag{8}$$

where $\beta'(t)$ equals $\beta(t)\gamma$.

Sediment discharge $Q(t)$ caused by the evolution of the stream net is related to the elongation rate of the streams:

$$Q(t) = m\,[\{C(t+\Delta t)l(t+\Delta t) - C(t)l(t)\}/\Delta t\,]^n \tag{9}$$

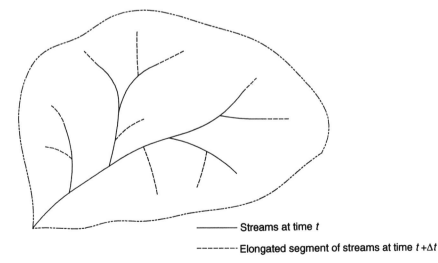

Streams at time $t$

Elongated segment of streams at time $t + \Delta t$

**Figure 4.2**  A model drainage basin

where $C(t)$ is the average cross-section area of the streams, $l(t)$ is the total length of the streams in a basin and $m$, $n$ are coefficients. If the cross-section area is assumed to be constant $(=C)$, then

$$Q(t) = m'\,[\{D(t+\Delta t) - D(t)\}/\Delta t\,]^n \tag{10}$$

$$m' = m(C/A)^n$$

where $D(t)$ is $l(t)/A$ and $A$ is the area of the basin or surface. That is

$$Q(t) = m'\,[\,\beta(t)\gamma\{1 - D(t)/\delta(t)\}D(t)\,]^n$$

$$= m'\,[\{\beta'(t)\int_0^t \beta'(\tau)\mathrm{d}\tau\,[\,\int_0^t \beta'(\tau)/\delta(\tau)\exp\{\int^\tau \beta'(x)\mathrm{d}x\}\mathrm{d}\tau + 1/D_0\,] - \exp\{\int_0^t \beta'(\tau)\mathrm{d}\tau\}$$

$$[\beta'(t)/\delta(t)\exp\{\int_0^t \beta'(\tau)\mathrm{d}\tau\}] / \{\int_0^t \beta'(\tau)/\delta(\tau)\exp\{\int^\tau \beta'(x)\mathrm{d}x\}\mathrm{d}\tau + 1/D_0\}^2]^n. \tag{11}$$

This equation also shows a close relationship between sediment discharge and external force;

$$\beta(t)' = \{Q(t)/m'\}^{1/n}/\{1 - D(t)/\delta(t)\}D(t). \tag{12}$$

When the maximum drainage density is infinite, i.e., $\delta(t \to) \infty$,

$$\beta'(t) = \{Q(t)/m'\}^{1/n}/D(t). \tag{13}$$

Temporal change of landforms $\{1 - D(t)/\delta(t)\}D(t)$ is, in general, much slower than change in mass transport $Q(t)$, and the following expression can be assumed;

$$\beta'(t) \propto Q(t)^{1/n}. \tag{14}$$

These estimations are supported with numerical calculations. Let us investigate temporal change in approximate functions of sediment discharge (equations (3) and (11)) and external force for two models by assuming that: $\beta'(t) = p\sin qt + r$, $V_c(t) = V_c\exp(-st)$, $\delta(t) = \delta_c$ (constant), $m' = 1$ and $n = 1$. Results of numerical calculations employing certain values for the coefficients are shown in Figure 4.3. These results indicate that variations in the external force sensitively reflect the variations in

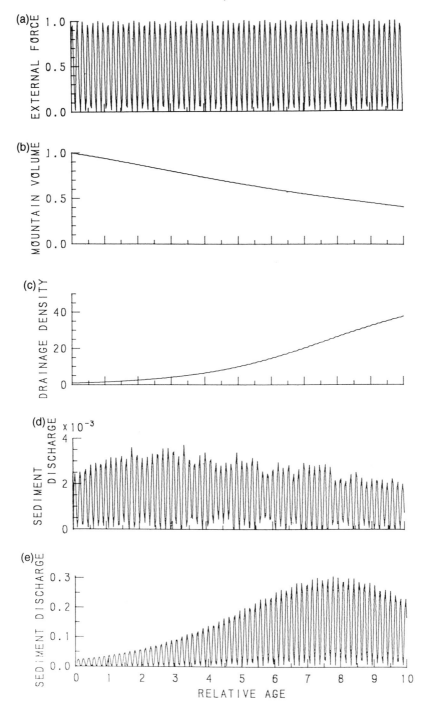

**Figure 4.3** (a) Change in external force: $\beta'(t) = p\sin qt + r$, $p = 0.5$, $q = 40.0$, $r = 0.5$, (b) mountain volume change: $V_c(t)/V_0 = V_{0c}/V_0 \exp(-st)$, $V_{0c}/V_0 = 0.9$, $s = 0.1$; (c) stream net change: $D_0 = 1.0$, $\delta_c = 50.0$, (d) sediment discharge from the mountain; and (e) sediment discharge from the stream net

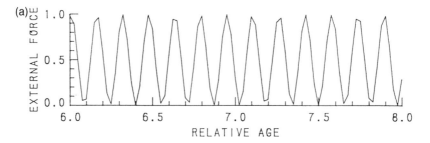

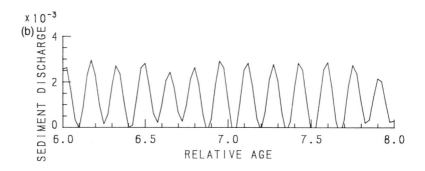

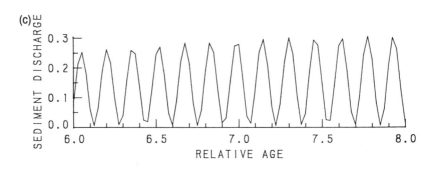

**Figure 4.4** (a) Change in external force in the 'developing' stage; (b) sediment discharge from the mountain in the 'developing' stage; and (c) sediment discharge from the stream net in the 'developing' stage

sediment discharge; the pattern of the variations clearly corresponds to the external force, especially in the 'developing' stage (Figure 4.4). This suggests that changes in external forces may be evaluated by searching for changes in the sediment discharge; by analysing the bottom sediments of closed or semi-closed lakes or seas. Sediment grain size may reflect the quantity of mass transfer, which is closely related to the magnitude of the external forces.

## GRAIN SIZE VARIATION AND EXTERNAL FORCE

The above equations indicate that we need mathematical formulations of the external forces to evaluate the processes properly. Accordingly, we shall discuss some approaches formulating the external force.

Some eroded material caused by landform change is kept on the slopes and in the river bed of the mountains. Other material is transported downstream and deposited in the mouths of the rivers. A part of this material is transported far away into deep lake bottom or sea bottom environments in the form of wash load. Wash load (suspended load) $Q_s$ is related to sediment discharge $Q$ and sediment discharge is related to water discharge $Q_w$;

$$Q_s \propto Q^a, \tag{15}$$

$$Q \propto Q_w^b. \tag{16}$$

As shear velocity $u_*$ becomes larger, larger grains can be transported because shear velocity is related to water discharge $(u_* \propto Q_w^p)$ and critical shear velocity $u_{*_c}$ is related to grain size $d$ $(u_{*_c} \propto d^q$, e.g., Ashida et al., 1982);

$$d \propto Q^r, \tag{17}$$

then

$$\beta' \propto Q^{1/n} \propto d^{1/nr} \tag{18}$$

or

$$\beta'(t) = \Phi\{d(t)\}. \tag{19}$$

Here, $\Phi(t)$ is a function of the grain size parameter and $a$, $b$, $c$, $p$, $q$ and $r$ are constants.

Changes in the grain size distribution of lake sediments are indicative of large fluctuations in sediment discharge, rapid increases in water flow (e.g., rainfall) or an abrupt fall in water level. Large changes in earth surface processes (landslides, mudflows, etc.) are related to changes in sedimentary conditions such as grain size (Taishi et al., 1987; Kashiwaya et al., 1988a). Therefore, it should be possible to trace changes in earth surface processes by examining the grain size of the coarse materials in bottom sediments, as is shown above as a working hypothesis.

Here, we employ grain size data from a 200-m sediment core obtained by Horie in 1971 from the central part of Lake Biwa (65.2 m deep) in central Japan, and analysed by Yamamoto (1984). The lake is semi-closed and is one of the oldest in the world. The sediments are composed of continuous clay, and sedimentary structures are not seen (Yokoyama & Horie, 1974). This indicates that the sediments obtained here can be used quantitatively to discuss the external forces operating in the surrounding areas. The core samples have been analysed by many researchers, clarifying the palaeoenvironment around the lake (Horie, 1984). The data investigated here have also been used in reconstructing palaeohydrological processes (Yamamoto et al., 1984, 1985; Kashiwaya

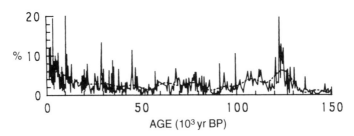

**Figure 4.5** Temporal changes in the percentage of materials coarser than 4.5 $\phi$ in sediment from Lake Biwa. The dotted line represents time-series smoothed by a numerical low-pass filter that removes components of periods shorter than 15 000 years from the original or raw data (cf. Kashiwaya *et al.*, 1991. Copyright by the American Geophysical Union)

*et al.*, 1987, 1988b, 1991). As one quantitative expression of external force, we consider variations in grain size coarser than 4.5 $\phi$ (Figure 4.5). In this lake, coarse sediment is said to reflect rainfall (Yamamoto, 1984). The heavy rainfall in this area causes landform deformation through landslides, rocky-mud flows, etc. Therefore, variations in this parameter are indicative of changes in the external force.

As a first approximation of the external force, we assume the following expression:

$$\beta'(t) \propto \Phi_{dc}(t). \tag{20}$$

In general, most equations can be expressed as a Fourier expansion; $\Phi_{dc}(t)$ can be set as follows:

$$\Phi_{dc}(t) = \sum_i A_i \cos\{2\pi/T_i(t - \phi_i)\} + \text{Ave.} \tag{21}$$

where $\Phi_{dc}(t)$ is a function of the content of fine sand coarser than 4.5 $\phi$, $A_i$ is the amplitude for period $T_i$, and $\phi_i$ is phase.

To obtain the dominant periods of grain size variation, we applied spectral analysis (Barrodale and Ericksson, 1980a,b) to the data. The results are shown in Figure 4.6. The 20 000-year and 40 000-year periods are considered to reflect astronomical forcing (Milankovitch, 1941): precessional parameters and obliquitical parameters, respectively (Kashiwaya *et al.*, 1991). Periods of 60 000 and 120 000 years show major variations. The 100 000-year period has been said to be strongest in long-term climatic change (Hays *et al.*, 1976; Barnola, 1988). However, it is pointed out that the period may be a combination of the 60 000–70 000-year period and the 120 000–130 000-year period. Hence, the periods of 60 000 and 120 000 years may also correspond to global climatic conditions.

Harmonic analysis was also used to calculate the amplitudes of the periods (Table 4.1). The results are employed to reconstruct a mathematical formulation of external force. That is, a synthesized curve using the periods is an approximation of the external force in this area (Figure 4.5). The dotted curve in the same figure represents the filtered

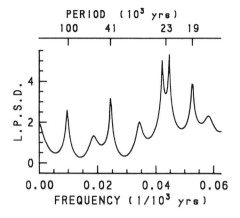

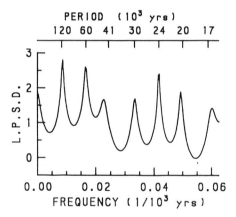

**Figure 4.6** Log power spectral density for average caloric summer isolation in the Northern Hemisphere (upper) and for the coarse fraction record of Lake Biwa (lower). (cf Kashiwaya *et al*, 1991. Copyright by the American Geophysical Union)

**Table 4.1** Amplitudes and phases of prevailing periods for the grain particles coarser than 4.5 $\phi$. Ave. = 2.716%

| Periods $T_i$ ($10^3$ yrs) | Amplitudes $A_i$ (%) | Phase $\phi_i$ ($10^3$ yrs) |
|---|---|---|
| 116.1 | 0.807 | 2.0 |
| 57.6 | 0.909 | 4.8 |
| 39.6 | 0.523 | 39.1 |
| 30.1 | 0.484 | 6.8 |
| 23.8 | 0.345 | 6.5 |
| 16.7 | 0.449 | 8.9 |

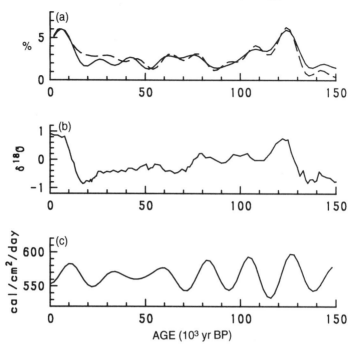

**Figure 4.7**  Temporal changes (a) in the coarse fraction; filtered (dotted) and synthesized (solid) curves using the periods (Table 4.1); (b) in the marine $\delta^{18}O$ record (Martinson *et al.*, 1987); and (c) insolation curve averaged over the Northern Hemisphere calculated from Berger's formulation (1978)

original curve obtained by using a numerical low-pass filter to remove components of periods shorter than 15 000 years (Ormsby, 1961). The trend of the curve is clarified by this operation. The agreement of the two curves shows that the synthesized curve is a good approximation of the original.

Next, we compare the results obtained here with data on global climatic changes. Variations in the marine $\delta^{18}O$ (Martinson *et al.*, 1987) and insolation curve averaged over the Northern Hemisphere (after Berger's formula, 1978) are shown in Figure 4.7. From these figures, the variation in the percentage of materials coarser than $4.5 \phi$ in sediment taken from Lake Biwa is closely related to global climatic changes, and is also linked with solar insolation. Long-term periods of 60 000 years and 120 000 years, as well as the precessional and obliquitical parameters are contained in the data of this $\delta^{18}O$ (Kashiwaya *et al.*, 1993). In this area, the coarse fraction increased during interglacial periods and decreased during glacial periods; the external force was strong in the interglacial periods and weak in the glacial periods in this area. The external force in this area seems to have responded to global climatic changes.

## REFERENCES

Ashida, K., Okabe, T. and Fujita, M. (1982). Threshold condition of particle suspension and the concentration of suspended sediments at a reference level. *Annuals, DPRI, Kyoto University*, **25B**, 401–416 (in Japanese with English abstract and illustrations).

Barnola, J. M. (1988). Vostok ice core provides 160 000-year record of atmospheric $CO_2$. *Nature*, **329**, 408–414.

Barrodale, I. and Ericksson, R. E. (1980a). Algorithm for least-square linear prediction and maximum entropy spectral analysis, part 1. Theory. *Geophysics*, **45**, 420–432.

Barrodale, I. and Ericksson, R. E. (1980b). Algorithm for least-square linear prediction and maximum entropy spectral analysis, part 2. Programme. *Geophysics*, **45**, 433–446.

Berger, A. (1978). Long-term variations of daily and Quaternary climatic changes. *Journal of Atmospheric Science*, **35**, 2362–2366.

Hays, J. D., Imbrie, J. and Shackleton, N. J. (1976). Variations in the Earth's orbit: pacemaker of the ice ages. *Science*, **194**, 1121–1132.

Horie, S. (ed.) (1984). *Lake Biwa*, Junk, Dordrecht, 654 pp.

Kashiwaya, K. (1979). On the stochastic model of rill development in a slope system. *Geographical Review of Japan*, **52**, 63–76 (in Japanese with English abstract and illustrations).

Kashiwaya, K. (1980). A study of rill development process based on field experiments. *Geographical Review of Japan*, **53**, 419–434 (in Japanese with English abstract and illustrations).

Kashiwaya, K. (1986). A mathematical model of the erosional process of a mountain. *Transactions, Japanese Geomorphological Union*, **7**, 69–77.

Kashiwaya, K. (1987). Theoretical investigation of the time variation of drainage density. *Earth Surface Processes and Landforms*, **12**, 39–46.

Kashiwaya, K., Fukuyama, K. and Yamamoto, A. (1993). Variations de la taille des grains de sédimonts lacustres pléistocenes, en relation avec l'environment de paléoérosion. *L'Anthropologie* **97** (in press).

Kashiwaya, K., Yamamoto, A. and Fukuyama, K. (1987). Time variations of erosional force and grain size in Pleistocene lake sediments. *Quaternary Research*, **28**, 61–68.

Kashiwaya, K., Taishi, H., Kawatani, T. and Okimura, T. (1988a). Grain size variation of pond sediments and landslide environment in Rokko Mountains. *Transactions, Japanese Geomorphological Union*, **9**, 193–200 (in Japanese with English abstract and illustrations).

Kashiwaya, K., Yamamoto, A. and Fukuyama, K. (1988b). Statistical analysis of grain size distribution in Pleistocene sediments from Lake Biwa, Japan. *Quaternary Research*, **30**, 61–68.

Kashiwaya, K., Fukuyama, K. and Yamamoto, A. (1991). Time variations in coarse materials from lake bottom sediments and secular paleoclimatic change. *Geophysical Research Letters*, **18**, 1245–1248.

Martinson, D. G., Pisias, N. G., Hays, J. D., Imbrie, J., Moore, T. C. Jr. and Shackleton, N. J. (1987). Age dating and orbital theory of the ice ages; development of a high-resolution 0 to 300 000-year chronostratigraphy. *Quaternary Research*, **27**, 1–29.

Milankovitch, M. (1941). *Kanon der Erdbestrahlung und seine Andwendung auf das Eiszeitenproblem*, Königlich Serbische Akademie, Bergrad, 633 pp.

Ormsby, J. F. A. (1961). Design of numerical filters with applications to missile data processing. *Journal of Association for Computer Machinery*, **8**, 440–466.

Schumm, S. A., Mosley, M. and Weaver, W. E. (1987). *Experimental Fluvial Geomorphology*, Wiley, New York, 413 pp.

Taishi, H., Okuda, S., Isozumu, Y. and Yokoyama, K. (1987). Estimations of sedimentation rates and sedimentary anomalies affected by abrupt meteorological events in the uppermost bottom sediments of Lake Biwa. *Research Report from Lake Biwa Research Institute*, **86-A04**, 91–97 (in Japanese with English abstract and illustrations).

Yammamoto, A. (1984). Grain size variation. In Horie, S. (ed.), *Lake Biwa*, Junk, Dordrecht, pp. 439–459.

Yamamoto, A., Kashiwaya, K. and Fukuyama, K. (1984). Periodic variations of grain size in the 200 meter core sample from Lake Biwa. *Transactions, Japanese Geomorphological Union*, **5**, 345–352.

Yamamoto, A., Kashiwaya, K. and Fukuyama, K. (1985). Periodic variations of grain size in Pleistocene sediments in Lake Biwa and earth-orbital cycles. *Geophysical Research Letters*, **12**, 585–588.

Yokoyama, T. and Horie, S. (1974). Lithofacies of a 200 m core sample from Lake Biwa. *Paleolimnology of Lake Biwa and Japanese Pleistocene*, **2**, 31–37.

# 5 An Evolutionary Model for the Coastal Range, Eastern Taiwan

JIUN-CHUAN LIN

*Department of Georgraphy, National Taiwan University*

## ABSTRACT

The aims of the study were to identify and interpret the neotectonic landforms of the Coastal Range of eastern Taiwan. A model for the geomorphological evolution of the Coastal Range is proposed based on geomorphological and linear information, and the calculated rate of plate motion.

The evolution of the Coastal Range can be divided into at least three stages from the north to the south beginning approximately 1.1 Ma ago. This view is consistent with recent sedimentological and palaeomagnetic studies and for the first time produces a model that is internally consistent with data from various disciplines.

Refer to the principal observations and lessons drawn – with respect to erosional processes and their relationship to tectonic deformation and other controls. The principle of the adjustment of processes to structure has been shown to be a valuable geological tool.

## INTRODUCTION

The collision between the Luzon Arc of the Philippine Sea Plate and the Eurasian continental margin is the most significant tectonic event in the evolution of the recent geomorphology of Taiwan. The collision probably began in the late Pliocene and reached its climax in the middle Pleistocene. The movement is called the Penglai Orogeny (Ho, 1975). There are several views as to the exact nature of the collision and emergence of Taiwan Island. These are summarised in Figure 5.1 A–E, although the differing viewpoints have not yet been resolved owing to incomplete information. It is a central aim of this chapter to contribute geomorphological evidence to this debate, if only on the local scale of the Coastal Range of the island.

## TECTONIC BACKGROUND

Arc–continent collision which first began only 4 Ma ago has already stopped in the northernmost quarter of Taiwan because of the southwestward propagation of the Ryukyu Trench subduction zone (Suppe, 1984). For the same reason the arc–continent collision is only just beginning in southernmost Taiwan (Chi *et al.*, 1981; Teng, 1982) (Figure 5.1 A–D).

*Process Models and Theoretical Geomorphology*. Edited by M. J. Kirkby
©1994 John Wiley & Sons Ltd

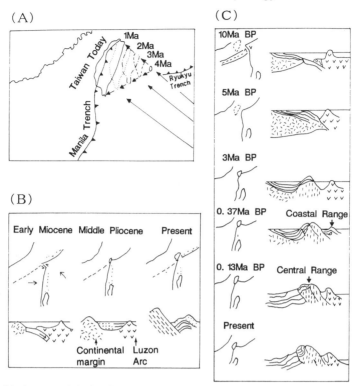

**Figure 5.1**   Various models for the convergence of Taiwan Island together with time scales and simple cross-sections. (A) Growth and motion of the island of Taiwan during arc–continent collision. (After Chi *et al.*, 1981. Reproduced by permission of The Geological Society of China.) (B) Tectonic evolution of the Coastal Range arc system. (After Teng & Wang, 1981. Reproduced by permission of The Geological Society of China.) (C) Evolutionary model of Taiwan since 10 Ma Ago. (After Teng, 1987. Reproduced by permission of The Geological Society of China.)

The current interpretations of plate motion suggest that the Philippine Plate close to Taiwan moves north. However, the collision of the Luzon Arc is oblique and the main stress trajectory is more northwestward (Lee & Wang, 1987). From palaeomagnetic studies of the Coastal Range, a clockwise rotation of about 30° occurred during the Pleistocene period. Prior to this there is no evidence for significant rotations for a period of at least 3 Ma. The rotation was very rapid. It first occurred in the northern part and then the middle part (2.3 to 1.9 Ma ago), and finally reached the southern part of the Coastal Range between 1.4 and 1 Ma ago. Thus, the collision is about 0.9 Ma younger in the south than in the middle of the Coastal Range (Lee, 1989).

During the collision, compressional deformation first occurred in the northern area of Taiwan and subsequently propagated southward. The indentation zone of the plate motion has been gradually enlarged through time (Lee & Wang, 1987) (Figure 5.1D).

The regional stress pattern derived from analysis of the focal mechanisms of principal stress directions of major earthquakes and composite focal mechanisms of micro-earthquakes shows predominant shear stresses at the western end of the Ryukyu Trench, with predominantly tensile stresses in the Central Mountain Range (Lin *et al.*, 1985).

(D)

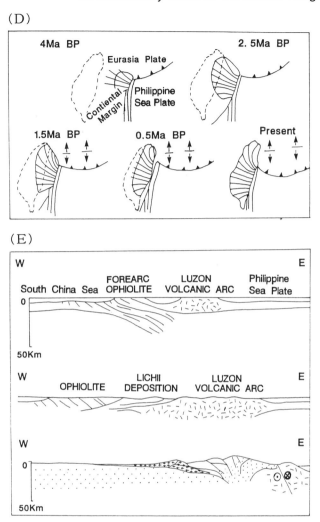

(E)

**Figure 5.1**(continued)   (D) Synthesis of arc–continent collision in Taiwan region. Open arrows indicate the direction of movement of the Philippine Sea Plate, and fan shapes indicate the trajectories of the maximum horizontal stress. (After Lee & Wang, 1987. Reproduced by permission of The Geological Society of China.) (E) Cross-section of Taiwan Island.⤜ = volcanic rock, dots = crystalline basement, + = molasse basins. (After Harris & Audley-Charles, 1987. Reproduced by permission of The Geological Society of China)

The rate of oblique collision has been calculated by Suppe (1981, 1984, 1987) and assumes the convergence rate of Seno (1977) of 70 km/Ma (7 cm/year) in a N55°W direction. The Luzon Arc is oriented N16°E and the continental margin N60°E. The angle between the arc and the compression is *c*. 71°, and between the Eurasian Plate margin and compression *c*. 65°. Solving the vector triangle yields a rate of propagation of the collision of 90 km/Ma with the arc and 95 km/Ma along the continental margin (Figure 5.2). When we consider the geomorphological evolution of the Coastal Range this fact becomes of fundamental importance.

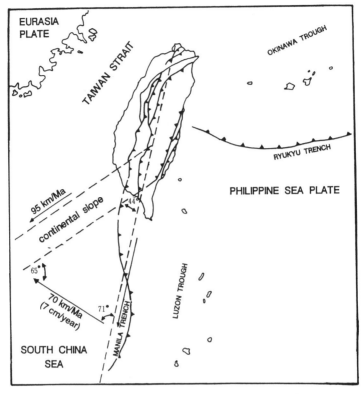

**Figure 5.2**  The nature of oblique collision in Taiwan calculated by Suppe (1981, 1984, 1987. Reproduced by permission of The Geological Society of China)

## METHODOLOGY

The proposed evolutionary model is mainly based on the nature of the tectonic movements together with the indirect archaeological and geodetic data. Such tectonic landforms are thought to be the product of both the endogenetic and exogenetic processes and the original geology, lithology and structure.

The fundamental principle behind this research is that all landforms reflect their geological structure. The processes pick out the weakest points of the system so that, through time, the landforms become adjusted to geological structure and reflect the tectonic hisotry. As a result, neotectonic landforms can be classified as (1) tectonically controlled, (2) process controlled and (3) structurally controlled landforms.

The difference between various forms is not always clear and is merely a matter of the relative dominance of the tectonic or process controls. It is important, however, to appreciate the subtle nature of the various forms.

The purposes of topographic interpretation were to record evidence for tectonic activity from contour maps, ortho-photo maps and aerial photographs, and to compare the results with known features on the published geological maps, papers and field data. The interpretation used standard techniques, often of qualitative nature and based on experience.

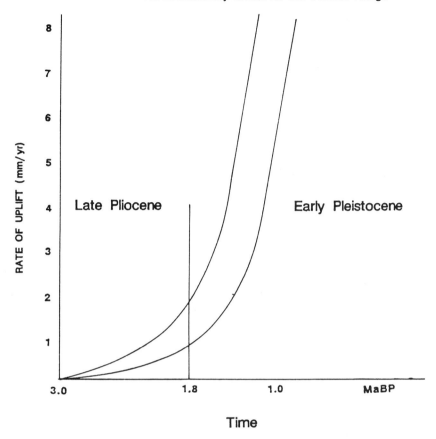

**Figure 5.3**   Rate of uplift during Late Pliocene and Early Pleistocene in west–central Taiwan. (After Lee, 1977. Reproduced by permission of The Geological Society of China)

## TECTONIC MOVEMENTS

Evidence of Quaternary uplift rates for Taiwan comes from the dating of uplifted Holocene reefs and marine terraces and from geodetic surveying. The data show that about 93% of the present mountain uplift was achieved during early Pleistocene time (Lee, 1977) (Figure 5.3). Several estimates of the Holocene uplift rate of Taiwan have been reported (Figure 5.4). According to the work of Lin (1969) and the data compiled by Bonilla (1975), the minimum uplift rate of the recent coastal terraces (less than 9000 year old) around the island ranges from 2 mm/year to 6.4 mm/year, averaging about 4 mm/year. The uplift rates in the north are 2 mm/year from 1500 years BP to 5500 years BP and 5.3 mm per year from 5500 years BP to 8500 years BP (Peng *et al.*, 1977). Konishi *et al.* (1968) and Lin (1989) estimate minimum uplift rates of *c.* 4.0–5.9 mm/year in the Coastal Range.

The characteristics of the neotectonics of the Coastal Range can be summarised as follows:

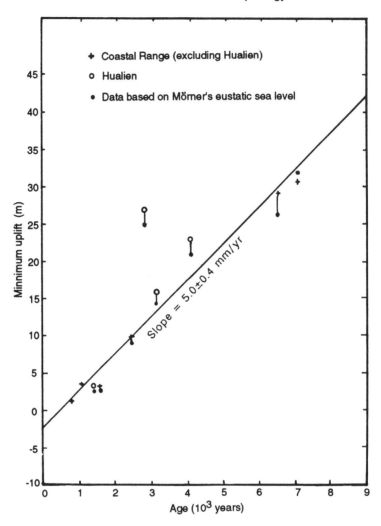

**Figure 5.4** Minimum uplift rate for the Coastal Range for the last 10 000 years. (After Peng *et al*, 1977. Reproduced by permission of The Geological Society of China)

1. Differential uplift within the Range is suggested from the first order [14]C dating (Lin, 1991). The results show that the uplift rate in Sanhseintai Island is 2–2.5 mm/year, Chengkung is 1 mm/year, and the minimum uplift rate in the Shihtiping area ranges from 1 to 4 mm/year. At the Sanyuan site it is 1.5 mm/year and at Shaoyehlieu it is 0.4–7.7 mm/year. The geodetic uplift rates are more variable and show a non-linear but dramatic difference from north to south (Lin, 1991). These first order data agree quite well with archaeological data (Peng *et al.*, 1977) and long-term rates. Nevertheless, there are other published [14]C dating data which give ranges of 4–7 mm/year, slightly higher than the first order dates, and which should therefore be treated with caution.

2. The drainage pattern shows close relationship with the geological structure and a similar pattern replicates itself from north to south. The drainage, alluvial fans, river terraces, raised fans, marine terraces and alluvial plains show different patterns within each tectonic unit.

The evolutionary model of the drainage pattern of the Coastal Range is thought to explain the present differential movements. The model is consistent with separate but integrated movements of the tectonic provinces (Lin, 1991). The replicated drainage patterns also suggest that the development of the drainage has taken place in sequential stages related to the movements of the tectonic blocks through time. However, the scale of the drainage basins remains similar throughout the area which may suggest that the formative tectonic events have taken place at approximately the same rate. This suggestion is similar to the conclusion from the palaeomagnetic data given by Lee (1989) and from the vertical displacement of marine deposits by Lundberg & Dorsey (1990).

3. The present seismicity deduced from seismic observations is concentrated in the Hualien, Tafu and Chengkung areas. These three areas are the most active at present.

4. The coastal landforms are mostly composed of a combination of steep slopes separating a series of uplifted marine terraces or residual smooth surfaces. These landforms vary within different structural units. This phenomenon means that the rapid uplift and mass deposition of colluvium on the surface can only be used as a guide for sea-level change, if the differential uplift rate along the coastal area is taken into account.

5. In the Longitudinal Valley area, the Wuho and Luyeh terraces were uplifted to 100–200 m above the alluvial plain. These features are important to an understanding of the neotectonics of the Range as these uplifted terraces should be internally consistent with other neotectonic features.

## IMAGE ANALYSIS

The aim of image interpretation is to look for the liner features which may be caused by tectonic movements or different geological structures on the remote sensing images and to find any relationship between the linear pattern and the shear field. In general it is not possible to indicate the nature of displacement from the images alone, due to the restriction of the resolution of the images, nor to classify the features in terms of precise structures or geological boundaries because the reflection characteristics of the electromagnetic waves may be similar in different geological strata. What the lineation pattern can reveal, however, is the stress field for a given tectonic unit.

Many definitions of 'lineaments' have been proposed. Zilioli & Antoninetti (1987) defined lineaments as 'simple or composite lines whose parts are aligned in a rectilinear relationship or composite lines whose parts are aligned in a rectilinear or slightly curvilinear relationship and which differ distinctly from the patterns of normal process features and presumably reflect a subsurface phenomenon'. This definition has been accepted for the purposes of this study.

In this research the 'fabric relief' features of neotectonic landforms were examined from remote sensing images by using lineament analysis. The main purpose of this lineament analysis was to record the lineament topographic patterns of the Coastal Range and from this 'fabric relief' to determine the characteristics of related neotectonic

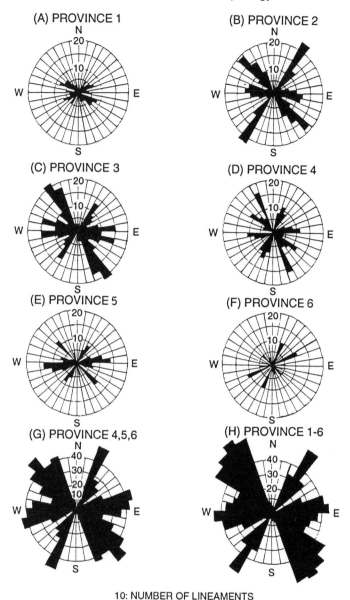

10: NUMBER OF LINEAMENTS

**Figure 5.5** Rose diagrams and neotectonic provinces of the Coastal Range. Provinces indicated in Figure 5.7

movements. The lineament data allow comparison with geological, geomorphological, geodetic and seismological data and form the basis of the tectonic provinces map (Figure 5.6). This zonal map then becomes the framework for the field study of coastal landforms and the reference for chronological correlation.

The rose diagram of lineaments (Figure 5.5) prepared from the basic lineament map shows that the orientations of the lineaments of the whole Coastal Range fall into

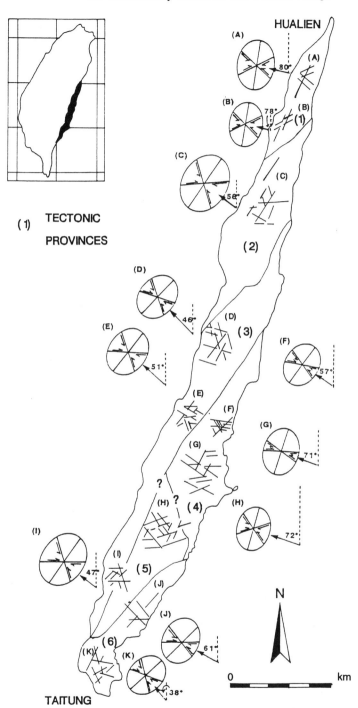

**Figure 5.6** Main conjugate sets of the Coastal Range

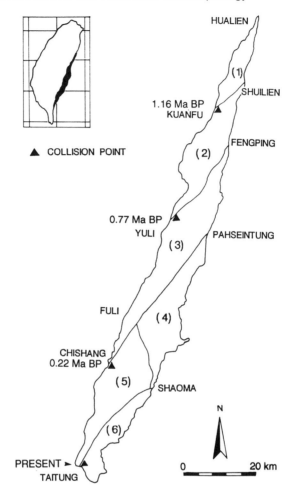

**Figure 5.7** Tectonic province-age diagram of the Coastal Range

three main groups: N20°–60°W, N20°–50°E and N70°–100°E. For the purposes of this research, the whole Coastal Range was subdivided into smaller provinces (Figure 5.7) based on the occurrence of major lineaments, important geological boundaries and known main faults. It is difficult to tell whether there is an anticlockwise or clockwise difference from the north to the south from the pattern shown on the combined rose diagram. From the conjugate sets (Figure 5.6) there is a variation within these provinces which suggests that some of the provinces, such as the Chengkung area, have more E–W to NW–SE direction lineaments numerically. The lineaments along the western side of the Longitudinal Valley of eastern Taiwan do not form a continuous straight line but consist of at least three main segments offset from one another: (1) Hualien to Juisui, (2) Juisui to Chishang and (3) Chishang to Taitung. These three lineament segments clearly form the boundary of the Central Range and Longitudinal Valley and indicate that the major Longitudinal Valley strike-slip fault may in fact not be a single tectonic feature or even of similar type or original fault system as is normally assumed in the literature (Hsu, 1954). The main lineaments of the Coastal Range actually continue across the floor of the Longitudinal Valley to terminate at the three boundary lineaments (Figure 5.7). The interpretation is confirmed by the sites of historical earthquakes and active faults.

Although there is no direct evidence to explain the relationship there is much co-incidence between the geodetic data, drainage patterns and lineaments to supplement our understanding of the tectonic movements. The tectonic provinces therefore form a good basis for further neotectonic and seismic research.

# EVOLUTIONARY MODEL FOR THE COASTAL RANGE

The evolution of the Coastal Range comprises at least three stages (Lin, 1991), see Figure 5.8.

## Stage 1

The arc–continent collision of the Philippine Sea Plate and the Eurasian Plate began in the northern part of the Range. The trend of the Coastal Range, in the northern part, formed an angle of c. 15° with that of the Central Range when they were colliding obliquely with each other. The Tafu area is the first point of the initial contact and today forms a distinct topographic rise and a watershed. This caused all the rivers north of Tafu and originating in the Central Range, to turn north and resulted in the Hualien Hsi and a series of alluvial fans along the northern Longitudinal Valley. The rivers Malankou Hsi and Hungyeh Hsi then formed an alluvial fan delta in the south. Subsequent uplift then formed the faulted (Hsu, 1976) Wuho raised fan terrace from this material. The observation is supported by the geodetic data (Figure 5.9).

## Stage 2

Following Stage 1, there is continued collision and uplift of the proto-Coastal Range. The area between Fuli and Chishang section forms the second place where the two plates collided. Chishang is the second divide within the Longitudinal Valley and all the streams from the Coastal Range were again forced to turn north when this segment of the Coastal Range collided with the Central Range. This second part of the Coastal Range is separated from those in the north by the Chimei Fault. During this second round of the collision the fan delta of the Malankou Hsi and Hungyeh Hsi was block-faulted and uplifted to form the Wuho terraces. The alluvial fans in the Hsiukuluan Valley are smaller and the valley is mainly composed of a broad sheet of alluvial deposits. Most of the active faults were found on the western side of the Range in this valley.

## Stage 3

Stage 3, involving the southern part of the Coastal Range, is more complex and may reflect a more prolonged period of collision with the Central Range. The alluvial plain was uplifted as a series of terraces from Kuanshan to Luyeh ranging from 2 to 6 km in

108

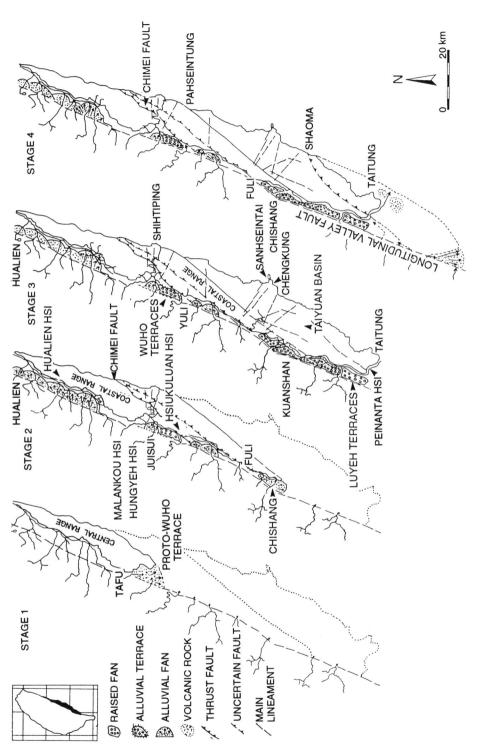

**Figure 5.8** Evolutionary model of the Coastal Range, and the predicted new alluvial fan at Stage 4 (400 000 years in the future)

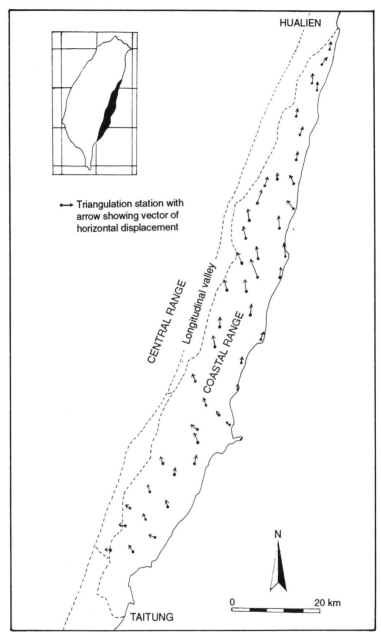

**Figure 5.9**  Horizontal displacements of triangulation stations in the Coastal Range of Taiwan from surveys of 1914–21 and 1976–79. (Redrawn from Biq, 1984. Reproduced by permission of The Geological Society of China)

width within the Longitudinal Valley. No large alluvial terraces are found on the western side of the Range. It is noteworthy that the original sedimentary model of Stage 1 is again repeated. A fan-delta formed at the southern exit of the Peinanta Hsi and has again been uplifted to form the raised, faulted, conglomerate terrace of Luyeh.

Stage 3 may be divided into substages of which the Taiyuan Basin is an example. The Taiyuan Basin is thought to have been formed by two different uplift episodes. The direction of compression is suggested by the arrangement of the axes of the anticlines and faults as well as the different drainage pattern. Stage 3 includes the most active movements up to the present day. There are many pronounced E–W lineaments and the known major NE–SW lineaments produced during this stage are much more complex than those from the earlier ones. It is quite likely that the basic accretion model proposed here will be modified as information on uplift and solid geology improves.

Within Stage 3, there is much variation in the coastal landforms. These variations, together with the differential uplift rates, concentrations of seismic activity and the occurrence of E–W main lineaments and fault lines, suggest that there may be several substages of accretion and landform development. Currently, the tectonic movements of the Coastal Range are also mainly on this part and a prediction of future accretion can easily be suggested (see Stage 4 of Figure 5.8).

## DISCUSSION

The evolutionary model for the Coastal Range has been proposed in a three stage model. Nevertheless, the discussions and conclusions can be summarised as follows:

1. Because accretion energy is being transferred from vertical to horizontal motion (Wu *et al.*, 1989) and because resistance is increasing as each slice accretes, it might be expected that the relief would vary from north to south and that the relief of the older blocks would be different from that of the younger units. The last blocks, which are still accreting, are still rising rapidly.

2. It should be noted that the rate of arc–continent collision means that the northern tip of the Coastal Range is older than the southern tip. The length of the Coastal Range collision boundary is 104.4 km. Dividing this by the collision rate of 90 km/million years means that collision took place 1.16 Ma ago. Units to the south are successively younger and the suggested ages are shown in Figure 5.7. To move north 10 km is equivalent to moving back in time by 0.1 Ma at the inner collision edge. The main tectonic units, moving north, therefore appear to have been produced by collisions at 0, 0.22, 0.77 and 1.16 Ma BP (Figure 5.7). This result is similar to the suggestion from Lundberg and Dorsey (1990) that parts of the Coastal Range have been uplifted for up to 1 Ma.

3. The drainage of the Coastal Range is mainly eastward or westward and the long axes (of the whole drainage area) are mainly in a NE–SW direction. The drainage pattern duplicated itself within the Coastal Range from the north to the south in a manner that is consistent with the main structural lines proposed by the linear analysis. However, it is interesting to note that there are different characteristics of landscape in the north compared with the south part of the Coastal Range because of different stages of tectonic movement. The drainage in the southern Coastal Range suggests that there are evolving from different movement patterns.

4. It is difficult to extrapolate the tectonic movements of the Coastal Range since the Pleistocene by using Holocene $^{14}$C dating data. However, the Holocene $^{14}$C

data and the marine terrace of the coastal area provide some information that the tectonic movements of the Coastal Range are in a differential movement and the boundaries of each tectonic unit could be the known main faults or important lineaments.

5.  Since there is no direct chronological evidence to help identify the age of movements within the Coastal Range, the zonation of the different stages and their subdivision needs the help of the geodetic uplift/horizontal movement data and the geological lithological/structural data. The model should be corrected whenever denudation/uplift data are obtained

## ACKNOWLEDGEMENTS

The author wishes to express thanks to all those who have assisted and advised during this research programme, with particular mention of Prof. D. Brunsden (Dept. of Geography, King's College, London).

## REFERENCES

Biq, C. C. (1984). Present-day manner of movement of the Coastal Range, eastern Taiwan, as reflected by triangulation changes. *Memoir of the Geological Society of China*, **6**, 35–40.

Bonilla, M. G. (1975). A review of recently active faults in Taiwan. *US Geological Survey Open File Report*, 41–75.

Chi, W. R., Namson, J. and Suppe, J. (1981). Stratigraphic record of plate interactions in the Coastal Range of eastern Taiwan. *Memoir of the Geological Society of China*, **4**, 155–194.

Harris, R. A. and Audley-Charles, M. G. (1987). Taiwan and Timor neotectonics: A comparative review. *Memoir of the Geological Society of China*, **9**, 45–61.

Ho, C. S. (1975). *An Introduction to the Geology of Taiwan*, explanatory text of the geologic map of Taiwan, The Ministry of Economic Affairs, Taipei, 153 pp.

Hsu, T. L. (1954). On the geomorphic features and the recent uplifting movement of the Coastal Range, eastern Taiwan. *Bulletin of the Geological Survey of Taiwan*, **7**, 51–57.

Hsu, T. L. (1976). Neotectonics of the Longitudinal Valley, eastern Taiwan. *Bulletin of the Geological Survey of Taiwan*, **25**, 53–62.

Konishi, K. Omura, A. and Kimura, T. (1968). $^{234}U$–$^{230}Th$ dating of some late Quaternary coralline limestones from southern Taiwan (Formosa). *Geology and Palaeontology of Southeast Asia*, **5**, 211–214.

Lee, C. T. and Wang, Y. (1987) Palaeostress change due to the Pliocene–Quaternary arc–continent collision in Taiwan. *Memoir of the Geological Society of China*, **9**, 63–86.

Lee, P. J. (1977). Rate of the early Pleistocene uplift in Taiwan. *Memoir of the Geological Society of China*, **9**, 63–86.

Lee, T. Q. (1989). Evolution tectonique et geodynamique Neogene et Quaternaire de la chaine cotiere de Taiwan: apport du Paleomagnetisme. These de doctorat de Univ. Paris 6.

Lin, C. C. (1969). Holocene geology of Taiwan. *Acta Geologica Taiwanica*, **13**, 83–126.

Lin, C. H., Yeh, Y. H. and Tsai, Y. B. (1985). Determination of regional principal stress directions in Taiwan from plate plane solutions. *Bulletin of the Institute of Earth Sciences, Academia Sinica*, **5**, 67–85.

Lin, J. C. (1991). Neotectonic landforms of the Coastal Range, Eastern Taiwan. Unpublished Ph.D. Thesis, University of London, 380 pp.

Lin, J. F. (1989). Uranium series dating of Holocene raised coral from Hengchun Peninsula and eastern coastal area, Taiwan. Unpublished Master Thesis, Institute of Geology, National Taiwan University.

Lundberg, N. and Dorsey, R. J. (1990). Rapid Quaternary emergency, uplift, and denudation of the Coastal Range, eastern Taiwan. *Geology*, **18**, 638–641.

Mörner, N.-A (1976). Eustasy and geoid changes. *Journal of Geology*, **84**, 123–151.

Peng, T. H., Li, Y. H. and Wu, F. T. (1977). Tectonic uplift rates of the Taiwan Island since the early Holocene. *Memoir of the Geological Society of China*, **2**, 57–69.

Seno, T. (1977). The instantaneous rotation vector of the Philippine Sea Plate relative to the Eurasian Plate. *Tectonophysics*, **42**, 209–226.

Suppe, J. (1981). Mechanics of mountain building and metamorphism in Taiwan. *Memoir of the Geological Society of China*, **4**, 67–90.

Suppe, J. (1984). Kinematics of arc–continent collision, flipping of subduction, and back–arc spreading near Taiwan. *Memoir of the Geological Society of China*, **6**, 21–33.

Suppe, J. (1987). The active mountain belt. In Schaer, S. P. and Rodgers, J. (eds), *The Anatomy of Mountain Ranges*, Princeton University Press, New Jersey, pp. 277–293.

Teng, L. S. (1982). Stratigraphy and sedimentation of the Suilien Conglomerate, northern Coastal Range, eastern Taiwan. *Acta Geologica Taiwanica*, **21**, 201–220.

Teng, L. S. (1987). Stratigraphic records of the Cenozoic Penglai Orogeny of Taiwan. *Acta Geologica Taiwanica*, **25**, 205–224.

Teng, L. S. and Wang, Y. (1981). Island arc system of the Coastal Range, eastern Taiwan. *Proceedings of the Geological Society of China*, **24**. 99–1120.

Wu, F. T., Chen, K. C., Wang, J. H., McCaffrey, R. and Salzberg, D. (1989). Focal mechanisms of recent large earthquakes and the nature of faulting in the Longitudinal Valley of eastern Taiwan. *Proceedings of the Geological Society of China*, **32**, 2, 157–177.

Zilioli, E. and Antoninetti, M. (1987). Geostructural evolution of the southern Alps: lineaments trends detected on Landsat images. *Remote Sensing of Environment*, **23**, 479–492.

# Part 2

## CHANNEL PROCESSES

# 6 The Physical Modelling of Braided Rivers and Deposition of Fine-grained Sediment

**P. J. ASHWORTH,[1] J. L. BEST,[2] J. O. LEDDY[1,2]**
[1]*School of Geography and* [2]*Department of Earth Sciences, University of Leeds, UK*

and

**G. W. GEEHAN**
*B. P. Research, Sunbury-on-Thames, Middlesex, TW16 7LN, UK*

## ABSTRACT

Correctly scaled physical models offer the potential to investigate the active processes and products of deposition within alluvial systems. Although the techniques of modelling have long been used by engineers, they have been underutilised within the Earth Sciences. This paper presents details of the scaled physical modelling of a gravel-bed braided river and an investigation of the depositional sites of fine-grained sediments within such systems. The flume model follows the principles of Froude scale modelling where the flow Reynolds number is relaxed in order to achieve model–prototype similarity. It is argued that scaled models can be operated at grain Reynolds numbers less than the critical value of 70 adopted in past studies; transitional and rough flow conditions prevail at grain Reynolds numbers greater than 15 and low grain Reynolds numbers are typical of many fine-grained sediments in the field. This Froude modelling results in an excellent hydraulic and morphological similarity between model and prototype.

Results from the flume model indicate six major sites for the deposition of fine-grained sediment within gravel-bed braided rivers: confluence scours, bar-top hollows, thalweg scours, abandoned channels, splays and backwater/leeside deposits. Details are presented of the morphology of these depositional niches together with the first results of experiments on the aggradation of braided rivers. These results are discussed in the light of database requirements on depositional morphology which are a necessary input within hydrocarbon reservoir simulation models.

## INTRODUCTION

Scale modelling has been used by scientists from many disciplines over the past two centuries to investigate complex natural phenomena in a controlled and simplified laboratory environment. Over the years, modelling experiments have become increasingly sophisticated so that they are now regarded as reliable and accurate predictive tools. Despite the widespread acceptance of scale modelling within engineering, little use has been made of its underlying principles in geomorphology and sedimentology. Early simple experiments by Reynolds (1883) & Gilbert (1914) demonstrated the potential for using

*Process Models and Theoretical Geomorphology.* Edited by M. J. Kirkby
©1994 John Wiley & Sons Ltd

flumes or stream tables for controlled experiments on flow structure and bedload transport. Later work by Leopold & Wolman (1957), Schumm & Khan (1972), Hooke (1975) and others highlighted the way in which straight and single channels can develop and helped to discriminate between different channel patterns. However, most of this work was analogue modelling of a 'typical' stream network (Schumm *et al.*, 1987), rather than the exact scaling of a known prototype, and usually involved the use of near-uniform bed material. With the move towards numerical models in the 1980s, the processes in single and meandering channels are now more completely understood (Bridge & Jarvis, 1982; Parker & Andrews, 1986; Dietrich & Whiting, 1989; Nelson & Smith, 1989) and scaled models only tend to be used as a test of numerical simulations, to support field observations, or to investigate applied engineering problems.

In comparison, scale modelling of braided systems has received much less attention despite the fact that the hydraulic and sedimentological interrelationships in braided streams are very complicated with substantial feedback (Ashworth & Ferguson, 1986), thus making numerical modelling problematic (particularly in three dimensions). Non-uniform flow (convergence and divergence around bars), mixed-sized bed material (ranging over several orders of magnitude in size) and the spatial and temporal variability in channel response to flood events result in difficulties in making detailed and representative quantitative measurements even under simplified laboratory conditions. In addition, there is the logistical constraint that few research establishments have flume facilities wide enough to cope with an extensive braided system and such rapidly changing conditions. Despite these problems, scale modelling has led to some major advances in our understanding of braided river behaviour. The pioneering work of Ashmore (1982, 1985) has shown how bars migrate and initiate braid development whilst Southard *et al.* (1984) showed the importance of the chute-lobe unit in controlling local sediment budgets and Hoey & Sutherland (1991) related 'pulses' in bedload transport to cycles of aggradation and degradation as bars are created and destroyed. More recently, Lisle *et al.* (1991) and Ashworth *et al.* (1992a) have started to tackle the complex problem of sediment sorting in alternate or single braid bar reaches and have highlighted the link between bar growth, flow structure and bedload size segregation. What all of these studies have shown is that it is possible to reproduce flow and sediment transport relations that closely correspond to the field braided prototype. Given the success in modelling general braided river behaviour, it is now appropriate to focus upon the reach-scale processes and in particular the type and geometry of depositional niches.

Information on the three-dimensional structure of braided river deposition has direct applications in the hydrocarbon industry. Large oil and gas reservoirs are found within ancient braided fluvial deposits (e.g. Geehan *et al.*, 1986; Atkinson *et al.*, 1991) which often form laterally extensive and highly permeable reservoirs. However, the intercalated fine-grained sediments within these reservoirs constitute effective permeability barriers which greatly affect hydrocarbon recovery (Haldorsen and Chang, 1986). These mudstones or shales can be divided into two types, either continuous (those which extend over two or more well spacings), or discontinuous. In braided rivers continuous shales may originate in overbank, floodplain or lacustrine environments whereas discontinuous shales are usually associated with abandoned channel fills, plugs or bar-top drapes. By controlling large-scale effective permeability, both shale types strongly influence the most expedient drilling methods and recovery strategies. Although information on shale geometry and distribution is fundamental to the success of oilfield development, even the most basic data are difficult to obtain, particularly for discontinuous shales. Cores provide only one dimension with limited internal structure. Outcrop studies of ancient analogues can expand observations into two dimensions but still cannot yield detailed information on shale three-dimensional geometry and spatial distribution. The development of Ground Penetrating Radar may help here (cf. Huggenberger, 1993), but currently the technique can only penetrate to depths up to 15 m and its resolution is highly susceptible to local groundwater and conductivity levels. Hence a new, or at least complementary technique is required which provides visualisation of the depositional system as it evolves over time and, with aggradation, can preserve it deposits which can be dissected and described.

In collaboration with British Petroleum, a study was undertaken to investigate the potential of applying the principles of scale modelling to a braided river typical of that which deposited the pebbly-gravel sequences of the Ivishak Formation, Prudhoe Bay, Alaska (see Geehan *et al.*, 1986 and Atkinson *et al.*, 1991 for more site information). The main objectives of this pilot study were (i) to validate the contention that a modern braided river can be scale modelled; (ii) to describe the geometry, spatial distribution and depositional environments of all the fine-grained depositional niches; and (iii) to test the feasibility of simulating aggradation and describing the preserved deposits. This paper highlights the preliminary results from a series of runs with scaled hydrographs using the modern braided Sunwapta River (Canada) as a field prototype.

## PRINCIPLES OF SCALE MODELLING

Scale modelling relies on the concept that a realistic miniature version (or 'model') of a field example (the 'prototype') can be achieved through consideration of geometric, kinematic and dynamic variables. A true model would require that all these groups were precisely scaled which would entail a 1:1 scaling. Therefore, in order to facilitate a reduction in scale, one or more of the controlling variables must be relaxed in order to ensure similarity but retain values within the range of those found in natural channels (Henderson, 1966; Yalin, 1971). Global parameters such as acceleration due to gravity, density of sediment and water, and fluid viscosity are often held constant between prototype and model so that it is some of the dynamic parameters that are relaxed. The variable which is normally altered is the flow Reynolds number (*Re*) defined by

$$Re = \frac{U.Y}{\nu} \tag{1}$$

where $U$ is the mean flow velocity, $Y$ the mean flow depth and $\nu$ the fluid kinematic viscosity. So long as the model Reynolds number is greater than 2000, fully turbulent flow is maintained and hence the model mimics conditions in natural rivers. This technique therefore makes use of relaxing the flow Reynolds number and making the velocity in the prototype scale to the square root of the model scale. It is known as 'Froude' scale modelling (Yalin, 1971) because the emphasis is on correctly modelling the Froude Number, *Fr*, defined as

$$Fr = \frac{U}{(g.Y)^{1/2}} \tag{2}$$

where $g$ is the acceleration due to gravity. A Froude number greater than 1 defines supercritical flow whilst that below 1 is termed subcritical flow. Braided rivers tend to have a Froude number close to unity and a regime transitional between super and subcritical flow (Ashmore, 1982). Froude similarity defines the relation between all other variables in the fluvial system (Henderson, 1966) so that some factors are unaltered from prototype to model, some are linearly scaled, whilst others are reduced by a fractional proportion (Table 6.1). These scaling principles can be taken a step further by 'distorting' some of the input variables, for example a vertical to horizontal length exaggeration (Yalin, 1971; Klaassen, 1991; Shen, 1991), but this can lead to unwarranted computation problems and is only really necessary when dealing with large fluvial systems or specific site or engineering structure applications.

The choice of scaling factor ($\lambda$) to be used is, to some extent, determined by the objectives of the research. However, in order to achieve dynamic similarity Yalin (1971) proposes

$$\lambda > \left(\frac{70}{Re_*}\right)^{2/3} \tag{3}$$

where $Re_*$ is the grain Reynolds number of the prototype

$$Re_* = \frac{U_* D}{\nu} \tag{4}$$

where $U_*$ is the shear velocity and $D$ the grain size that represents the roughness of the bed, usually taken as the $D_{90}$ of the bed (the subscript 90 meaning the size that 90% of the bed is finer than). This choice of $\lambda$ for this project is discussed under the prototype details later.

This scaling ratio given by equation (3) is, in part, derived from the desire to keep the grain Reynolds number, $Re_*$, above a value of 70. This value is normally recommended to ensure fully rough flow where grains do not lie within the viscous sublayer and is derived from the Shields graph where dimensionless shear stress at entrainment above a value of $Re_* = 70$ is assumed constant (see Yalin, 1971,

**Table 6.1**  The Froude scaling laws of the basic hydraulic and geometric variables encountered in physical scale modelling. A more comprehensive list of variables is given in French (1985). Note that $\lambda$ is the scaling factor selected (see equation (3) in text) and the subscripts p and m refer to the prototype and model respectively

| Variable | Scaling ratio |
|---|---|
| Grain size | $D_p/D_m = (\lambda)$ |
| Water discharge | $Q_p/Q_m = (\lambda)^{2.5}$ |
| Sediment discharge | $Qs_p/Qs_m = (\lambda)^{1.5}$ |
| Velocity | $U_p/U_m = (\lambda)^{0.5}$ |
| Time | $T_p/T_m = (\lambda)^{0.5}$ |
| Slope | $S_p/S_m = 1$ |

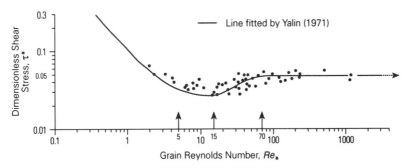

**Figure 6.1** A tracing of the Shields curve as plotted by Yalin (1971, Fig. 6.2, p. 154). Note the high degree of scatter and hence the difficulty in precisely determining the shape and position of the interpretative line. The critical upper value of $Re_*$ for hydraulically smooth flow is at 5. Transitional rough flows occur at higher $Re_*$ values but because of the inherent scatter it is difficult to pinpoint exactly when the flow becomes fully rough (i.e. the Shields line becomes horizontal)

Figs 3.3 and 6.2), although this point may not be reached until $Re_*$ exceeds values of over 250 (e.g. Richards, 1982, Fig. 3.8d). Yalin's (1971) figure is reproduced for discussion purposes in Figure 6.1. The discrepancy in the critical value of $Re_*$ which represents fully rough flow stems from the different interpretation by workers of the shape of the curve that bisects the Shields data set (note that the Shields curve is usually plotted on logarithmic axes, hence reducing the visual impact of the scatter, see Figure 6.1). One common theme that all plots of the Shields curve contain is the division of the data into three components representing smooth, transitional and fully rough flows. Close examination of all plots shows that the upper limit of hydraulically smooth flow occurs somewhere near a $Re_*$ value of 5. Beyond this there may either be a 'dip' in the curve (see Figure 6.1) before a constant line is approached, or a horizontal line stretching from $Re_* = 5$ onwards. Regardless of how the original data is interpreted there seems no doubt that there is at least an upper limit to the significance of $Re_*$ upon grain movement which occurs at a much smaller value than $Re_* = 70$, possibly around 5–15. Hence, the requirement of the minimum grain Reynolds number may also be relaxed somewhat as long as a $Re_*$ value of approximately 15 is exceeded (Parker, 1979). If we assume that shear velocity is a function of the flow depth and bed slope, then a range of $Re_*$ values may be calculated for corresponding combinations of grain size, slope and water depth (Figure 6.2). In braided rivers, non-uniform flow often results in reach average depth–slope products that overestimate the shear velocity (and hence $Re_*$) in shallow water depths (e.g. channel and bar margins), and vice versa in the deepest thalwegs. Nonetheless, these graphs indicate the realms of $Re_*$ that are possible within a model and may be of direct use to future workers who must consider $Re_*$ effects within scale modelling. Figure 6.2 also shows that $Re_*$ values for fine-grained sediment may fall below the critical $Re_*$ value of 15 outlined above; the significance of this within natural channels is discussed later.

Depending on how strictly the laws of Froude similarity are followed, a scaled model can range from being an exact replica of the prototype (de Vries & van der Zwaard, 1975; Francis, 1975), to an approximate analogue or generic type (Southard *et al.*, 1984; Ashmore, 1991; Young & Davies, 1991a), to a smaller system with 'similarity of process'

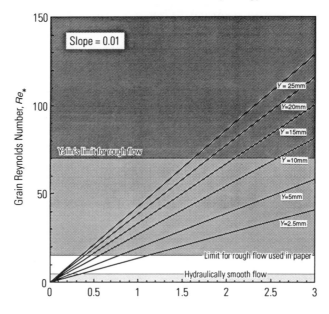

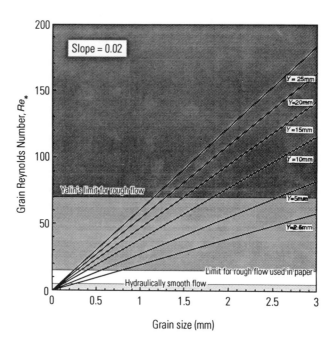

**Figure 6.2** Plots of grain Reynolds number as a function of slope ($S$), flow depth ($Y$) and grain size ($D$). Shear velocity, $U_*$, has been calculated from $U_* = (gYS)^{0.5}$. $Re_*$ values assume a kinematic viscosity of $0.00000114 \, m^2 \, s^{-1}$ at 15 °C. The graphs illustrate the range of $Re_*$ that may be obtained within the flume model and where they lie in terms of the hydraulic limits of smooth and rough flow (highlighted by shading, see text for more details)

to the field prototype (Hooke, 1968; Schumm *et al.*, 1987). Because this project needed to investigate general trends in braided river behaviour, modelling followed generic principles whereby the initial scaling criteria were based on a known prototype but the results did not necessarily simulate a known reach.

## THE FLUME FACILITY AND MEASUREMENT TECHNIQUES

The British Petroleum (BP) flume was specially constructed to simulate braided river deposition and aggradation. The flume measures $5.5 \times 3.7 \times 0.5$ m with a 0.4 m wide $\times 2.3$ m long feeder channel (Figure 6.3). Dry sediment is fed into the channel via a motor-controlled conveyor belt capable of feed rates up to $50 \text{ g s}^{-1}$. Sediment falls off the belt onto a conical plate which ensures dispersion across the whole feeder channel width. Water is pumped over a double 'V' notch weir, under the conical plate and then transports the sediment into the flume ensuring a steady bedload supply. Sediment is not recirculated but trapped at the tail of the flume. Observations and measurements of the braided channel planform are taken by two colour video cameras sited directly

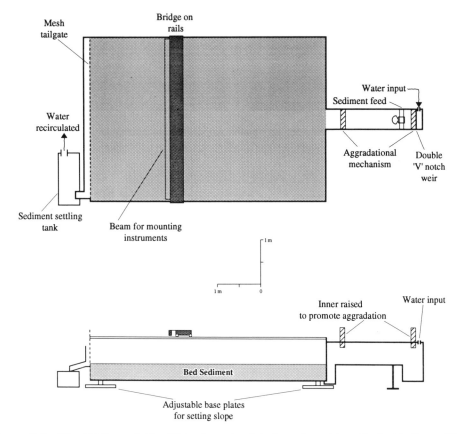

**Figure 6.3** The BP flume table. Water is recirculated, sediment is collected in the settling tank at the end of the flume. A sealed inner cradle in the feeder channel can be raised to promote aggradation

above the centre of the flume and obliquely over the left wall. Bed and water surface elevations are measured using a point gauge and instrument trolley mounted on the movable bridge spanning the width of the flume (Figure 6.3).

Water discharge is controlled manually by a gate valve near the feeder channel inlet and measured by calibration of water height above the 'V' notch weir and discharge collected at the tail of the flume. The non-aggradational set of experiments used five different flood hydrograph types (see later). These were modelled by changing the discharge at between 0.5 and 2-minute intervals depending on the 'flashiness' of the hydrograph.

In order to check the scaling parameters, measurements were taken of individual channel cross-section topography (to the nearest 0.1 mm using the point gauge during and after runs), mean flow velocities (using surface floats traced on the video and converted to mean velocity through multiplication by 0.7), water and bed surface slopes (during and after runs), mean water depths (using the point gauge whilst measuring the cross-sections) and water temperature. A photographic record of channel change was kept for each run from a fixed site. After each run the flume was allowed to dry and then measurements taken of the $x$, $y$ and $z$ coordinates and dimensions of each fine-grained morphological element (as defined later). The upper and lower 0.5 m of the flume were not surveyed in order to avoid any distortion from entrance or exit effects.

## THE SUNWAPTA RIVER FIELD PROTOTYPE

In order to simulate a gravel-bed river with characteristics similar to, although not identical with, those of the Ivishak Formation (see below) a gravel-bed river prototype was selected which fined rapidly downstream but still maintained its fine-gravel component. The Sunwapta River in Jasper National Park, Alberta (Figure 6.4), suited this need although detailed grain size analysis showed that it was slightly too coarse for an exact comparison with the Ivishak Formation (see later). The Sunwapta River is a proglacial stream flowing from the Athabasca glacier through a steep, confined valley with occasional tributary alluvial fans. A study reach was selected 17 km from its source where the discharge was concentrated into one main channel which then began braiding into three distinct distributaries (Figure 6.4). The flume planform mimicked this field situation whereby a single feeder channel braided into three or four distributaries. Extensive background information on the Sunwapta River can be found in Rice (1982), Dawson (1988) and Ferguson et al. (1992).

The study reach has a braiding index value of 3.14 (Brice, 1960) and average sinuosity of 1.16. Of the wetted area 52% is occupied by gravel bars (72% of which are longitudinal forms) with mean aspect ratio (intermediate/long axis ratio) of 0.37. In order to permit comparisons with the flume model several sets of hydraulic and topographic measurements were made:

(i) 37 cross-sections to obtain mean bankfull depth and width for the channel zone (Figure 6.4);
(ii) 13 bed slopes over a cumulative distance of 1005 m;
(iii) 19 velocity profiles in the thalwegs of alternate cross-section sites to obtain mean velocity, bed shear stress, $\tau$ (using the 'law of the wall') and maximum water depth; and

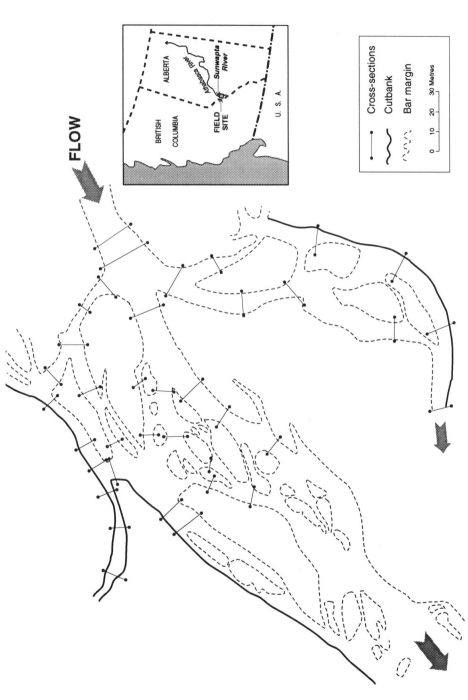

**Figure 6.4** Location of the Sunwapta River field prototype and surveyed planimetric map of the study reach where hydraulic and geometric information was collected for comparison with the flume model. Bar margins were defined as the water's edge at low flow

**Table 6.2** Range of Sunwapta prototype channel dimensions as measured in the reach and cross-sections shown in Figure 6.4.

|  |  | Mean | Range |
|---|---|---|---|
| Mean bankfull depth | $\langle d \rangle$ | 0.28 m | 0.12–0.51 m |
| Width | $w$ | 11.4 m | 5.01–27.0 m |
| Slope | $S$ | 0.00680 | 0.000160–0.0162 |
| Width/depth ratio | $w/\langle d \rangle$ | 51.3 | 14.8–166 |

(iv) volumetric 'bulk' bed surface grain size at the same 19 sites to enable calculation of $D_{50}$ and $D_{90}$ (for input into equation (4)). Sampling of the bed surface layer (by scooping into a mesh bag) was chosen in preference to the subsurface because it was assumed that it was this texture that influenced channel and bar geometry.

Because the hydraulic data were obtained at low discharge ($1.96 \text{ m}^3 \text{ s}^{-1}$ in the single input channel), computed values of $Re$, $Fr$ and $Re_*$ (equations (2–4)) and $Y$ and $\tau$ were not typical of the hydraulics during channel-forming conditions. Thus these data were supplemented by a larger data set obtained for a braided reach with similar bed grain size (average $D_{50} = 26$ mm) 3 km upstream (Ashworth *et al.*, 1992b; Ferguson *et al.*, 1992). These measurements were made during summer snowmelt floods using the techniques described above.

Summaries of the channel geometry statistics (Table 6.2) and flow hydraulics for the two low and high discharge sampling periods (Table 6.3) illustrate the major characteristics of the Sunwapta River. As expected for a gravel-bed river, flow was always fully turbulent and largely subcritical, although some values approached the supercritical transition. The range of field $Fr$ is in line with that reported by other workers (Ashmore, 1985; Young & Davies, 1991b). The flow was always fully rough ($Re_* < 70$), thus viscous effects are unimportant for the coarser fractions of the grain size distribution.

Figure 6.5 shows the pooled grain size distribution of the bulk channel bed surface samples. The average bed $D_{50}$ grain size of the study reach is 37 mm which is much coarser than the Ivishak conglomerate braided unit which has a $D_{50}$ in the range 2–5.6 mm (Ashworth & Best, 1992). Because one of the overriding objectives of the flume modelling was to test the feasibility of scaling a grain size distribution similar to that in the Ivishak, the Sunwapta grain size was used only to compute a 'typical shape' grain size curve for a coarse-grained braided river (see below). Differences in bed grain size between the Sunwapta and Ivishak may cause a change in the hydraulic geometry of scaled-up equivalent anabranches but since the modelling strategy is only looking at a portion of the braided system it is still valid to make general comparisons between flume and Sunwapta channel geometries. Unfortunately this does not resolve the question as to what extent a portion of one braid system can be faithfully modelled by a portion (possibly a different part) of another although this is not so important in the generic modelling principles followed here.

## MODELLING CRITERIA AND EXPERIMENTAL RUNS

If it is assumed that part of the braided system typical of the Ivishak Formation has the same range of channel geometries as the Sunwapta (mean depth = 0.28 m, Table 6.2)

**Table 6.3** Hydraulic data from the Sunwapta prototype. Data were obtained for two different magnitudes of discharge as explained in the text. The range and mean values of water depth are near equal at low and high discharge because in the former case velocity profiles were taken only in the thalwegs, whilst in the latter they are a mix of thalweg and barhead sites. Note that the coarser bed material in the low-discharge study ready results in the high $Re_*$.

|  |  | Mean | Range |
| --- | --- | --- | --- |
| **Low discharge** |  |  |  |
| Mean water depth $Y$ (m) |  | 0.26 | 0.12–0.51 |
| Shear stress $\tau$ (Nm$^{-2}$) |  | 12.3 | 0.01–42.0 |
| Mean velocity $U$ (ms$^{-1}$) |  | 0.75 | 0.19–1.38 |
| Froude No. $Fr$ |  | 0.49 | 0.10–0.78 |
| Flow Reynolds No. $Re$ |  | 170 000 | 58 100–446 000 |
| Grain Reynolds No. $Re_*$ |  | 4649 | 2347–8388 |
|  |  |  |  |
| **High discharge** |  |  |  |
| Mean water depth $Y$ (m) |  | 0.26 | 0.10–0.65 |
| Shear stress $\tau$ (Nm$^{-2}$) |  | 20.2 | 3.00–57.1 |
| Mean velocity $U$ (ms$^{-1}$) |  | 0.89 | 0.43–1.44 |
| Froude No. $Fr$ |  | 0.58 | 0.33–1.07 |
| Flow Reynolds No. $Re$ |  | 156 000 | 30 300–585 000 |
| Grain Reynlds No. $Re_*$ |  | 3480 | 1370–6100 |
| Bedload transport rate (g m$^{-1}$s$^{-1}$) |  | 24.9 | 0.00–294 |

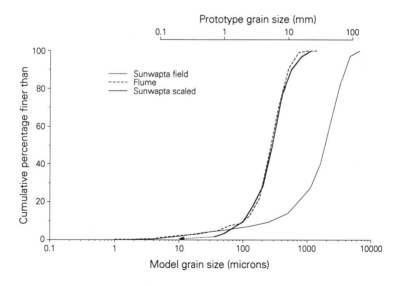

**Figure 6.5** Grain size distributions of the Sunwapta study reach (an average of 19 surface channel bed samples), the scaled Sunwapta grain size that mimics a hypothetical Ivishak size distribution with $D_{50}$ of 5.6 mm, and the final flume grain size used in the experiments

but steeper bed slopes (because it was probably most extensively braided in its upstream reaches, cf. Atkinson *et al.*, 1991) then equation (3) can be used to calculate the maximum scaling ratio ($\lambda$) of the model. Using $S = 0.02$, an Ivishak $D_{90}$ of 10 mm and following the previous arguments, a critical Reynolds Number of 15, yields $\lambda = 0.0471$ or 1:21. Thus a scaling ratio of 1:20 was adopted so that in the field the 0.4 m wide feeder channel represents 8 m while the flume's total area is equivalent to $110 \times 74$ m. As Figure 6.4 and Table 6.2 illustrate, these sizes represent a portion of the Sunwapta channel network and therefore it must be remembered that the flume model is scaling only second and third order channels.

In order to derive a full grain size distribution that can be taken as typical of the Ivishak conglomerate, the grain size curve from the Sunwapta was shifted leftwards in Figure 6.5 until its $D_{50}$ matched an assumed Ivishak $D_{50}$ of 5.6 mm. Although this transformation may not strictly be valid (because the skewness may change as the overall grain size gets finer) keeping a log-normal distribution is probably reasonable. Five tonnes of sediment were machine mixed using a scaling ratio of 1:20. Because of the limited number of sediment size populations commercially available, the final grain size distribution for the model deviated slightly from the Ivishak (although the $D_{50}$ was identical, see Figure 6.5). Of the Ivishak grain size distribution 99% was scaled with a lower truncation limit of 0.02 mm (flume = 1 $\mu$m). Use of silica flour (1–20 $\mu$m) avoided any problems of cohesion and had the added advantage that it was pure white as compared to the various orange/red colours of the sand. This proved an invaluable help in the later visualisation of fine-grained depositional areas. It is worth noting that previous flume studies have truncated their model grain size distributions at levels much coarser than those shown in Figure 6.5, for example the lower limit of Ashmore (1985) was about 0.25 mm and Young & Davies (1991b) was 0.1 mm. However, so long as the onset of cohesion and particle suspension does not change with the scaled hydraulics and grain size mix (there is no evidence to suggest that it does) then there is no reason why Froude scaling principles cannot be extended to finer grain sizes.

Once mixed, the scaled sediment was placed in the flume to a depth of 60 mm (the deepest anticipated scour) and a template dragged over the surface to smooth the floodplain and cut a straight rectangular channel 0.4 m wide and 15 mm deep (field = 0.3 m deep) down the centre of the flume. In order to model sediment depositional conditions realistically it is necessary to run flood hydrographs so that fines can both be transported out-of-channel and deposited during the recession limbs of the hydrograph. Five separate hydrographs were run (termed HY1–HY5) with intervening periods for the measurement of the dry bed surface. Figure 6.6 shows the shape of HY5; HY1–HY4 were similar in shape but had a less rapid rise to peak discharge. The magnitude of peak and low flow discharge was scaled to snowmelt hydrographs from the Sunwapta using the Froude law in Table 6.1. Sediment feed rates were also varied during the hydrograph to mirror the scaled equivalent of the measured values for the Sunwapta (Table 6.3) and to lag the flood peak as is typical of sediment delivery ratios in upland streams. A continuous change in sediment feed rate and water discharge was not possible so their supply curves contain several steps (see Figure 6.6). Although somewhat subjective in terms of temporal variability, the use of hydrographs and changing sediment discharge can only improve the realism of the scaled model, especially if the maximum and minimum values are based on prototype field data.

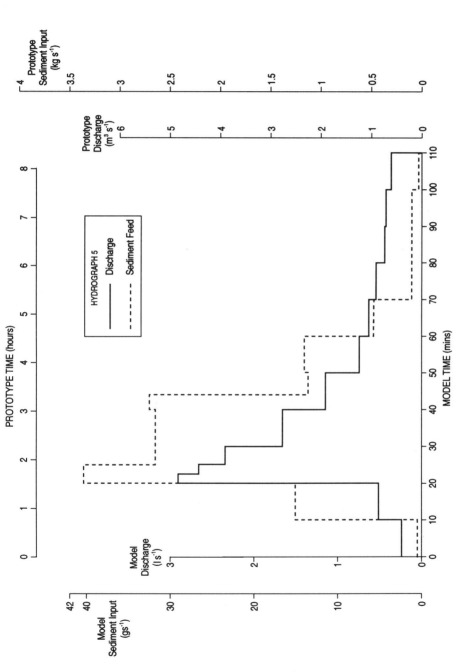

**Figure 6.6** Hydrograph 5 showing the rapid rise to the flood peak and a more gentle falling limb. Note the magnitude of discharge and sediment transport rate are based on Sunwapta River values. Sediment delivery is lagged after the flood peak as is commonly found in the field

Before commencing the five hydrograph experiments the model was run to allow the development of a braided pattern from the initial straight channel. This required approximately six hours at a constant bankfull discharge. The discharge was then dropped and each hydrograph run sequentially (thus each run inherited the planform from the previous hydrograph).

## COMPARISONS BETWEEN FIELD AND FLUME

A comparison between the hydraulic and morphological characteristics of field and flume (Table 6.4) shows a close correspondence. Measurements of surface velocity and mean channel depth were recorded for the largest channels in the centre of the flume (usually three or four discrete channels). Model flows were usually fully turbulent and subcritical although supercritical flows were evidenced by standing waves in some of the smaller, shallower channels. The grain Reynolds numbers show a range of values from 18 to 56. The $Re_*$ distribution for the grain size of the model using the bed slope (0.02) and mean water depth (11.6 mm) is plotted in Figure 6.7. Bearing in mind that this plot does not include the deepest thalwegs that are responsible for the largest $Re_*$, the $D_{90}$ produces a $Re_*$ value of 25 with 56% of the grain size distribution having a $Re_*$ below 15. It is argued here that the coarsest fractions of the bed material make the channel hydraulically rough and therefore the model is similar to the prototype. Clearly there is still scope for a large proportion of the finer model grain sizes to be affected by the viscous forces associated with hydraulically smooth flow. However, this is also true in the field prototype. For example, using the *mean* values of $\tau$ measured in the Sunwapta *thalwegs* at low and high discharge (Table 6.3), for a sand particle of 0.1 mm, gives $Re_*$ of 6.9 and 8.9, respectively. In portions of the braid system where the shear stresses are considerably lower (and by implication the areas where fines are likely to be deposited) the grain Reynolds numbers may be even lower. Thus the flume model matches the field situation whereby the coarsest fractions are in fully rough flow whilst the finer grades are often in smooth or transitional flows. The influence of grain Reynolds number on the mobility and deposition of fine sediment is a little researched topic,

**Table 6.4** Comparison between the flume model and Sunwapta field prototype for purposes of checking hydraulic and geometric similarity. Explanations for deviation from similarity are detailed in the text

| Variable | Sunwapta River[†] | | Flume model | | Comments |
|---|---|---|---|---|---|
| | Mean | Range | Mean | Range | |
| $Re$ | 170 000 | 58 100–446 000 | 2400 | 1885–2632 | Fully turbulent |
| $Fr$ | 0.49 | 0.10–0.78 | 0.50 | 0.43–0.61 | Subcritical flows |
| $Re_*$ | 4649 | 2347–8388 | 35.6 | 18.1–56.4 | Transitional flows |
| $Y/D_{50}$ | 69.9 | 32.4–138 | 41.7 | 39.1–43.4 | Similarity OK[§] |
| $w/<d>$ | 51.3 | 14.8–166 | 50.5 | 20.5–90.5 | Similarity OK |
| $\rho_s/\rho_f$ | 2.72 | – | 2.65 | – | Similarity OK |

[†]For the October fieldwork period in low discharge, see Tables 6.2 and 6.3.
[§]Similarity between flume and field is coincident and no relationship should necessarily be expected (see text for more details).

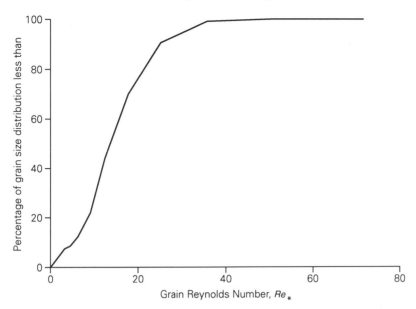

**Figure 6.7** Grain Reynolds number distribution calculated for the model grain size using the mean water depth of 11.6 mm and mean $U_*$ of 0.0476 m s$^{-1}$. Note the maximum $Re_*$ is 71 which corresponds to the $D_{99}$ flume grain size of 1.42 mm. Of the grain size distribution 56% has a $Re_*$ below 15 (see text for discussion)

but similarity with the field prototype will only be achieved if part of the grain size distribution is hydraulically smooth.

The geometric scaling of the flume channels can also be seen to be in good agreement with the field prototype and the model therefore satisfies the similarity of form and process as required if the model is to successfully replicate the field (Hooke, 1968). Figure 6.8

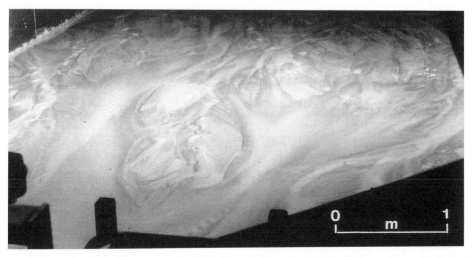

**Figure 6.8** Oblique view of model during low flow in hydrograph 5. The planform is characteristic of many proglacial outwash streams. Flow is away from the camera

shows the typical multichannel planform observed during model runs, which even at the gross scale is clearly characteristic of an extensively braided outwash stream. This similarity in form is also borne out by examination of the planform and types of depositional niche for the fine-grained sediment (see below).

## CLASSIFICATION AND GEOMETRY OF FINE-GRAINED DEPOSITIONAL NICHES

The finest grades of sediment were easily recognisable by their sharply contrasting white colour and represented the field equivalent of silts and very fine sands. Observations of cores from the Ivishak Formation suggests that all preserved shales are predominantly silts with only a low percentage clay. Thus by recording the depositional position of the silica flour, information can be gained on the modes of shale deposition and their typical geometries. Because the initial experiments were conducted in a non-aggrading setting, the preservation potential of some of these fine-grained depositional niches is low. Hence the morphology of the depositional sites was defined from maximum dimensions before aggradation and possible migration or reworking. Thus in the case of scours, for example, if fines were being deposited, the size of the entire scour hole was measured rather than just the patch of fines draped in the hollow.

The modes of fine-grained deposition were grouped into six categories. These were common to all five hydrographs.

1. *Confluence scours*: present at most channel junctions (cf. Best, 1988), often possessing steep, oblique, avalanche faces which can migrate rapidly into the confluence scour and infill it when one confluent flow becomes dominant over the other (Krigström, 1962; Reid *et al.*, 1989).
2. *Bar-top hollows*: found randomly distributed on the tops of large mid-channel bars. These may owe their origin to either the non-uniform fill of abandoned junction scours which have been incorporated into a compound bar, the topographic low that naturally exists between two inwardly migrating lateral accretion surfaces of a braid bar (cf. Ashworth *et al.*, 1992a), or due to secondary circulation/eddies generated in the non-uniform, shallow flow over the bar-tops. Similar morphological elements have been described by Williams and Rust (1969) and Gustavson (1974) but their exact origin is uncertain; ongoing time-lapse video camera work may help solve this issue. Once formed, the bar-top hollows become sites of dead water which may then become zones of fine-grained deposition in the time between the maximum overbank flows.
3. *Thalweg scours*: an all-embracing term that includes scours not associated with convergent channels or bar-tops. These may originate as scours on the outside of channel bends which then form areas of flow separation near the banks and substantial within-channel fine-grained deposition (cf. the 'concave benches' of Woodyer (1975)), or be part of the pool-riffle sequence associated with most alluvial rivers which become abandoned and infilled as the active channel shifts its planform position (usually through avulsion).
4. *Abandoned channels*: a common depositional feature of braided rivers when the active channels frequently and rapidly switch positions leaving an entire channel abandoned

which is then infilled with fines from slow-moving or stagnant water. Abandoned channels are often at a wide angle from the dominant flow direction and are found across the entire range of channel scales.

5. *Splays/Overbanks*: laterally extensive, but often thin, drapes on the tops of mid-channel bars and overbanks on the floodplain. These usually form a continuous deposit with local thicker accumulations in bar and floodplain depressions.

6. *Backwater and bar leeside*: normally associated with flow separation at different scales. Found attached to the banks immediately downstream of deep confluences (Best, 1988), behind obstructions and bank discontinuities (Lisle, 1986), and in the lee of bar avalanche faces (especially transverse bars with their long axis perpendicular to the main flow direction).

At the end of each hydrograph all fine-grained depositional sites were located and their three axes (maxima) and deviation angle from the main downstream flow direction measured. From this database the aspect ratio for each niche was calculated which can be taken as an approximate indication of planform shape (Figure 6.9). Although the overall mean aspect ratio was 0.47 there are marked variations from this according to the depositional type (Figure 6.9). All scours have similar shapes with ratios near 0.6. Because splays are commonly found on bar-tops they often mirror the shape of the bar morphology which is near to a ratio of 0.4 (see Figure 6.4 and earlier description of the Sunwapta River). The abandoned channels and backwater/leeside deposits were

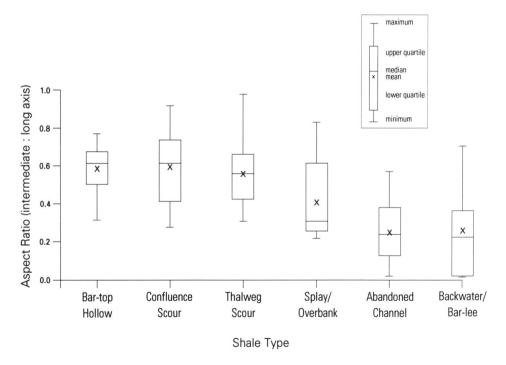

**Figure 6.9**  A comparison of aspect ratios for different fine-grained depositional niches. Note the different types of scours have similar values but are markedly different from the splays, abandoned channels and backwater/bar-lees. This has implications for hydrocarbon reservoir heterogeneity modelling (see text)

markedly different with mean ratios near 0.25 reflecting their long and narrow shape. No systematic change in the aspect ratio was observed for the range of sizes in each individual depositional type.

These measurements of aspect ratio for each category of fine-grained deposition can be compared to the Sunwapta field prototype to help validate the sensitivity of the scale model. Only the geometries of abandoned channels, backwater/bar-lees and splays were measured in the field (most scours were either active and hence contained few fines so were omitted, or amalgamated with abandoned channels and so were merged into the abandoned channel category). Mean aspect ratios for these fine-grained deposits were 0.22, 0.43 and 0.37, respectively, with an overall mean of 0.35. This compares favourably with the aspect ratios found in the flume model with the exception of the backwater/bar-lee category (which is much higher, 0.43 compared to 0.25). Closer inspection of the Sunwapta field data shows this can be explained by the amalgamation of two different fine-grained families – the bar-lee deposits are more elongate (mean $B/A = 0.26$, $n = 25$), similar to the flume results, but the backwater deposits are more equant (mean $B/A = 0.64$, $n = 20$). Because almost twice as many bar-lee deposits were included in the backwater/bar-lee category in the flume compared to the field, the mean class aspect ratio for the flume is biased towards lower values. This points to the need to ensure the same proportion of each niche is sampled in the flume and field or more importantly that this fine-grained category is divided into two different subunits.

The fines were not always deposited parallel to the dominant flow direction (expressed as the difference between the long axes of the fine-grained units and the downstream flume direction) as may be expected in a braided environment with highly non-uniform flow and flow convergence and divergence around bars. A histogram of deviation angles for all the fine-grained depositional niches (Figure 6.10) shows that the maximum deviation angle was 65° (an abandoned channel) with a mean of 12°. These deviation

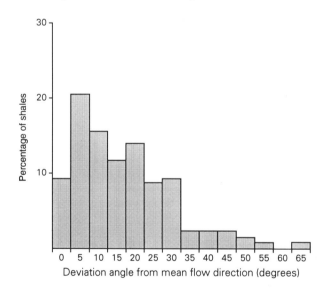

**Figure 6.10** Deviation in orientation of the fine-grained depositional niches from the mean flow direction (data is for all the fine-grained niches recorded, $n = 116$). Deviation angles were taken to the nearest 5° so that each histogram class represents the $x$ axis value $\pm 2.5°$

angles are generally less than the angles measured for the fine-grained deposits in the Sunwapta River (mean = 23°) and smaller than the 22° recorded by Williams & Rust (1969) for bedforms and bars in the braided Donjek River, Canada. This suggests the flume model is less sinuous than the field prototype or that some of the channels may have been confined by the flume walls (though Williams and Rust (1969) did record a mean stream channel orientation of 5° from the downvalley river trend).

What is clear from Figure 6.9 is that the six different fine-grained depositional types each have their own morphological and geometric characteristics. This is important to recognise and quantify because previous hydrocarbon reservoir heterogeneity models have used single geometric values for all shale types in a reservoir (although they may often be divided into continuous and discontinuous groups). For example, the three-dimensional numerical simulation model run by Wadman *et al.* (1979) for the Prudhoe Bay Field assumed an average shale aspect ratio of 0.33. As Figure 6.9 shows this may be a close approximation for abandoned channels and even splays but cannot be taken as representative for scour and poolhead fills. Although the stochastic model of Wadman *et al.* (1979) 'does a remarkable job of estimating reasonable effective megascopic vertical permeabilities' (Haldorsen *et al.*, 1987), the crucial factor affecting predictive success is the shale frequency and size. In particular Haldorsen *et al.* (1987) note that 'the most important information to be specified by the geologist is shale length and shale width statistics and some guidance as to their interdependence'. Figure 6.9 shows preliminary modelling results of such information and highlights the need for quantifying different geometrical parameters for each depositional shale family as an input into numerical reservoir simulation models. If the predictive models are stochastic then an assumed frequency distribution of different shale families can be used (based on flume or field observations) to generate the shale distribution and sizes throughout the reservoir. If a deterministic model is being used (usually based on inter-well correlation and core description), then careful consideration has to be given to matching the correct shale family geometry to the likely environment of deposition. This requires detailed core logging to determine whether the shales are scours, splays, etc., and then representation by their respective aspect ratios (possibly scaled according to likely channel size) and orientations. In this manner, a picture of the vertical variation in shale type and size can be assembled for each core which can be extrapolated through the field by assuming a spatially varying distribution of shale types. Such an approach has been followed by Begg *et al.* (1989) with a high degree of success although they clearly state that assigning similar shale dimensions to different facies is one of the main weaknesses of their model.

## AGGRADATION AND THREE-DIMENSIONAL ARCHITECTURE

The aforementioned experiments were all undertaken on a constant bed level with no restrictions on channel migration and network development. As discussed, the degree of braiding was very similar between flume and field with three or four main braid channels and frequent dissection of mid-channel bars. The majority of erosion and deposition was enacted by the gradual migration of sinuous thalwegs and the downstream accumulation of tongue-shaped lobes (cf. Rundle (1985) and Ferguson *et al.* (1992)). Major avulsion was rare, but when it did occur, it led to abandonment of approximately 30% of the active floodplain. Since the depositional system had no opportunity to

134

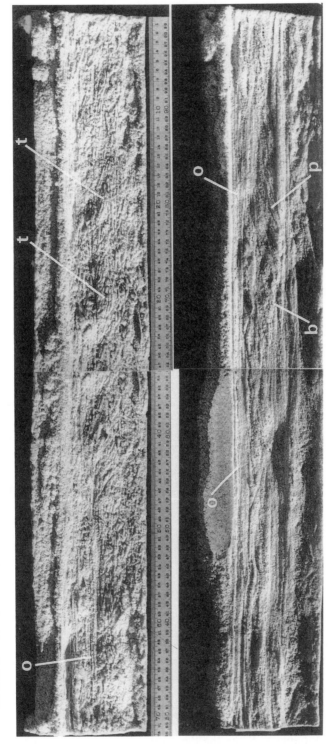

*Core 6*: **Across Stream**

*Core 4*: **Downstream**

**Figure 6.11** Photographs of two box cores taken during a 55 mm aggradation run. The top core is perpendicular to the mean flow direction whilst the lower core is parallel to flow (from right to left). The cross-stream core shows several arcuate channel fills with trough cross-bedding (arrowed t) and parallel-laminated overbank fines (labelled o). The downstream core shows evidence of planar cross-bedding preserved through accretion at the avalanche face of mid-channel bars (p), basal erosion surfaces (b) and overbank fines (o)

aggrade, the fine-grained units shown in Figure 6.9 represent surface morphologies which may not necessarily have the same geometry when preserved. Repeated dissection, reworking and erosion by channels migrating across the floodplain over time may lead to preservation of only a small proportion of the original depositional feature (and sometimes none at all). The absolute values of shale dimensions used to construct Figure 6.9, therefore, must be viewed as maxima, although dimensionless variables such as the aspect ratio may stay constant even after a degree of reworking.

In order to address this problem, and again with the aim of providing the reservoir engineer with realistic shale dimensions, the flume design has been modified to permit aggradation of up to 0.5 m. By raising an inner cradle in the inlet channel which is controlled by two stepper motors (Figure 6.3), the channel and bed are forced to aggrade. The rate of aggradation is not continuous but can be set to uplift at any time interval (or amount). In nature, aggradation is probably not a smooth steady progression and therefore this is not too far from reality. One problem that does arise, however, is what aggradation rate to impose since clearly at higher rates there is an increased likelihood of preservation (Salter, 1993). Figures of 0.58 m/1000 years have been estimated for large aggrading fluvial systems such as the Mississippi (Schumm & Brakenridge, 1987). Although time is scaled in the flume model (see Table 6.1), such slow aggradation/subsidence rates are impossible to simulate; however, if it is assumed that much of the aggradation actually occurs during rare, large floods then the rate reduces to more manageable levels.

A preliminary aggradational experiment was run for 36 hours after HY5. The rate of aggradation was arbitrarily set at 1 mm/hour for the first 17.5 hours and 2 mm/hour thereafter. This built up 55 mm of sediment spread fairly evenly over the flume. After completion, several cores were taken down to the flume base in a 0.6 m$^2$ area in the centre of the flume using a 0.7 m long, 0.07 m wide, 0.2 m deep metal box corer. The cores were preserved using Araldite epoxy resin and hardener. Figure 6.11 shows an example of two cores taken across and downstream. Both show remarkably clear internal preservation structures including planar and trough cross-bedding produced by bar migration, channel fills, basal erosion surfaces and overbank deposits. The success of this preliminary aggradation experiment will permit future quantification of the three-dimensional fluvial architecture and investigation of the links between surficial processes (recorded via the video) and preserved form.

## CONCLUSIONS

The potential of dimensionally correct scaled models of fluvial systems has not yet been fully realised in the Earth Sciences. Only a handful of workers have used scaled models to investigate the relationship between geomorphological form and sedimentological product in alluvial systems. Additionally, most of these studies have been confined to single thread channels. As part of a collaborative research project with British Petroleum, a large and versatile flume has been constructed at Leeds which allows experiments on scaled braided river depositional mechanisms. These experiments have direct applications in the hydrocarbon industry as many reservoirs contain substantial ancient braided river deposits which, because of their lateral extent and connectivity, may form excellent reservoir rocks. The reservoir engineer has to predict the size, shape and spatial distribution of the fine-grained shales which often form impermeable barriers and therefore impede the passage of fluids. Current numerical simulation models of reservoir performance use a 'minimum, maximum and most likely' fit for 'typical' shale geometries (cf. Begg *et al.*, 1989) based on one- and two-dimensional information from core and outcrop analogues. A series of scaled flume experiments has been undertaken to improve this existing database by providing information in the missing third dimension. Using a correctly scaled model and allowing some relaxation in the flow and grain Reynolds numbers produces an excellent hydraulic and geometric analogue of the field prototype. Not only can such scale models of braided rivers accurately reproduce the prototype morphology but they also hold considerable promise for quantifying the various depositional niches within such systems. This has been demonstrated in the present experiments through quantification of the depositional sites of fine-grained sediment. Future work upon the aggradation and preservation potential of such niches holds considerable scope for evaluating the architecture of such sedimentary suites and isolating and investigating the autocyclic (e.g. bedload calibre, channel size) and allocyclic (e.g. hydrograph type, climatic, tectonic) controls upon the preservation style of alluvial systems.

## ACKNOWLEDGEMENTS

We gratefully acknowledge the financial support of British Petroleum PLC (Exploration) and thank them for permission to publish this paper. Many BP scientists have helped us with this project and provided constructive criticism and ideas. In alphabetical order we thank Steve Begg, Mike Bowman, Andrew Brayshaw, Martin Dawson, Mark Eller, Gus Gustason, Phil Hirst and Marcus Richards. JOL thanks BP for financial support as a Research Assistant to this project which enabled him to undertake an M.Phil. A Royal Society grant to PJA and JLB enabled purchase of the cameras and video. The Environment Canada Parks Service kindly allowed us to work in Jasper National Park and Environment Canada provided the Sunwapta discharge records. The Canadian Hostelling Association is thanked for giving us a welcome stay at Beauty Creek Hostel. We are grateful for the detailed review comments of Mike Church, Trevor Hoey and Keith Richards which gave us some new ideas and helped improve the clarity of this paper.

## REFERENCES

Ashmore, P. E. (1982). Laboratory modelling of gravel, braided stream morphology. *Earth Surface Processes and Landforms*, **7**, 201–225.

Ashmore, P. E. (1985). Process and form in gravel braided streams: laboratory modelling and field observations. Unpublished Ph.D. Thesis, University of Alberta, 414 pp.

Ashmore, P. E. (1991). Channel morphology and bed load pulses in braided, gravel-bed streams. *Geografiska Annaler*, **73A**, 37–52.

Ashworth, P. J. and Best, J. L. (1992). The scale modelling of braided rivers of the Ivishak Formation, Prudhoe Bay. *Internal Report to British Petroleum*, 5 March 1992, 9 pp.

Ashworth, P. J. and Ferguson, R. I. (1986). Interrelationships of channel processes, changes and sediments in a proglacial river. *Geografiska Annaler*, **68**, 361–371.

Ashworth, P. J., Ferguson, R. I. and Powell, D. M. (1992a) Bedload transport and sorting in braided channels. In Billi, P., Hey, R. D., Thorne, C. R. and Tacconi, P. (eds), *Dynamics of Gravel-Bed Rivers*, Wiley, Chichester, pp. 497–513.

Ashworth, P. J., Ferguson, R. I., Ashmore, P. E., Paola, C., Powell, M. D. and Prestegaard, K. L. (1992b). Measurements in a braided river chute and lobe: II Sorting of bedload during entrainment, transport and deposition. *Water Resources Research*, **28**, 1887–1896.

Atkinson, C. D., McGowen, J. H., Bloch, S., Lundell, L. L. and Trumbly, P. N. (1991). Braidplain and deltaic reservoir, Prudhoe Bay Field, Alaska. In Barwis, J. H., McPherson, J. G. and Studlick, J. R. J. (eds), *Sandstone Petroleum Reservoirs*, Springer-Verlag, New York, pp. 7–29.

Begg, S. H., Carter, R. R. and Dranfield, P. (1989). Assigning effective values to simulator gridblock parameters for heterogeneous reservoirs. *Society of Petroleum Engineers, Reservoir Engineering*, **Nov.**, 455–463.

Best, J. L. (1988). Sediment transport and bed morphology at river channel confluences. *Sedimentology*, **35**, 481–498.

Brice, J. C. (1960). Index for description of channel braiding. *Geological Society of America, Bulletin (Abstracts)*, **71**, 1833.

Bridge, J. S. and Jarvis, J. (1982). The dynamics of a river bend: a study in flow and sedimentary processes. *Sedimentology*, **29**, 499–541.

Dawson, M. (1988). Sediment size variations in a braided reach of the Sunwapta River, Alberta, Canada. *Earth Surface Processes and Landforms*, **13**, 599–618.

de Vries, M. and van der Zwaard, J. (1975). Movable bed river models. *Proceedings of International Association of Hydraulic Research, Conference on Hydraulic Modelling*, pp. 540–559.

Dietrich, W. E. and Whiting, P. (1989). Boundary shear stress and sediment transport in river meanders of sand and gravel. In Ikeda, S. and Parker, G. (eds), *River Meandering, American Geophysical Union Water Resources Monograph*, **12**, 1–50.

Ferguson, R. I., Ashmore, P. E., Ashworth, P. J., Paola, C. and Prestegaard, K. L. (1992). Measurements in a braided river chute and lobe: I. Flow pattern, sediment transport and channel change. *Water Resources Research*, **28**, 1877–1886.

Francis, J. R. D. (1975). *Fluid Mechanics for Engineering Students*, 4th edition, Edward Arnold, 370 pp.

French, R. H. (1985). *Open Channel Hydraulics*, McGraw-Hill, New York, 705 pp.

Geehan, G. W., Lawton, T. F., Sakurai, S., Klob, H., Clifton, T. R., Inman, K. F. and Nitzberg, K. E. (1986). Geologic prediction of shale continuity, Prudhoe Bay field. In Lake, L. W. and Carroll, H. B. Jr (eds), *Reservoir Characterization*, Academic Press, New York, pp. 63–82.

Gilbert, G. K. (1914) The transportation of debris by running water. *US Geological Survey Professional Paper*, **86**, 259 pp.

Gustavson, T. C. (1974). Sedimentation on gravel outwash fans, Malaspina Glacier Foreland, Alaska. *Journal of Sedimentary Petrology*, **44**, 374–389.

Haldorsen, H. H. and Chang, D. M. (1986). Notes on stochastic shales; from outcrop to simulation model. In Lake, L. W. and Carroll, H. B. Jr (eds), *Reservoir Characterization*, Academic Press, New York, pp. 445–487.

Haldorsen, H. H., Chang, D. M. and Begg, S. H. (1987). Discontinuous vertical permeability barriers: a challenge to engineers and geologists. In Graham, G. and Troutman, F. (eds), *North Sea Oil and Gas Reservoirs*, Norwegian Institute of Technology, pp. 127–151.

Henderson, F. M. (1966). *Open Channel Flow*, Macmillan, New York, 552 pp.

Hoey, T. B. and Sutherland, A. J. (1991). Channel morphology and bedload pulses in braided rivers: a laboratory study. *Earth Surface Processes and Landforms*, **16**, 447–462.

Hooke, R., Le B. (1968). Model geology: Prototype and laboratory streams: Discussion. *Geological Society of America, Bulletin*, **79**, 391–394.

Hooke, R., Le B. (1975). Distribution of sediment transport and shear stress in a meander bend. *Journal of Geology*, **83**, 543–565.

Huggenberger, P. (1993). Radar facies: recognition of facies patterns and heterogeneity estimation (Pleistocene Rhine Gravel, N.E. Switzerland). In Best, J. L. and Bristow, C. S. (eds), *Braided Rivers: Forms, Special Publication, Geological Society*, **75**, London, 432 pp.

Klaassen, G. J. (1991). On the scaling of braided sand-bed rivers. In Shen, H. W. (ed.), *Movable Bed Physical Models*, Kluwer Academic, New York, pp. 59–72.

Krigström, A. (1962). Geomorphological studies of sandur plains and their braided rivers in Iceland. *Geografiska Annaler*, **44**, 328–346.

Leopold, L. and Wolman, M. G. (1957). River channel patterns: braided, meandering, and straight US Geological Survey Professional Paper, **262-B**, 85 pp.

Lisle, T. E. (1986). Stabilization of a gravel channel by large streamside obstructions and bedrock bends, Jacoby Creek, northwestern California. *Geological Society of America Bulletin*, **97**, 999–1011.

Lisle, T. E., Ikeda, H. and Iseya, F. (1991). Formation of stationary alternate bars in a steep channel with mixed-size sediment: a flume experiment. *Earth Surface Processes and Landforms*, **16**, 463–469.

Nelson, J. M. and Smith, J. D. (1989). Evolution and stability of erodible channel beds. In Ikeda, S. and Parker, G. (eds), *River Meandering, American Geophysical Union Water Resources Monograph*, **12**, 321–377.

Parker, G. (1979) Hydraulic geometry of active gravel rivers. *Journal Hydraulics Division, American Society of Civil Engineering*, **105 (HY9)**, 1185–1201.

Parker, G. and Andrews, E. D. (1986). On the time development of meander bends. *Journal of Fluid Mechanics*, **162**, 139–156.

Reid, I., Best, J. L. and Frostick, L. E. (1989). Floods and flood sediments at river confluences. In Beven, K. and Carling, P. A. (eds), *Floods: Hydrological, Sedimentological and Geomorphological Implications*, Wiley, Chichester, pp. 135–150.

Reynolds, O. (1883). An experimental investigation of the circumstances which determine whether the motion of water shall be direct or sinuous, and of the laws of resistance in parallel channels. *Royal Society of London, Philosophical Transactions*, **174**, 935–982.

Rice, R. J. (1982). The hydraulic geometry of the lower portion of the Sunwapta valley train, Jasper National Park, Alberta. In Davidson-Arnott, R., Nickling, W. and Fahey, D. B. (eds), *Research in Glacial, Glaciofluvial and Glaciolacustrine Systems*, Geo-Books, Norwich, pp. 151–174.

Richards, K. S. (1982). *Rivers: Form and Process in Alluvial Channels, Methuen*, London, 358 pp.

Rundle, A. (1985). Mechanisms of braiding. *Zeitschrift für Geomorphologie Supplement*, **55**, 1–14.

Salter, T. (1993) Fluvial scour and incision; models for the development of realistic reservoir geometries. In North, C. P. and Prosser, D. J. (eds), *Characterisation of Fluvial and Aeolian Reservoirs, Special Publication, Geological Society*, **73**, London, 33–51.

Schumm, S. A. and Brakenridge, G. R. (1987). River responses in North America and adjacent oceans during the last glaciation. In Ruddiman, W. F. and Wright, H. E. (eds), *The Geology of North America, Vol. K-3*, Geological Society of America, Boulder, Colorado, pp. 221–240.

Schumm, S. A. and Khan, H. R. (1972). Experimental study of channel patterns. *Geological Society of America Bulletin*, **93** 1755–1770.

Schumm, S. A., Mosley, M. P. and Weaver, W. E. (1987). *Experimental Fluvial Geomorphology*, Wiley, New York, 413 pp.

Shen, H. W. (1991). Introductory remarks for the NATO workshop on movable bed physical models. In Shen, M. W. (ed.), *Movable Bed Physical Models*, Kluwer Academic, New York, pp. 1–15.

Southard, J. B., Smith, N. D. and Kuhnle, R. A. (1984). Chutes and lobes: newly identified elements in braiding in shallow gravelly streams. In Koster, E. H. and Steel, R. J. (eds), *Sedimentology of Gravels and Conglomerates, Canadian Society of Petroleum Geologists Memoir*, **10**, 51–59.

Wadman, D. H., Lamprecht, D. E. and Morosovsky, I. (1979). Joint geologic/engineering analysis of the Sadlerochit Reservoir, Prudhoe Bay Field. *Journal Petroleum Technology*, July 1979, 933–940.

Williams, P. F. and Rust, B. R. (1969). The sedimentology of a braided river. *Journal of Sedimentary Petrology*, **39**, 649–679.

Woodyer, K. D. (1975). Concave-bank benches on the Barwon River, New South Wales. *Australian Geographer*, **13**, 36–40.

Yalin, M. S. (1971). *Theory of Hydraulic Models*, Macmillan, London, 266 pp.

Young, W. J. and Davies, T. R. H. (1991a). Bedload transport processes in a braided gravel-bed river model. *Earth Surface Processes and Landforms*, **16**, 499–511.

Young, W. J. and Davis, T. R. H. (1991b). Prediction of bedload transport rates in braided rivers: a hydraulic model study. *Journal of Hydrology (NZ)*, **29**, 75–92.

# 7  Interrelationships between Bedload Transfer and River-bed Adjustment in Mountain Rivers: An Example from Squaw Creek, Montana

**P. E. ERGENZINGER, C. de JONG**
*Institut für Geographische Wissenschaften, Freie Universität Berlin, Germany*

**and**

**G. CHRISTALLER**
*Labor für Sensortechnik, Technische Fachhochschule Berlin, Germany*

## ABSTRACT

River-bed adjustment in coarse-grained mountain streams is part of a more complex site-interactive system than previously thought with a number of positive and negative feedbacks. An interpretation of river-bed adjustment solely in terms of the forces created by flowing water is inadequate. Only when detailed analyses of bedload transfer are incorporated can the reaction of the river bed be understood. At Squaw Creek, Montana, 35 m of river bed were monitored with an electronic magnetic detection system, giving an exact knowledge of amounts of incoming and outgoing bedload. Changes in geometry and roughness are closely related to discharge and amount of bedload transfer. Maximum change occurred in conjunction with maximum bedload transfer. A first-time opportunity was provided for measuring the parameters required for the FAST (Fluid and Sediment Transfer) theory. The FAST theory establishes that there exists an intensive interrelationship between river-bed adjustment and bedload transfer and, further, that during unsteady flood flow both river-bed geometry and bedload transfer are ruled by the development of flow cells. Only during large bedload pulses can these vortex cells be replaced by a more chaotic two-layered flow.

## INTRODUCTION

In order to bring light to the complex but presently poorly understood interrelationships between bedload transport and river-bed adjustment, fluvial experiments under natural conditions are a necessity. It has long been realized that in order to link near-bed fluid dynamics, bed structures and sediment transport in predictive modelling, an improvement in the often inadequate measuring techniques is required. Bedload transport and dynamically changing bed conditions need to be reviewed under a much higher spatial and temporal resolution than considered up to now. Continual measurements of bedload

*Process Models and Theoretical Geomorphology.* Edited by M. J. Kirkby
©1994 John Wiley & Sons Ltd

transport and adjustments in river geometry pose innumerable problems under natural conditions. Nevertheless, there is an urgent need for an empirical rather than theoretical approach in order to study the complex system interactions between river hydraulics, sediment transport and the related river-bed conditions.

The electro-magnetic pebble detector system at Squaw Creek, Montana, has been applied as one solution towards the problem of continual bedload measurement (Ergenzinger and Custer, 1983; Custer *et al.*, 1987). Under the coarse grain size spectrum available in rivers of this kind, bedload transport progresses in a very unsteady and pulsed manner (Reid *et al.*, 1985; Tacconi and Billi, 1987; Bänziger and Bursch, 1991; Bunte, 1991). Since bedload transport has been observed not to behave as a continuous process but analogous to single bedload pulses, there also have to be individual phases of changes in bed geometry. Up till the present, viable explanations for these bedload pulses which differ in magnitude and time over various temporal and spatial scales in coarse sediment mixtures in mountain regions are not satisfactory. At the Squaw Creek research site, pulses can be subdivided into micro-pulses over hecto-seconds, meso-pulses over minutes and macro-pulses over hours (Ergenzinger, 1992), but in this paper only macro-pulses are considered in view of more general interactive processes.

Reid *et al.* (1985) have explained pulses, with a mean periodicity of 1.7 hours, to occur as a result of kinematic waves of particles travelling in a slow-moving traction carpet. This is dependent on the availability of sediment and resistance of the bed to bedload movement. In other cases, bedform migration has also been correlated with the bedload transport rate (Kuhnle and Southard, 1988; Whiting *et al.*, 1988). Other pulses have been observed to occur without bedform migration (Klingeman and Emmett, 1982; Bunte *et al.*, 1987). Billi (1987) put forward an explanation where the amount of entrained material increases progressively up to a maximum, after which the sediment supply is exhausted. Although not directly comparable with gravel-bed rivers, Hoey and Sutherland (1991) identifies strong links between phases of adjustment in three-dimensional channel morphology and bedload pulses in sand-bed flume experiments. Ashmore (1988) explains pulses over several hours in terms of cycles of aggradation and degradation caused by migration of agglomerates of bars.

It is important to differentiate between processes operating in sand and gravel-bed rivers, braided and single channel systems as well as laboratory and natural river studies. Thus while dune and sheet transport is the dominant process of sediment transfer in sand-bed rivers, very different processes operate on coarser river beds. The macro-pulses have to be explained in terms of hydraulics and river-bed adjustment distinguishable into separate phases. According to most experiences there is no direct relationship between shear stress or shear velocity and bedload pulses. Using FAST (Fluid and Sediment Transfer) theory an attempt is made to overcome some of the shortcomings of the existing models and to stress the relationship between channel adjustment and bedload transport.

The aim of the paper is to outline an approach which sets out to theorize the dynamical system existing in mountain streams under very unstable fluid and bed transferring conditions into a theoretical framework. Three hypotheses are formulated; the first labels the interactive conditions as chaotic, with no apparent pattern or regularity in hydraulic conditions, bedload transport and bed adjustment. The second hypothesis states that a self-regulatory system exists between discharge and river-bed reactions that adjust to particular states in time. In the third, and most plausible hypothesis, it is suggested that we are observing an ambivalent process consisting of phases with a high amount of

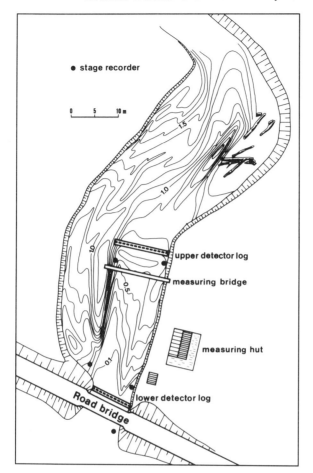

**Figure 7.1**  The measuring site at Squaw Creek in 1991 showing the location of the upper and lower detector logs, the measuring bridge with Tausendfüssler, and stage recorders. Cars pass the bridge infrequently and have no influence on the detector logs. Flow covers 5–10 cm of the right gravel bar only during very high flow events and therefore bedload transported will constitute <5% of total transport. Contours are in dm, unit numbers relative to height above lower detector log

bedload transfer alternating with regulated and transitional phases the behaviour of which can be explained more deterministically.

## FIELD SITE

Squaw Creek is a mountain stream that has its source in the Gallatin range, Bozeman County, Montana. It drains over a distance of 22 km before joining the Gallatin River which forms the headwaters of the Missouri. The study reach, which lies 42 km south of Bozeman, has a catchment area of 106 km² with Devils Point forming the highest peak at 3154 m. The central and upper parts of the study area consist of Tertiary volcanic rocks, mainly basic andesites containing up to 7% magnetite. The rest of the basin consists primarily of sedimentary and metamorphic rocks. Glacial and glacio-fluvial deposits can be found in the central valley. Average annual precipitation values lie around 810 mm. Floods occur most frequently during snowmelt events from mid-May to mid-June (Bunte, 1991). Flood events usually peak around 23:00 at the site, due to the time lag involved in the afternoon snowmelt, with an average discharge of $6\,m^3\,s^{-1}$. The study site is located along a straight reach below a river bend (Figure 7.1). Bed material at the study site ranges from gravel to cobble-sized clasts with a $D_{50}$ of 24 mm in the subsurface of the bar and 128 mm on the surface of the channel.

## MEASURING TECHNIQUES

### Electronic Measurements of Bedload Transfer

Individual grain transport was measured over two magnetically sensitive sills (Figure 7.1). This technique can detect all particles with $b$ axes $>30\,mm$, i.e., approximately 70% of all material in motion. The measuring technique, as described below, is based on the induction law which is usually described in terms of the second Maxwell equation:

$$\text{rot}\ E = -\,dB/dt \tag{1}$$

where $E$ is the electrical field strength and $B$ is the magnetic induction. Then, by integration:

$$\int E\ ds = -(d/dt)_A \int B_N dA \tag{2}$$

where $B_N$ is the normal component, $s$ is the length of a conductor and $A$ is the area enclosed by the inductor. With a magnetic flux of

$$\Phi = {}_A\!\int B_N dA \tag{3}$$

it follows that:

$$\phi E\ dS = -d\Phi/dt = u_{1q} = \text{source voltage} \tag{4}$$

If the coil has $N$ windings and the ferrite core used has a relative permeability of $\mu_r$, then:

$$u_{Nq} = -N\ d\Phi/dt = -N\mu_0\mu_r dH/dt. \tag{5}$$

At Squaw Creek up to 70% of the bedload material consists of magnetic andesites with a magnetic induction of between $10^{-8}$ and $10^{-6}$ tesla $(10nT–1\mu T)$. Depending on its course of motion, a signal voltage, $u_s$, is calculated by:

$$10\,nV < u_s < 10\,\mu V. \tag{6}$$

In order to obtain the required signal magnitude for the electronics an amplification of $10^3$ to $10^6$ is necessary.

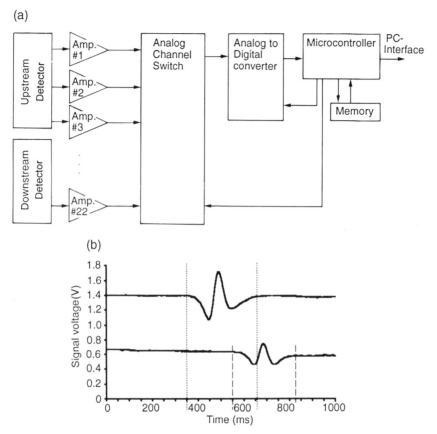

**Figure 7.2**    (a) Schematic diagram of the electronic bedload measuring system. (b) An example of the signal produced by a magnetic pebble passing the upper and lower detector logs

The electronics consist of four main parts: a double auto zero amplifier (DAZA), an analog multiplexer (MUX), a data acquisition controller (DAC) with digital converter (ADC) and some auxiliary electronics with a microcontroller (Figure 7.2a). The sensors consist of individual segments of ferrite-cored coils connected to the detector logs. Both the upstream and downstream detector logs contain these segments and have them arranged in double rows. The signals are fed through 22 twin wires to the amplifier. An inbuilt filter serves to reduce low-frequency noise produced by the vibration of the detector beam itself and the signal wires.

The data-processing system consists of a microcontroller with a control program and a PC (a 80386 SX-portable) with corresponding software. The microcontroller software is responsible for digitizing the amplifier output signals and transferring these values to the PC (Figure 7.2b). This includes controlling the analog multiplexers, the analog digital converter and serving the serial interface.

## River-Bed and Water-Surface Adjustment

Geometrical changes of the river bed were determined at hourly intervals over a 24 hour flood event in a cross-section using the 'Tausendfüssler' device attached to the measuring

(a)

## Upper Sill

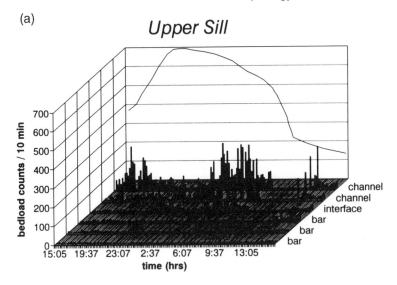

(b)

## Lower Sill

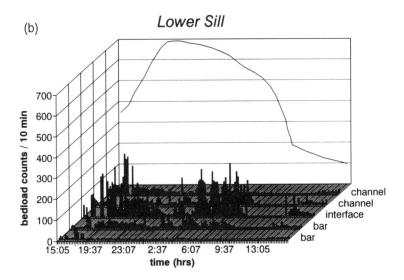

**Figure 7.3**  Bedload transfer versus flow stage during the flood event of 5–6 June 1991 for (a) upper sill and (b) lower sill 30 m further down. Individual bars represent bedload counts summed over 10 minutes. The detector logs are subdivided cross-sectionally into five or six separate segments representing bar, interface and channel

bridge (see Figure 7.1 and de Jong and Ergenzinger, 1992; Ergenzinger, 1992). This instrument measures the bed micro-topography and water-surface topography from a straight, fixed horizontal beam at 10 cm intervals with an accuracy of ±1 cm. Water-surface level was measured every 50 cm simultaneously with the river-bed measurement.

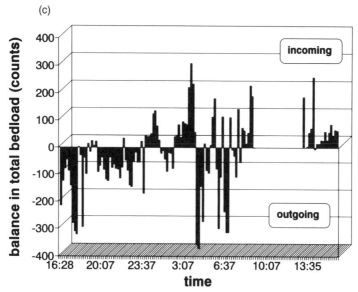

(c)

**Figure 7.3**(continued) (c) Net balance in bedload discharge between upper and lower sills. At the beginning as well as the end of the flood event, there is a large amount of erosion within the measuring reach, i.e. material is leaving the lower detector log without being registered over the upper logs

## RESULTS

### Hydrology

The selected flood event of 5–6 June 1991 was atypical, since it maintained an attenuated flood peak and an asymmetrical pattern of bedload transfer. The discharge hydrograph started as a typical snowmelt pattern, rising rapidly between 15:00 and 18:00 with an unusually broad discharge plateau resulting from prolonged rainfall and lasting from 20:00 to 23:00. The extended recession continued until 14:00 the next afternoon.

### The Nature of Bedload Transport

Bedload transport occurred in a series of pulses with no direct relationship to discharge (Figure 7.3a and 7.3b). According to previous observations, maximum bedload transport should be related to maximum changes in discharge, i.e. on the rising limb and in the recession. This has been verified by continual measurement of magnetic pebbles on the upper and lower sills.

The bedload transport counts registered approximately 70% of the material transported at Squaw Creek. The dataset generated was faulty in only two short instances. Neither on the upper sill nor on the lower sill did the pulses occur synchronously across the channel. On the upper sill, the bar was subject to their maximum pulses on the rising limb of the flood, but it was not until the flood recession, between 04:00 and 08:00 hours, that the main channel indicated the passage of its largest pulses. Only minor pulses (4–12 stones/minute over 1 hour) occurred during the rising limb of the flood whereas major pulses (> 14 stones/minute over 2–3 hours) occurred during the recession. Phases of steady transport intensities occurred over half an hour. The balance in total bedload (Figure 7.3c) demonstrates the striking asymmetry between the upper and lower sills.

148

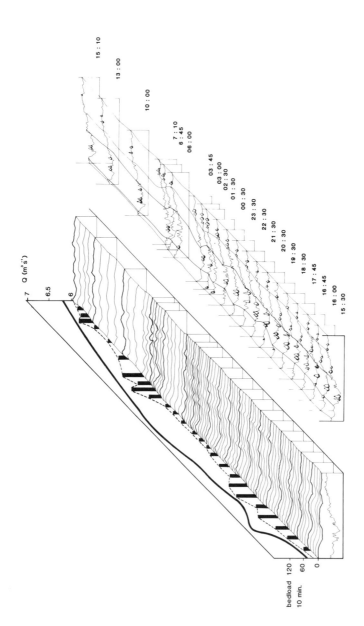

**Figure 7.4** Time-sequential illustration of river-bed (lower right) and water-surface (upper left) topographical adjustments across same cross-section over the 24 hour event reproduced from detailed Tausendfüssler measurements. Obstacles on the river bed represent bed elements that remained unchanged between one measurement period and the next. Longitudinal lines indicate changing subdivision between bar, interface and channel. Thick lines on the water surface represent measured cross-sections while thin lines were interpolated. Notice abrupt changes in bed geometry at 20:30 and 06:45 during the two major bedload pulses, as well as transition from even water surface to highly irregular surface at 06:45

During the rising limb and even during the large pulses of the falling limb there was more magnetic material leaving the reach. Pronounced accretion occurred only during the first phases of recession and towards the end of the flood. Under the prevailing conditions there is no direct relationship between hydraulic conditions and the amount of bed load transport.

## River-bed Adjustment

The river bed can be described as the level mobile boundary between the fluid flow, bedload and bed material zone. In order to study river-bed interactions, changes have to be investigated over short intervals of space and time.

River-bed adjustment was measured on an hourly time basis from the Tausendfüssler cross-profiles (Figure 7.4). The long interactive slope between the inner channel and the bar was defined as the interface. In principle this interface always existed between the main bar on the left and the channel but it was only poorly developed along the short right bar. The width of the inner channel changed rhythmically during the duration

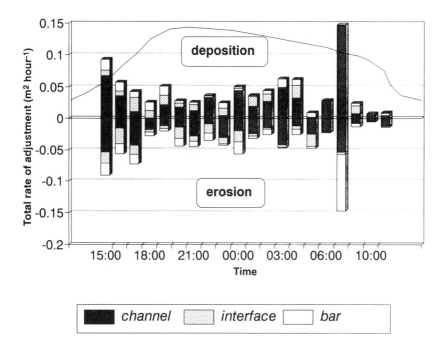

**Figure 7.5**  Hourly rates of erosion and deposition during the flood event, derived from comparisons between one hourly Tausendfüssler measurement series and the next. Note the marked total channel readjustment at the beginning of the flood and the overall shift to accretion during the major bedload pulse at 20:00. Most readjustment occurred during and after the main bedload pulse at 06:45 during the flood recession as heavy accretion in the main channel and heavy erosion on the bar

of the flattened peak. However, during the time of transit of the major bedload pulse in the early morning, maximum width occurred together with minimum depth and there was a shift of the inner channel towards the right bank.

If we compare the erosion and deposition rates as well as the overall rates of adjustment in $m^2$/hour for the channel, interface and bar, it is obvious that the channel remains most active (Figure 7.5). The channel is being eroded during several phases approximately every 2–4 hours. The time of peak discharge coincided with the period of minimum rate of adjustment. At the very beginning of the flood and during the major bedload pulse at 07:00, when there was maximum bedload transport, the highest rates of erosion and accretion occurred. Processes of erosion and accretion mostly remained in phase. The most stable area was invariably presented by the interface. It is important to remember that peaks in total rates of adjustment were built up during the flood wave in the inner channel up to their final stage at 06:00–07:00 hours. Changes increased in magnitude over several waves during the time of recession, right up to 07:00, ending abruptly after a secondary peak. With a starvation of sediment supply, rates of change declined definitively even in the region of the inner channel. As will be demonstrated in the context of roughness, river-bed adjustment is ultimately dependent on the development of flow cells and/or the passage of long-lasting bedload.

## Roughness

Since roughness has to be considered under a high spatial resolution, the Darcy-Weissbach approach using average roughness co-efficients could not be considered and the $K_3$

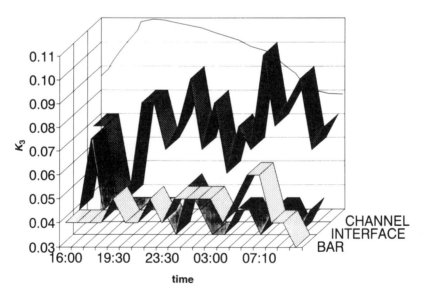

**Figure 7.6** Average $K_3$ roughness values varying over the 24 hours (5–6 June 1991) over the channel, bar and interface. Notice that the channel maintains the highest roughness values throughout, which build up and disintegrate cyclically. As soon as a bedload pulse is initiated, at 18:00 and 04:00, roughness decreases cross-sectionally to a minimum

roughness coefficient was applied instead (de Jong and Ergenzinger, 1992; Ergenzinger, 1992). Since energy dissipation by roughness is mainly due to turbulence created along the lower boundary, a description of local roughness in terms of the local difference in height is appropriate. Therefore the $K_3$ coefficient is calculated from the maximum vertical difference between three adjacent Tausendfüssler points, each 10 cm apart. The next $K_3$ value is calculated from the second last, last and new point, etc., as a gliding value of maximum differences across the entire section. When average $K_3$ values are calculated for the entire cross-section they can be divided into two phases, one reaching from the rising flood limb up to the end of the peak and the other one lasting throughout the recession (Figure 7.6). The $K_3$ values in the channel and interface drop sharply during the first phase of active sediment transport with the rising stage, but increase during non-transit bedload phases, only to drop to a minimum again at the beginning of the major bedload pulse at 05:00. Once bedload transport has ceased, roughness resumes its former state. These differences are also reflected in the pattern of maximum $K_3$ values. When the cross-section is differentiated into bar, channel and interface, average $K_3$ values are, however, far more dynamical indicators of changing roughness.

The $K_3$ values can be calculated in detail for every part of the cross-section (Figure 7.7). Again the time of maximum bedload transfer poses the most interesting situation. At 6:45 when there was a small lapse in bedload counts a very pronounced pattern of $K_3$ values appeared. Maximum $K_3$ values lying above 0.10 m occurred very regularly at 1 m intervals in the channel area. The interface area showed low roughness values whereas on the left bar maximum local values were associated with the distribution of large blocks. Half an hour later the roughness situation was similar on the interface and bar whereas the pattern in the area of the inner channel had changed extensively. In combination with extreme amounts of bedload transport, the regular pattern described above was replaced by roughness values of up to twice those previously and there was no longer any regular spacing. It required several hours before a situation as regular as that described previously could be re-established.

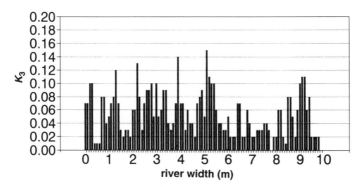

**Figure 7.7** Distribution of $K_3$ roughness values over the entire river cross-section during bedload pulse at 06:45, 6 June. Roughness peaks are spaced regularly at 1 m intervals in the channel/interface area

## Water-surface Dynamics and the Development of Flow Cells

Due to the relatively shallow water depths during flood peaks at Squaw Creek, the interrelationship between hydraulics and river bed is reflected in the upper boundary, i.e. the water surface. Discharge is distributed over a channel with a high width to depth ratio (40:1). Water levels never remain horizontal in cross-section – instead a relative height difference of up to 0.2 m (lateral inclination of up to 4%) occurred. Lateral inclinations are often higher than their longitudinal counterparts.

During the entire flood, two bulges were constantly maintained next to the boundary of the inner channel (Figure 7.4) of the Tausendfüssler cross-section. Waves were restricted to the inner channel and interface. Especially after the time of maximum bedload transfer, the number of waves increased from 3 to 7. It became obvious that the 20 cm depth recorded over the bar was insufficient to create such surface waves.

Waves of the type described at Squaw Creek can only exist in the presence of secondary circulation. Below the upper detector log with its hydraulic jump, in the straight channel reach, these waves could not possibly have been induced by momentum transfer through curvature. It has been shown that even if there is a slight influence from the curve above the measuring reach, any secondary circulation induced by this is destroyed by the disturbing effects of the upper detector sill. The secondary circulation could therefore either be caused by the channel banks and/or by large particles, large roughness elements or where there are large vertical differences in altitude.

The vortex flow cells formed particularly after periods of high bedload transfer and showed a distinct pattern of distribution. Water-level adjustment between 06:45 and 07:10 shows that distinct waves built up at 2 m and 4.5 m distance whereas troughs developed next to these waves at 0.5 m, 3.3 m and 5.3 m (Figure 7.8a). When these results are compared with the river bed across the same time interval, waves build up where there was a tendency towards accretion and troughs where there is erosion (Figure 7.8b). These steady dynamics were maintained over longer phases, not only shorter pulses.

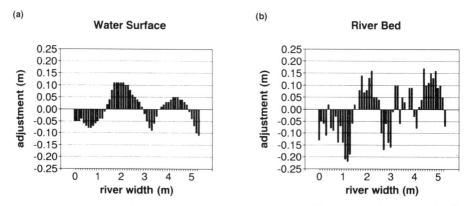

**Figure 7.8** (a) Water surface adjustment across channel and interface during major bedload pulse between 06:45 and 07:10, 6 June. Note regular 1 m wave amplitudes. (b) River-bed adjustment across channel and interface during major bedload pulse between 06:45 and 07:10, 6 June

The misfit between these measurements is due to the reluctance of the river bed to adjust to the hydraulics as fast as the water level.

## INTERPRETATION

### Causes of Bedload Pulsing

The cross-sectional adjustment of the river bed profile in response to bedload pulses will have to be explained in terms of the capacity and supply of sediment transport, which is ultimately limited by the bed structure and presence or absence of flow cells. Bedload pulses will be preferentially generated during smaller floods (average discharge of about $6 \, \text{m}^3 \, \text{s}^{-1}$) as is represented by the flood described in this paper when sediment that has accumulated in local deposits such as on a bar or in a river bend is set into motion again. The development of flow cells plays an important role in such processes. Due to the large spatial and temporal variation in the magnitude of these pulses, it will be necessary in future to describe their nature in terms of a probability approach. The probability of a bedload pulse occurring wil be ultimately dependent on the magnitude of discharge and time period involved and the organization of secondary flow. The impact of coarse particles in the uppermost layer on the river bed is particularly important. This positive feedback mechanism is mainly responsible for the enlargement of pulses. The kinematic wave theory proposed by Langbein and Leopold (1968) is not applicable for short transport paths. It is doubtful whether the changing shape of a bedload pulse will maintain a constant volume during its journey downstream when bed material is as highly interchangeable as observed during transport.

Roughness elements that cause large vertical differences are responsible for locally increasing vertical velocities which are in turn the driving mechanism of the flow cells. The differentiation between accretion, erosion and transport in cross-section is related to the development of flow cells which are enlarged and displaced locally within the wave system through a positive feedback mechanism. These minor oscillations within the hydraulic system are responsible for the creation and magnification of bedload pulses in coarse-grained material.

### Flow Cells and River-Bed Adjustment

A strong coupling effect was observed between vortex flow and river-bed adjustment. During phases of very low bedload transport surface waves develop. In order to maintain the waves elevated above the general level, there has to be a surplus in flowing water at certain locations. In contrast to these upwellings there has to be downwelling in the trough sites. In terms of mass balance this means that there has to be divergent flow below the trough and convergent flow below the upwelling waves. This is related to a strong tendency towards erosion on the one hand and accretion on the other. Where erosion proceeds, there is a strong tendency towards low $K_3$ roughness values, and where there is accretion roughness values tend to become higher. At the same time flow cell boundaries and bedload transfer streets occur in those localities where flow is divergent over the bed (Figure 7.9).

So far it has been assumed that within a reach as short as that presented at Squaw Creek, the input and output parameters of water and sediment discharge during an event have to be in equilibrium and that the river bed adjusts accordingly. If this is the case, then over a given period of time the sediment discharge should be calculable as a direct function of water discharge. This hypothesis could not be supported during the flood event of 5–6 June 1991. A mere distance of 35 m was sufficient to generate the main bedload pulse within the reach during the rising flood limb as recorded over the low sill, i.e. the reach was clearly experiencing an erosive phase. During the recessional stages, contrary to model predictions, the main pulse did cross the upper sill but elapsed before reaching the lower sill. In this situation the reach acted as a sediment storage depot. This indicates the importance of the buffering action of the bar and channel through river-bed adjustment. Since the bedload pulse travels along a local path and does not mobilize the entire river bed, the bed responds only locally.

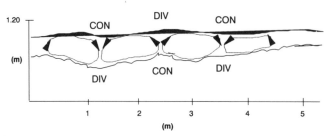

**Figure 7.9** The development of flow cells at 07:10, after the major bedload pulse, in converging and diverging zones over and between major roughness elements in the channel and interface. It is suggested that where flow is convergent at the river bed, the forces are erosive, thus lowering roughness values, whereas where flow is divergent at the river bed, high roughness values occur. This pattern is reflected at the water surface

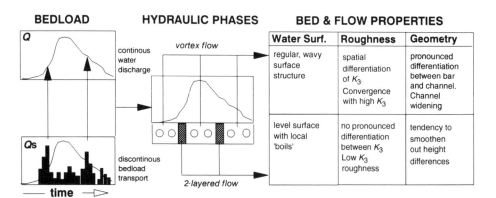

**Figure 7.10** Flow diagram showing FAST theory with interactions between discharge and bedload transport during a flood event. Stage is considered as a continuous variable while bedload transfer is of a discontinuous, pulsed nature and independent of discharge. Distinct hydraulic phases are proposed where two-layered 'chaotic' flow exists during bedload pulses, i.e. during the rising and falling flood stage, interchanging with phases of vortex flow development after bedload pulses. At Squaw Creek this pattern recurs for various flood events in the sequence modelled. During the vortex flow phases, the water surface is wavy, $K_3$ roughness values are high and there is a clear difference between channel and bar. During the two-layered flow phases the water surface becomes virtually level and roughness as well as geometrical differences within the channel diminish

## FAST Theory

The complexity of the interrelationships between hydraulic conditions, bedload and river-bed adjustment has been mentioned ubiquitously throughout the subject-related literature. Channel adjustment both in geometry and in roughness depends not only on the hydraulics of unsteady flow but also on the transfer of bedload and cannot be derived only from average shear stress values.

The FAST theory can be used to relate observed changes in geometry and roughness with changes in discharge and coarse bedload transport. According to the observations and measurements at Squaw Creek (Figure 7.10), it can be stated that:

- Discharge is only a general prerequisite for bedload transport and channel adjustment. When a threshold shear stress value of 50 Pa is surpassed during increasing water depth, it is with some probability that a resulting reaction will ensue.
- The erosion and mobilization of bed material is closely linked to phases of high bedload input into the local system. Due to the impact of coarse colliding bed particles there is a higher destabilization of the river bed and a positive feedback will develop with the particle exchange. These conditions must be taken into account. Erosion cannot be related only to the surpassing of critical shear.
- During phases of very intensive bedload transport, the water surface often has the appearance of 'boiling' water. The related turbulence creates a rather chaotic hydraulic situation which destroys the secondary flow cells and creates a far more level water surface. Roughness is lowered and the mobile bed layer will result in a 'two-layered' flow system (Wang and Larsen, 1993). In the two-layered system the lower layer is assumed to be the mobile, bedload transporting layer while the upper layer is the fluid transferring layer.
- When water depth is sufficient (deeper than about 20 cm) there is a strong tendency for flow cells to develop under high roughness and this is not only the case in the inner channel. This development is associated with very distinct surface and river-bed features. On the water surface, zones of upwelling raise the water level and zones of downwelling lower the water level. On the river bed there is a divergence of flow below the troughs with a strong tendency towards erosion and/or preferred longitudinal sediment transfer but at a much lower rate than during the 'two-layered' flow and with relatively low roughness. In areas of convergence, relatively high roughness is coupled with low bedload transfer and a strong tendency towards accretion.
- River-bed adjustments are strongly connected to changing conditions, not only of water discharge but also of bedload transfer and related differences of hydraulics. Models of river-bed adjustments and coarse material bedload transport must take the impact of moving particles into consideration.
- The spacing of roughness elements is such that flow resistance is minimized during phases of maximum bedload transport and maximized during phases of minimum bedload transport. Similar results were generated in a flume (Kuhnle and Southard, 1988). When the general flow structure organizes itself into individual flow cells after the passage of a bedload wave, the morphology of the river bed is no longer required to encourage the smooth and efficient transfer of sediment. Separate roughness elements and a special bed geometry may develop.

The flood investigated above presents a case where only part of the river bed is mobile over long intervals and where there are only local transport paths. If only partial mobility of the channel exists, then the kinematic wave approach cannot be applied for the description of coarse material bedload transport. Indeed 35 m of differentiated river bed are sufficient to change the pattern of bedload pulses. Maximum adjustment of geometry and roughness has a very close relationship to moving particles. River-bed erosion is not only dependent on water transfer and shear stress but also on the impact of moving particles.

According to these results, the hypothesis that the interactive relationships between river-bed conditions, bedload transfer and water discharge are simply chaotic can be refuted (hypothesis 1). There exists a highly self-regulatory system (hypothesis 2) which is mainly reflected by its different hydraulic phases. A close relationship is in existence between the different types of runoff conditions (vortex flow or two-layered flow) and the related boundary conditions.

In the third hypothesis it was stated that during the more regulated phases the characteristics of river-bed roughness and morphology depend on the evolution of secondary flow circulation, i.e. development and decay of flow cells during and after bedload pulses. Nakagawa *et al.* (1991) also suggest that in a roughness sublayer of small relative submergence, an organized flow pattern exists under the influence of individual roughness elements. The development of such flow cells even in straight reaches allows particular sections of the river bed to approach equilibrium. It is further suggested that the river bed is adjusting so as to accommodate an optimal roughness that will facilitate bedload and water transfer.

Due to the difficulties encountered in measuring three-dimensional velocity components, the hydraulics of vortex and of 'two-layered' flow could only be derived from their indirect effects. An improved understanding of the interrelationships between bedload transfer and river-bed adjustment remains a challenge in terms of hydraulic measurements.

## CONCLUSIONS

Attempts to derive processes of river-bed adjustment from the dynamics of flowing water alone fail under boundary conditions posed by coarse-bedded mountain rivers. Experiences from Squaw Creek indicate that these processes are a function of both hydraulic conditions and dynamics of bedload transfer. These interrelationships need to be analysed in future on the basis of the FAST theory.

Using the relatively low magnitude flood of 5–6 June 1991 as a selected example, two distinct hydraulic phases can be differentiated during the period of bedload transfer. On the whole there is a strong tendency towards the development of flow cells, but during phases of intensive bed-material transport this situation is replaced by more chaotic flow. Simultaneously these phases respond to the most active periods of river-bed adjustment. It is suggested that under these conditions the flow cells are replaced by a turmoil of two-layered flow. At the same time roughness and water surface slope react quite discretely. In the channel water slope is declining and $K_3$ roughness values increase up to 10 cm, whereas on the bar water slope is inclining and $K_3$ roughness is decreasing. Our current model suppositions are insufficient to explain these situations in detail. In future it will be necessary to widen our scope of understanding and modelling by increasing our knowledge on the:

- development of flow cells,
- development of two-layered flows,
- critical conditions for bed-material erosion through particle impact.

At present there are only a limited number of measuring sites and coarse-material flumes where a solution to these problems can be approached.

## ACKNOWLEDGEMENTS

The authors wish to thank their co-worker K. Bunte for field assistance and stage curves, J. Bruggemann and U. Achter from the Technische Fachhochschule Berlin for the electronics, J. Palmer and J. Hogin for the woodwork and hardware, and S. Custer, C. Riley, S. Lohr and J. Cassano from Montana State University for their assistance in the field. We are also grateful to Dirk Wiydenes and two anonymous referees for helpful reviews. The research was funded by the DFG (German Research Council) and by the European Research of the US Army, and Standardisation Group in cooperation with Paul Carling, Freshwater Biological Institute, Windermere, UK. CDJ is in receipt of a Studienstiftung scholarship.

## REFERENCES

Ashmore, P. E. (1988). Bed load transport in braided gravel-bed stream models. *Earth Surface Processes and Landforms*, **13**, 677–695.

Bänziger, R. and Bursch, H. (1990). Geschiebetransport in Wildbächen. *Schweizer Ingenieur und Architekt*, Part **24**.

Billi, P. (1987). Sediment storage, bed fabric and particle features of two mountain streams at Plynlimon (mid-Wales). University of Florence. *Institute of Hydrology, Report no. 97*.

Bunte, K. (1991). Untersuchungen zur zeitlichen Variation des Grobgeschiebetransportes und seiner Korngrößenzusammensetzung (Squaw Creek, Montana, USA). Ph.D. Thesis, Freie Universität, Berlin.

Bunte, K., Custer, S. G., Ergenzinger, P. and Spieker, R. (1987). Messungen des Grobgeschiebe-transportes mit der Magnettracertechnik. *Deutsche Gewässerkundliche Mitteilungen*, **31** (2/3), 60–67.

Custer, S. G., Ergenzinger, P., Anderson, B. C. and Bugosh, N. (1987). Electro-magnetic detection of pebble transport in streams: A method for measurement of sediment-transport waves. In Ethridge, F. and Flores, R. (eds), *Recent Developments in Fluvial Sedimentology, Society of Economic Paleontologists and Mineralogists Special Publication*, **39**, pp. 21–26.

de Jong, C. and Ergenzinger, P. (1992). Unsteady flow, bedload transport and bed geometry responses in steep mountain torrents. In Larsen, P. and Eisenhauer, N. (eds), *Proceedings of the 5th International Symposium on River Sedimentation, Sediment Management*. Vol. 1, Karlsruhe, pp. 181–193.

Ergenzinger, P. (1988). The nature of coarse material bedload transport. In Walling, D. E. and Bordas, M. (eds), *Sediment Budgets*, IAHS, Porto Allegro, pp. 207–216.

Ergenzinger, P. (1992). River bed adjustments in a step-pool system: Lainbach, Upper Bavaria. In Hey, R. D., Billi, P., Thorne, C. R. and Tacconi, P. (eds), *Dynamics of Gravel Bed Rivers*, Wiley, Chichester, pp. 415–430.

Ergenzinger, P. (in press, 1993). Approaching problems of bed and bedload dynamics – F.A.S.T.! In Ergenzinger, P. and Schmidt, K. H. (eds) *Dynamics and Geomorphology of Mountain Rivers*, Springer-Verlag, Benediktbeuern.

Ergenzinger, P. and Custer, S. G. (1983). Determination of bedload transport using naturally magnetic tracers. First experiences at Squaw Creek, Gallatin County, Montana. *Water Resources Research*, **19** (1), 187–193.

Hoey, T. B. and Sutherland, A. J. (1991). Channel morphology and bedload pulses in braided rivers: a laboratory study. *Earth Surface Processes and Landforms*, **16**, 447–462.

Klingeman, P. C. and Emmet W. W. (1982). Gravel bedload transport processes. In Hey, R. D., Bathurst, J. C. and Thorne, C. R. (eds), *Gravel-Bed Rivers*, Wiley, Chichester, pp. 141–179.

Kuhnle, R. A. and Southard, J. B. (1988). Bedload transport fluctuations in gravel bed laboratory channel. *Water Resources Research*, **24**, 247–260.

Langbein, W. B., Leopold, L. B. (1968). River channels, bars and dunes-theory of kinematic waves. *U.S. Geological Survey Professional Paper*. 122L, L1–20.

Nakagawa, H., Tsujimoto T. and Shimizu Y. (1991). Turbulent flow with small relative submergence. In Armanini, A. and Di Silvio, G. (eds), *Fluvial Hydraulics of Mountain Regions*, Springer-Verlag, Berlin, 33–44.

Reid, I. and Frostick, L. E. (1986). Dynamics of bedload transport in Turkey Brook, a coarse-grained alluvial channel. *Earth Surface Processes and Landforms*, **11**, 143–155.

Reid, I., Frostick, L. E. and Layman J. T. (1985). The incidence and nature of bedload transport during flood flows in coarse-grained alluvial channels. *Earth Surface Processes and Landforms*, **10** (1), 33–44.

Tacconi, P. and Billi, P. (1987). Bedload transport measurements by the vortex-tube trap on Virginio Creek, Italy. In Hey, R. D., Bathurst, J. C. and Thorne, C. R. (eds), *Gravel-bed Rivers*, Wiley, Chichester, pp. 583–616.

Wang, Z. and Larsen, P. (in press, 1993). Turbulent structure of flows of water and clay suspensions with bedload motion. In Ergenzinger, P. and Schmidt, K. H. (eds) *Dynamics and Geomorphology of Mountain Rivers*, Springer-Verlag, Benediktbeuern.

Whiting, P. J., Dietrich, W. E., Leopold, L. B., Drake, T. G. and Shreve, R. L. (1988). Bedload sheets in heterogeneous sediment. *Geology*, **16**, 105–108.

# 8 Towards a Model of Changes in Bed Material Texture at the Drainage Basin Scale

STEPHEN RICE

*Department of Geography, University of British Columbia, Canada*

## ABSTRACT

A negative exponential function is often used to describe the downstream diminution of grain size parameters along a single drainage line. As a model of grain size change it is unsophisticated and provides little predictive capability at the drainage basin scale. A preliminary model is presented in which the drainage basin is considered as a network of interdependent hillslope–channel links which differ in their sediment supply and storage characteristics. Particular emphasis is placed on the drainage network, mixing of discrete sediment populations at confluences, and the role of non-alluvial sediment sources, channel boundaries and obstructions. Empirical data collected in the Queen Charlotte Islands, British Columbia, illustrate the nature of grain size changes in hillslope-channel coupled links, and confirm the model's *a priori* expectation that bed material texture varies stochastically in headwater areas.

## INTRODUCTION

This paper introduces a preliminary model of variations in grain size at the drainage basin scale. A negative exponential function is commonly used to describe the undifferentiated effects of both abrasion and selective transport in causing grain size diminution in a downstream direction (Church and Kellerhals, 1978; Knighton, 1980). When calibrated using empirical data it is generally consistent with the overall trend of the observations, but scatter about the relation is usually significant. Even as a functional, if not a physically based model, it is unsophisticated and is applicable only to a single drainage line. It therefore provides little predictive capability at the basin scale.

In the approach outlined here particular emphasis is placed on the network of drainage links, the combination of discrete sediment populations at network confluences, and the importance of non-alluvial sediment sources, channel boundaries and obstructions. A general discussion of these key elements is followed by a description of the model currently under development. While the relevance of sediment inputs at tributary junctions has been widely recognised (Miller, 1958 *et seq.*), if not thoroughly investigated, non-alluvial factors have not been explicitly considered. In the final section of this paper, field data from the Queen Charlotte Islands, BC, are therefore used to illustrate their significance.

*Process Models and Theoretical Geomorphology.* Edited by M. J. Kirkby
© 1994 John Wiley & Sons Ltd

**Networks and Tributaries**

Discrete grain size discontinuities in an underlying systematic trend are often observed at tributary junctions (Miller, 1958; Church & Kellerhals, 1978; Knighton, 1980; Shaw & Kellerhals, 1982; Dawson, 1988). Knighton (1984) suggests that inputs of distinct material at confluences are superimposed on the systematic modification of main-stem sediments by sorting and abrasion processes.

The implications of understanding the tributary effect go beyond the ability to predict textural changes along a main-stem channel. If tributaries are a source of variation then sediment texture locally is intimately linked to drainage network topology. In turn an understanding of the tributary effect opens up the possibility of predicting basin-wide variations largely on the basis of cartographically derived network information and some headward data. Pizzuto (1991) explains grain size changes in a small Pennsylvanian catchment in terms of the pattern of tributary inputs rather than specific geomorphic processes.

**Non-alluvial Sediment Storage and Supply**

Within a drainage network the composition of the bed material supplied to any point may be primarily alluvial (delivered by flowing water) or colluvial (delivered by gravity) or it may be lag material from some prior regime. The degree of channel–hillslope interaction is important in this respect, and is reflected by the frequency of non-fluvial sediment inputs. In the headward areas of a drainage basin such sediment sources are likely to be common and to be active often (Church, 1983). Consequently the stream may be described as 'strongly coupled' to the adjacent land surface.

Associated with strong coupling in headward reaches is the presence, in the channel, of non-alluvial boundaries and obstructions. Bedrock is generally close to the surface, large boulders of colluvial origin are common and, in forested areas, large organic debris (LOD) is supplied in abundance by debris torrents, debris slides, bank erosion and blowdown. Fans and other depositional forms associated with mass movements and steep tributaries may also be present in the valley bottoms. Sediment storage is often controlled by the trap efficiency of these elements and their effect on local water surface slope. Since sediment storage is not then determined by the reciprocal relations between flow hydraulics and a mobile channel boundary, these elements invalidate the basic assumption of channel self-formation (Lisle, 1987).

In general, the impact of coupling and non-alluvial storage controls can be expected to decline downstream as a consequence of the downstream increase in the magnitude of fluvial activity relative to hillslope activity. As discharge increases the river has increasing power to remove and assimilate non-alluvial sediments and obstructions. As sediment discharge increases the relative impact of colluvial contamination diminishes and flood-plain deposits accumulate. These buffer the active channel from colluvial inputs and reduce the occurrence of bedrock controls. As the channel widens and supply is limited to flotation from upstream, large organic debris becomes less abundant (Keller & Swanson, 1979) and less significant. Depositional forms become less common where tributary gradients decline and the main-stem channel is competent to rework tributary sediment yield.

The rate at which these factors become less important downstream, and the spatial distribution of these phenomena depend on basin morphometry, biophysical conditions and geomorphological history. Relatively strong coupling is evident in the headwaters of all drainage basins which transport sediment, and may continue to be important for some distance downstream.

Although not discussed in such terms, coupling and non-alluvial storage controls are a source of unexplained variation in a number of downstream fining studies which utilise the exponential decline function. MacPherson (1971) reports an overall downstream decline in grain size along Two O'Clock Creek, a short, steep tributary of the North Saskatchewan River in the Canadian Rockies. Scatter about the relation is very high, however, and the model yields an $r^2$ value of 0.24. As MacPherson (1971) points out, the persistent delivery of fresh colluvium and till, which mantle the steep slopes of the basin, is probably responsible for this variance. Miller (1958), Bradley *et al.* (1972) and Brewer & Lewin (1990), comment on the entrainment of glacial material along their study channels, and Dawson (1988) illustrates how tributary fans cause local variations in competence. A comprehensive model of grain size trends at the drainage basin scale must then recognise the significance of non-alluvial sediment sources, boundaries and obstructions.

## MODEL DEVELOPMENT

### Link Subsystems

The drainage basin consists of interdependent hillslope and channel components. The channel component can be subdivided into progressively smaller networks and ultimately described as a set of channel reach subsystems, each of which is positioned between two hillslopes. Considering the drainage basin in this way allows one to isolate within-reach variation from between-reach variation and so study downstream trends and tributary effects independently.

A link is here defined as a portion of channel connecting adjacent network junctions (equivalent to a channel link in network geometry). Each link can be conceptualised as a discrete subsystem which consists of various sediment and water inputs, storage elements, and sediment and water outputs. Storage of sediment can occur in response to hydraulic relations or in association with non-alluvial controls. In the former case short-term stores (the active bed material and bar deposits), and long-term stores (the floodplain adjacent to the channel, and any vegetated islands), can be expected to exhibit predictable longitudinal structure. In contrast storage associated with non-alluvial controls does not exhibit such structure.

The nature of sediment input to individual links varies as a function of the degree of coupling within the link. One can identify two extreme cases which represent the ends of a continuum: strongly coupled links, and completely buffered links. An appreciation of their characteristics and the modelling implications is necessary before it becomes clear how links of intermediate character might behave, and how they can be handled within the model.

### Strongly Coupled Links

In the absence of a buffering alluvial floodplain material is transferred directly from the adjacent hillslopes to the channel by creep and episodic mass-movement processes.

The spatial and temporal frequency of these transfers ensure an almost continuous supply of non-fluvial sediment along the length of the channel. The colluvial inputs consist of a wide range of grain sizes and are often poorly sorted. Non-alluvial storage controls are common.

It is hypothesised that the combined effect is to preclude the development of systematic downstream variation in texture which fluvial sorting and abrasion processes might otherwise produce. Grain size in the channel fluctuates in response to hillslope inputs and local sorting around non-alluvial storage controls. Tributary inputs may be indistinguishable as discontinuities within the resulting signal. Such links may be regarded simply as suppliers of sediment to the downstream system. Within these links grain size is essentially independent of systematic fluvial processes and needs to be determined independently.

## Completely Buffered Links

Floodplain deposits are sufficiently extensive to prevent colluvial material from reaching the channel so that the only sediment entering the link is that transported fluvially from the upstream channels. Sediment storage is characterised by regularly spaced, hydraulically controlled bars and non-alluvial storage controls are not present. Within-link exchanges of sediment involve material of fluvial provenance only; that is, the material in the floodplain is alluvial and reflects the existing transport regime.

Abrasion and sorting (selective entrainment and selective deposition) modify the sediment supplied at the head of the link. Neither process is well understood and despite a long history of interest (see Werrity, 1992, for a review), and recent attempts to isolate their roles (Parker, 1991; Werrity, 1992), the general nature of the modification process remains enigmatic. However, identification and parameterisation of the modification process is of secondary importance to the model introduced here. The key point is that within completely buffered links alluvial control provides a basis for computing grain size changes.

Presently the modification process is best approached via empirical assessment of the relation between sediment texture and distance downstream *within links*. Observations suggest that a negative exponential model describes this relation (Church & Kellerhals, 1978; Knighton, 1980):

$$D = D_0 e^{-\alpha L}$$

where $D$ is grain size, $D_0$ is grain size at $L = 0$, $L$ is distance downstream and $\alpha$ is a calibration coefficient $> 0$.

Whatever model is forthcoming it is reasonable to expect that the rate of textural modification with distance is determined by the hydraulic conditions in the link and the abrasion characteristics of the sediment. One can expect both the rate of abrasion and sorting to increase with increasing discharge as sediment mobility and grain-to-grain interactions increase (at least above the sand–gravel boundary). It is therefore supposed that for a given lithological type there is some relation between hydraulic conditions and the calibration coefficient (e.g. $\alpha$), in the modification model.

In most published calibrations of the negative exponential model, 'grain size' has been taken to be a particular percentile or moment of the cumulative size distribution.

These studies do not therefore constitute studies of the physical processes which cause diminution of particular grains nor do the coefficients represent the true rates of the processes involved. Rather they report the empirical rate of change of a statistical parameter. In this sense they are purely descriptive and reported values of $\alpha$ are in part statistical artifacts of the sediment distributions investigated. In applied situations modelling the behaviour of distribution characteristics may be preferable for predictive purposes, though it fails to add to our physical understanding.

## Changes in Texture and Hydraulic Conditions at Confluences

Where any two links meet, the sediment outputs of each link are mixed to produce the sediment input to the downstream link. The texture of the mixed sediment depends on the textures of the two inputs and their relative volumes. The texture of these will usually differ because the upstream basins from which they are derived are likely to vary in size, morphometry, and possibly dominant lithology. Even where initial inputs are similar the material will have moved through a different number of links of different sizes and will be modified according to these specific circumstances. Summation of the respective grain size fractions in proportion to relative sediment discharge will determine the texture of the mix. As long as the mix is not affected by the complex hydraulic conditions in the confluence (for example preferential removal of certain fractions in the flow separation zones), then the mixed material provides the sediment input to the modification process in the downstream link.

The hydraulic conditions within the downstream link are determined by the combination of the flows at the upstream confluence. It is assumed that major changes in hydraulic variables occur at confluences in response to the significant change in water volume, and that within-link variations are small in comparison. Thus, although conditions will vary locally on a pool-riffle scale, departures from the mean link condition are assumed to be insignificant (Richards, 1980).

The sum of the two upstream discharges must equal that in the channel below. With appropriate hydraulic geometry relations it is possible to determine changes in width, depth and other derived conditions at a confluence using discharge information:

$$x_c = x_1 [ (Q_1 + Q_2)/Q_1 ]^f$$

where $x$ is a hydraulic parameter, e.g. depth, $Q$ is discharge, f is the appropriate hydraulic geometry exponent, and c, 1, 2 refer to the confluent channel and the two upstream channels, respectively. This is one of several approaches. Miller (1958), Richards (1980) and Roy & Roy (1988) offer alternatives.

Changes in both sediment texture and hydraulic characteristics are assumed to occur essentially instantaneously at confluences. Input parameters for the modification process in the downstream link are reset, and can be computed from the output parameters of the two upstream links.

## Network Propagation

Consider a hypothetical basin which consists only of links which are either strongly coupled or completely buffered.

Theoretically the output texture of any buffered link can be determined. If the sediment output from any two links and the discharge within those links is known then, with an appropriate relation between hydraulic characteristic and modification model exponent (e.g. $\alpha$), the texture of the output from the downstream link can be computed. The volume of any input can be determined as the sum of all the outputs from the strongly coupled basins upstream of this point. If one knows the texture of the material leaving the strongly coupled links in the headwaters and one can estimate the volume of material leaving each one, one can work through the network using the modification model within links and a mixing function at confluences to estimate texture at any point in the basin. Where selective transport is dominant, such that a net zero change in sediment throughput cannot be assumed, this approach would be buttressed by within-link sediment transport considerations. In order to apply the modification model it is also necessary to determine the hydraulic characteristics in any link. If one can measure or somehow estimate the discharge of the fingertip links (half of the links in the basin) then this is possible using the continuity relation for discharge.

### Intermittently Coupled Links

Links which are neither strongly coupled nor strongly buffered but which are intermittently coupled to varying degrees are most common in the landscape. Point sediment sources may occur where active hillslope processes abut the channel or where the floodplain is discontinuous and bank erosion undercuts the toe of the valley side. At such points non-fluvial material is introduced to the channel and is expected to perturb the systematic sorting–abrasion modification. If such sources can be identified they can be incorporated into the framework described above by treating them as dry tributaries; that is as tributaries which introduce sediment but not water. The non-fluvial input will be mixed with the modified link input to produce a new texture. The same mixing function can be used as is used at confluences as long as the texture and volume of the new material can be estimated. The sediment parameter of the modification function is reset but the new mixture is modified according to the same hydraulic conditions.

As it is presented here a distinction is made between the strongly coupled links and these intermittently coupled links. In fact the former are an extreme case of the latter, and could be considered as such in practice. This becomes inexpedient when the frequencies of 'dry tributaries' and non-alluvial stores preclude any between-input systematic modification. At this point the link involved must be regarded as a strongly coupled link. The sediment which it supplies to the downstream system must be identified independently as an information requirement of the predictive model.

### Operationalising the Model

Sketching such a basin-wide model of grain size change conceptually is valuable since it reveals the issues which need to be addressed. The following issues are being investigated in a suite of rivers in western Canada:

1. Can a general within-link sediment texture modification function be isolated?
2. Can a general relation between within-link hydraulic characteristics and modification function exponents be identified? (It is anticipated that lithological control will be necessary in order to achieve this.)

3.  Is sediment mixing at confluences a straightforward summation of corresponding grain size fractions or is selective deposition of certain fractions evident?
4.  Are within-link variations of hydraulic characteristics significantly less than between-link variations?

Assuming that these issues are resolved the model described above may ultimately provide a feasible method of predicting local grain size throughout a drainage basin. Information requirements within a particular drainage would be as follows.

1.  Delineation of the network.
2.  Identification of sediment sources, their texture and volume.
3.  Identification of fingertip discharges and a set of hydraulic geometry relations appropriate for the river involved.

While these are by no means trivial, they are not prohibitive. Much of this information can be garnered from maps and aerial photographs, but field measurements would be unavoidable. Based on limited gauging, relations between basin area and runoff, and basin area and sediment discharge, may prove to be expedient and adequate in this respect. Given the inadequacy of the extant model and the costly, time-consuming nature of bed material sampling, a predictive model with limited input requirements is desirable in a variety of applied contexts.

## TEXTURAL VARIATION IN STRONGLY COUPLED BASINS ON THE QUEEN CHARLOTTE ISLANDS, BRITISH COLUMBIA

Existing empirical work is largely restricted to intermittently coupled reaches. Least is known about the nature of grain size changes in strongly coupled links, which are prevalent in basin headwaters. The *a priori* ideas about these links therefore require empirical investigation. The major issue is whether or not systematic downstream structure can develop given the supposed dominance of colluvial inputs and non-alluvial storage controls. To this end a study of textural variation in strongly coupled basins was conducted on the Queen Charlotte Islands, British Columbia.

The Islands lie 130 km off the coast of British Columbia (Figure 8.1). Climatic, geologic, topographic and tectonic conditions promote high rates of sediment production and rapid mass wasting (Roberts & Church, 1986). Small ($< 50$ km$^2$), linear, steep basins which lack a buffering alluvial floodplain are typical, and material is transferred directly from the adjacent hillslopes to the channel. Gimbarzevsky (1986) estimated an average major slope failure frequency of 1.0 per square kilometre over the entire landmass. Rood (1984) estimated that 43% of the total volume of sediment mobilised by mass wasting directly enters the stream system. Additional colluvial material is delivered to the streams by bank erosion (Roberts & Church, 1986) and creep.

Debris torrents and debris slides introduce substantial amounts of large organic debris (LOD: logs, limbs and root boles greater than 10 cm in diameter). Additional LOD is supplied by bank erosion, windthrow and snowloading. The volume of LOD per unit channel area varies from 0.03 to 0.06 m$^3$m$^{-2}$ (Hogan, 1986). Channel stability and morphology are strongly dependent on the spatial and temporal dynamics of LOD accumulations (log jams) which occur at a variety of scales (Hogan, 1989).

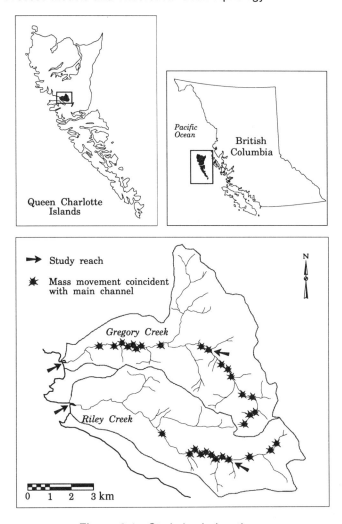

**Figure 8.1**   Study basin location

The variations of bed material texture were examined in two study creeks (Figure 8.1). Details of the sampling and analysis can be found in Rice (1990, 1993). Mass-movement events coincident with the main channel are indicated on Figure 8.1 following mapping by J. Schwab & D. Hogan (pers. comm., 1990). For the purposes of this discussion the behaviour of the median surface grain size ($D_{50}$) is considered.

A longitudinal survey was conducted within each creek. In each main-stem link between three and ten sample sites were located at regular intervals irrespective of local LOD and hillslope inputs (Figure 8.2). Consistent use of riffle-pool breaks as sampling positions at each site ensured that local-scale variability was minimised (Wolman, 1954; Church & Kellerhals, 1978). $D_{50}$ was established on the basis of 100-stone Wolman samples. Significant between-site variations are apparent within several links (Table 8.1). However, only in link 2 in Gregory Creek is a significant, systematic, clearly exponential trend observed (Figure 8.3). In general the data behave erratically and in one case there is systematic downstream coarsening. Although a perturbed exponential signal might be interpreted in several links it is clear that there is no systematic modification at this resolution.

Higher resolution photographic survey data provide a detailed picture of grain size variation within four links over distances of up to 1.9 km (Figure 8.4). The technique used is based on establishing an empirical relation between median grain size and the number of particles exposed on the surface in a given area. In general grain size fluctuates erratically over short distances and shows no systematic downstream modification.

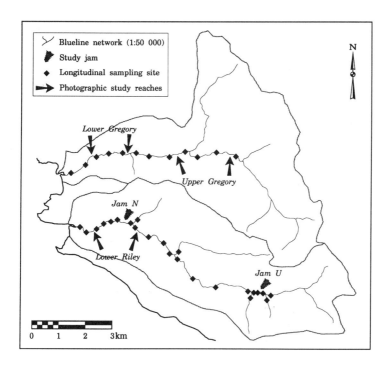

**Figure 8.2**  Sampling sites and study jams

**Table 8.1**  Within-link variance of longitudinal survey surface $D_{50}$

| Link | $\sigma^2 B$ | $n_1$ | $\sigma^2 W$ | $n_2$ | $F$ | $F$crit | Signif. |
|------|------|------|------|------|------|------|------|
| Surface $D_{50}$ Riley | | | | | | | |
| 1 | 0.17 | 6 | 0.02 | 8 | 7.62 | 3.97 | Yes |
| 2 | 0.05 | 4 | | | 2.21 | 4.35 | No |
| 3 | 1.24 | 4 | | | 56.19 | 4.35 | Yes |
| 4 | 0.02 | 3 | | | 0.70 | 4.74 | No |
| Surface $D_{50}$ Gregory | | | | | | | |
| 1 | 0.52 | 6 | 0.02 | 8 | 23.47 | 3.97 | Yes |
| 2 | 0.34 | 3 | | | 15.43 | 4.74 | Yes |
| 3 | 2.21 | 3 | | | 99.86 | 4.74 | Yes |

*Note*: $F$ is the ratio of between-site variance ($\sigma^2 B$) to within-site variance ($\sigma W$). $F$ crit is for $\alpha = 0.05$, ($n_1 - 1$, $n_2 - 1$) degrees of freedom

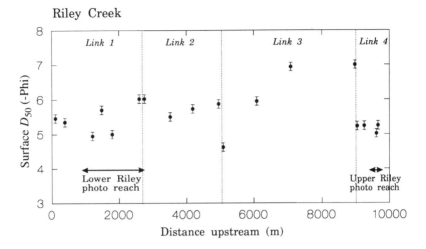

Riley Creek

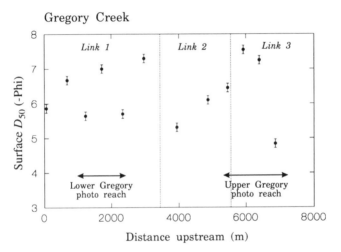

Gregory Creek

**Figure 8.3**  Longitudinal survey surface $D_{50}$. Replications provide an estimate of within-site variance and error bars ($\alpha = 0.05$) are shown. Distances are from the ocean

In a few places relatively systematic grain size declines over limited distances are observed. Discontinuities are apparent but these are not associated with tributary inputs.

Sediment sources (mass movements, eroding banks, gullies) are so numerous that they are difficult to isolate, but evidence of their importance is available in the photographic survey data. The variance of surface $D_{50}$ is markedly less in Lower Riley than it is in the other study reaches (Figure 8.4). While all types of input are common in the latter, a limited valley flat hinders mass movement inputs in the former (Figure 8.1).

Individual log jams can have a significant impact on local sediment texture. In the most simple case a log jam acts as a dam which is impermeable to sediment. This barrier leads to aggradation upstream and local channel gradient is reduced. Entrapment of material prevents resupply of the mobile fractions to the bed downstream and, supposing selective transport is in operation, relative coarsening will occur there. Upstream the

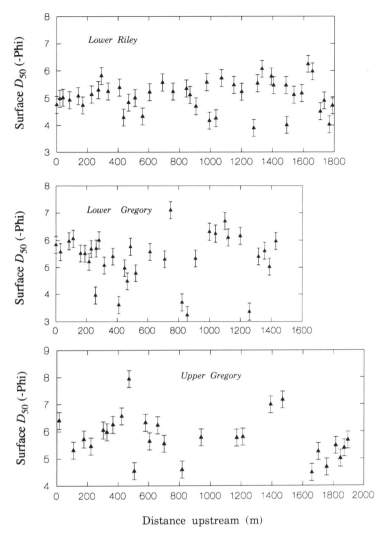

**Figure 8.4**  Photographic survey surface $D_{50}$. A within-site variance estimate and confidence limits on the number – $D_{50}$ calibration curve, provide an error bar ($\alpha = 0.05$)

most mobile fractions will constitute the largest portion of the material being trapped, and consequently, relative fining will occur here (Figure 8.5). The ambient rate of sediment transport and the stability and longevity of the jam are important determinants of the magnitude of the upstream/downstream contrast in grain size.

Log jams are temporally dynamic, commonly forming and disintegrating on time scales of up to 60 years in the Queen Charlotte Islands (Hogan, 1989). An initially impermeable jam gradually becomes more permeable as it gets older. The integrity (strength) of the jam and the proportion of the channel width which it occupies decrease, while the number of channels which flow through it increase. In turn the upstream sediment accumulation is gradually released downstream and the upstream/downstream grain size disparity is moderated until it is no longer evident (Figure 8.5).

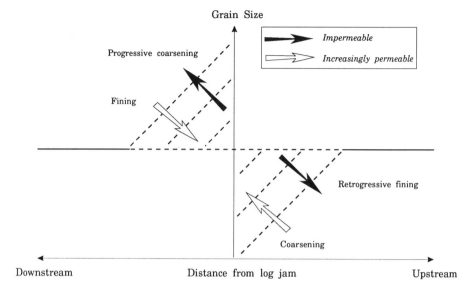

**Figure 8.5**  Schematic diagram of changes in sediment texture in the vicinity of a log jam

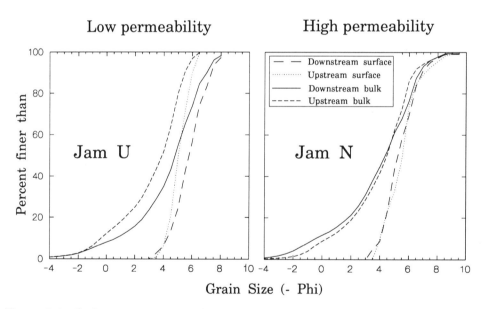

**Figure 8.6**  Sediment texture in the vicinity of two log jams with differing permeability characteristics. See text for details

Measurements of sediment throughput (permeability) using magnetically tagged tracer particles and scour chains, detailed bed material sampling in the vicinity of jams, semi-quantitative descriptions of jams in relation to photographic survey sites, and informal observations all support this hypothesis (Rice, 1990, in press). Figure 8.6 shows how sediment characteristics vary in the vicinity of two jams with different characteristics. Jam U is young, of high integrity, is breached by two small channels, and is associated with a lengthy upstream sediment accumulation. In contrast Jam N is old, weak and breached by three large channels. An estimated 75% of the upstream wedge has been degraded.

The large-scale impact of log jams is ensured by their frequency, which at an average of one per 90 metres approaches the frequency of sediment accumulations in the photographic study reaches. While isolated jams show the relative grain size variability expected, adjacent jams interact in a complex way. Together with hillslope inputs they prevent the development of a simple pattern of textural variation in the basins studied. A test of randomness using runs up and down reveals that variations in surface median grain size are in fact stochastic.

In summary, median surface grain size in these strongly coupled channels exhibits no systematic relation with distance downstream and, in general, fluctuates stochastically.

## CONCLUSIONS

A novel approach to the question of grain size variations at the drainage basin scale has been outlined. The drainage basin is considered as a network of links which differ in terms of sediment supply and sediment storage characteristics. In strongly coupled links, hillslope inputs and non-alluvial storage controls severely disrupt abrasion and sorting effects and mask the systematic modification of bed material texture. In completely buffered links, fluvial inputs and alluvial storage ensure that abrasion and sorting processes dominate, which provides a basis for computing changes in texture. Either empirical relations or theory might be used to describe the modification effect within such links. Mixing of sediment and changes in hydraulic parameters at confluences can also be computed. By considering hillslope inputs as dry tributaries, grain size changes in intermittently coupled links, the most common type, can also be managed.

The conceptual framework presented is accompanied by a number of operational problems, for which solutions are only partially develcped. Many of these are interesting questions in their own right and are currently under investigation by various groups. Initial work has confirmed the *a priori* ideas about strongly coupled links, and work on testing and developing other aspects of the model is in progress in British Columbia. It is hoped that the model will ultimately provide a geomorphological framework for examining grain size change in the fluvial system.

## ACKNOWLEDGEMENTS

The ideas developed in this paper benefited substantially from conversations with Dr Michael Church. I would also like to thank Dr Marwan Hassan and Dr Olav Slaymaker for their advice and encouragement. The Queen Charlotte work was funded by the Canada/British Columbia Fish Forestry Interaction Program and I would like to thank Steve Chatwin and Dan Hogan for their interest and support.

## REFERENCES

Bradley, W. C., Fahnestock, R. K. and Rowekamp, E. T. (1972). Coarse sediment transport by flood flow on Knik River, Alaska. *Geological Society of America Bulletin*, **83**, 1261–1284.

Brewer, P. A. and Lewin, J. (1990). Transport modification of alluvial sedment: field evidence and laboratory experiments. *Manuscript presented at the Fourth International Conference on Fluvial Sedimentology*, Barcelona, Spain (October 1989), 35 pp.

Church, M. (1983). Concepts of sediment transfer and transport on the Queen Charlotte Islands. *Fish/Forestry Interaction Program, Working Paper 2/83*, Vancouver, BC, 30 pp.

Church, M. and Kellerhals, R. (1978). On the statistics of grain size variation along a gravel river. *Canadian Journal of Earth Science*, 15, 1151–1160.

Dawson, M. (1988) Sediment size variation in a braided reach of the Sunwapta River, Alberta, Canada. *Earth Surface Processes and Landforms*, 13, 599–618.

Gimbarzevsky, P. (1986). Regional inventory of mass wasting on the Queen Charlotte Islands. *British Columbia Ministry of Forests, Land Management Report 29*, 96 pp.

Hogan, D. J. (1986). Channel morphology of unlogged, logged, and debris torrented streams in the Queen Charlotte Islands, British Columbia. *British Columbia Minstry of Forests, Land Management Report 49*, 94 pp.

Hogan, D. L. (1989). Channel responses to mass wasting inputs, Queen Charlotte Islands, British Columbia. *Fish/Forestry Interaction Program Phase II; Watershed 89; A conference on the Stewardship of Soil, Air, and Water Resources, Juneau, Alaska*, 22 pp.

Keller, E. A. and Swanson, F. J. (1979). Effects of large organic material on channel form and fluvial processes. *Earth Surface Processes*, 4, 361–380.

Knighton, A. D. (1980). Longitudinal changes in size and sorting of stream-bed material in four English rivers. *Geological Society of America Bulletin*, 91, 55–62.

Knighton, A. D. (1984). *Fluvial Forms and Processes*. Edward Arnold, London, pp. 78–80.

Lisle, T. F. (1987). Overview: Channel morphology and sediment transport in steepland streams. *Erosion and Sedimentation in the Pacific Rim (Proceedings of the Corvallis Symposium)*, IAHS Publ. No. 165, pp. 287–297.

MacPherson, H. J. (1971). Downstream changes in a sediment character in a high energy mountain stream channel. *Arctic and Alpine Research*, 3 (1), 65–79.

Miller, J. P. (1958). High mountain streams: Effects of geology on channel characteristics and bed material. *New Mexico State Bureau of Mines and Mineral Resources, Memoir*, 4, 23–76.

Parker, G. (1991). Selective sorting and abrasion of river gravels I: Theory. *Journal of Hydraulic Engineering, ASCE*, 117, 131–149.

Pizzuto, J. E. (1991). Interpreting the pattern of downstream fining in a network: Why process is insignificant. *EOS Supplement, AGU Spring Meeting Program and Abstracts 1991*, p. 134.

Rice, S. P. (1990). The spatial variation of bed material texture in coupled basins on the Queen Charlotte Islands. M.Sc. Thesis, University of British Columbia, Vancouver, 127 pp.

Rice, S. P. (in press). The spatial variation and routine sampling of spawning gravels in small coastal streams. *British Columbia Ministry of Forests, Land Management report*.

Richards, K. S. (1980). A note on changes in channel geometry at tributary junctions. *Water Resources Research*, 16 (1), 241–244.

Roberts, R. G. and Church, M. (1986). The sediment budget in severely disturbed watersheds, Queen Charlotte Ranges, British Columbia. *Canadian Journal of Forest Research*, 16 (5), 1092–1106.

Rood, K. M. (1984). An aerial photograph inventory of the frequency and yield of mass wasting on the Queen Charlotte Islands, British Columbia. *British Columbia Ministry of Forests, Land Management Report 35*, 55 pp.

Roy, A. G. and Roy, R. (1988). Changes in channel size at river confluences with coarse bed material. *Earth Surface Processes and Landforms*, 13, 77–84.

Shaw, J. and Kellerhals, R. (1982). The composition of recent alluvial gravels in Alberta. *Alberta Research Council Memoir*, 43, 151 pp.

Werrity, A. (1992). Downstream fining in a gravel bed river in Southern Poland: Lithological controls and the role of abrasion. In Billi, P. *et al.* (eds), *Dynamics in Gravel Bed Rivers*, Wiley, Chichester, pp. 333–346.

Wolman, M. G. (1954). A method of sampling coarse river bed materials. *American Geophysical Union Transactions*, 35, 951–956.

# 9 A Theory of Channel and Floodplain Responses to Alternating Regimes and its Application to Actual Adjustments in the Hawkesbury River, Australia

ROBIN F. WARNER
*Department of Geography, University of Sydney, Australia*

## ABSTRACT

Alternating flood- (FDRs) and drought-dominated regimes (DDRs), as well as human-induced changes to catchments and channels, have caused considerable instability in New South Wales coastal rivers. This study concentrates on developing a theory to explain channel changes and modification to floodplain stores caused by FDRs and DDRs. The theory is developed to help explain natural complex responses, as part of channel and floodplain metamorphosis. It is then tested in the Hawkesbury–Nepean system northwest of Sydney, considering also additional modifications superimposed by human-induced impacts to water and sediment discharges in both catchment and channel. This is possible by considering channel and floodplain changes derived from repeated surveys, photogrammetry and by observations on remnant sedimentary forms.

## INTRODUCTION

In much of New South Wales, it has been possible to show that significant shifts in hydrological regimes have occurred since European settlement. Periods with frequent, high magnitude floods have been called flood-dominated regimes (FDRs). Those with fewer, low magnitude floods have been designated drought-dominated regimes (DDRs). Such shifts may last up to 50 years and in the former (FDRs) channel-forming discharges may be two to four times greater than in the latter (DDRs). Such variations induce channel and floodplain adjustments, with the rate of change dependent on local energy factors and perimeter resistance conditions. Superimposed on these natural modifications are those associated with human interactions with the catchment and channel through time.

So far most of the research on this topic has been concerned with channel change through time. In FDRs, for instance, it is common to find widths increasing ($w^+$) and depths decreasing ($d^-$) in compliance with Schumm's (1977) notions of channel metamorphosis, where both water ($Q^+$) and sediment ($Q_s^+$) discharges are increased. If channel perimeters are very resistant to change, floodplain stripping (Nanson, 1986) or chute cutting are alternative forms of adjustment. In DDRs the main floodplain is seldom surcharged and width reductions are accompanied by the development of sandy,

*Process Models and Theoretical Geomorphology*. Edited by M. J. Kirkby
©1994 John Wiley & Sons Ltd

inset incipient floodplains, as well as the deepening of the main channel. Again, such adjustments accord with changes indicated by metamorphosis theory.

In this paper a theory of channel changes and floodplain modification associated with alternating regimes is developed and discussed for the first time. This is presented with the aid of simple conceptual diagrams, minus the human impacts in the first instance, and for potentially mobile perimeters. This is then tested with the evidence available in the Hawkesbury River (Figure 9.1) for the 25 km of channel between Wilberforce and Sackville (Figure 9.2), which includes the impacts of numerous human activities upstream.

Thus the major aim of this paper is to develop the theoretical basis for channel and floodplain changes associated with FDRs and DDRs. This provides a framework for understanding channel responses to what has been called cyclic equilibrium (Nanson & Erskine, 1988). It is then possible to separate out additional impacts to channels and floodplains caused by human activity.

Following an elaboration of this theoretical framework, background information is presented for this reach of the Hawkesbury–Nepean. This details the flood records, the nature of the shifting regimes and the human occupancy of this area. In the next section evidence of sedimentation and erosion is presented for the present FDR and the last DDR, together with some aspects of erosion and sedimentation associated with human activities. In the discussion which follows the theory is tested in light of changing regimes, the superimposition of human impacts and the complexities added by differing perimeter conditions.

## THE THEORETICAL BASIS FOR CHANNEL AND FLOODPLAIN CHANGES IN ALTERNATING REGIMES

In landscapes subject to cyclic equilibrium (Nanson & Erskine, 1988), changing hydrological regimes induce thresholds which condition adjustments in the channel and on the floodplain, as rivers seek to attain some new form of equilibrium every few decades. Such channel and floodplain modifications involve both erosion and sedimentation. They promote an instability which could not be explained in terms of simple natural or human-induced changes.

The NSW rivers involved have annual regimes which are highly variable and this further complicates the sequence and magnitude of power introduced to the channel perimeter and floodplain surface by high flow events. Catastrophic floods, such as those which occurred in 1949 and 1950 in several valleys, or a series of floods can effect considerable change, such as alluvial stripping (Nanson, 1986), and the effects of such events persist for a long time.

The hydrologic records allow the definition of flood- and drought-dominated regimes (FDRs and DDRs) (Erskine & Warner, 1988). Their effect in terms of channel and floodplain modifications, their occurrence in time, and factors inducing delayed responses have been considered in numerous studies of channel changes and floodplains (Warner, 1983, 1984, 1987a, 1990, 1992; Erskine & Warner, 1988).

In the following discussion, changes from DDR to FDR and from FDR to DDR are considered in theoretical terms for several types of channels, prior to the brief elaboration of the sequences of changes and their consequences.

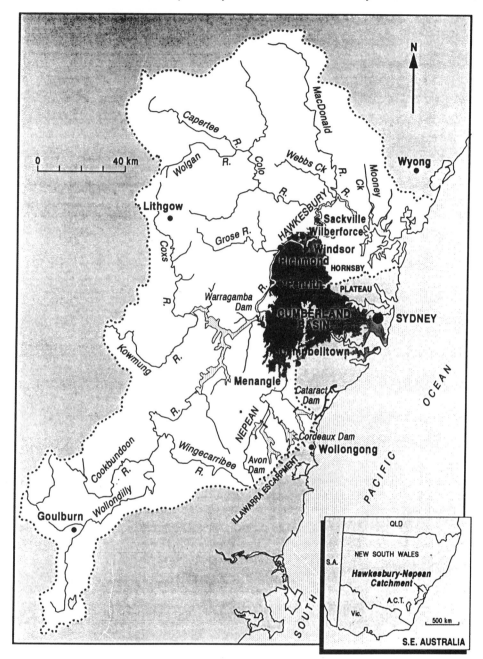

**Figure 9.1**  The Hawkesbury–Nepean catchment

## Changes from DDR to FDR

Such a change last occurred in the late 1940s. It was initially discussed in geomorphical terms by Pickup (1974). Since this is the most recent change of regime, there is more evidence available to help document change. Impacts were immediately obvious when

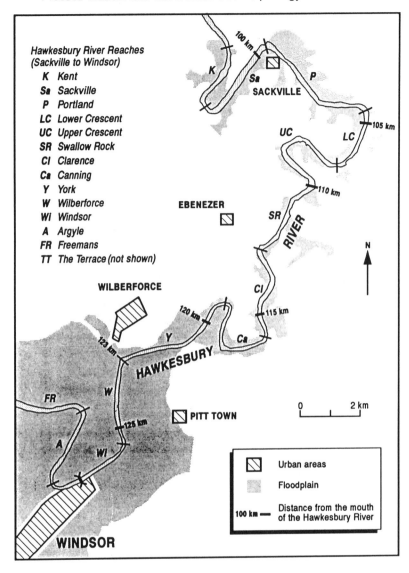

**Figure 9.2** Map to show reaches between Windsor and Sackville

comparing 1940s air photographs with those taken in the 1950s and 1960s (Warner, 1971).

From such sources the most obvious changes were channel widening. Earlier channels were narrow with deep pools separated by gravel riffles. The channel was often flanked by dense riparian tree vegetation. Only 10 years later much of that had changed and the channel was wider, the vegetation often removed and pools had in many cases been filled in. Such changes have been noted by Simon (1992) who indicated that 'channel widening means a much more important role in the high-energy environment (FDR) than in the low-energy environment where vertical processes dominate'.

Thus, in metamorphosis terms, the changes induced by the onset of a FDR were those associated with increases in water and sediment discharge:

$$Q^+ Q_S^+ \approx w^+ d^-$$  (1)

where $Q$ is discharge, $Q_S$ is sediment discharge, $w$ is channel width and $d$ is channel depth.

However, such changes are restricted to those parts of the channel where banks readily fail and where the bedload components are not readily moved out of the reach. Where the banks are cohesive, as in lower estuarine, backwater reaches (Pickup, 1984) or in well protected densely vegetated reaches, width changes are not so obvious. In these reaches, where high energy prevails, lateral or parallel chute channels may be cut in the floodplain to provide additional flow capacity. These have followed European settlement where floodplain roughness has been considerably reduced by the removal of dense forests. In estuarine and deltaic reaches, where lower energy prevails in floods, the incidence of flooding has increased, levees have been raised, and floods utilise much greater overbank storage capacities. These are not available upstream where valley-floor zones are narrow and overbank storage limited (Warner, 1992).

In such cases:

$$Q^+ Q_S^+ \approx w^0 d^0 \text{ FP chutes}^+ \text{(high energy)}$$  (2)

$$Q^+ Q_S^+ \approx w^{0+} d^0 \text{ FP inundation} + \text{Levees}^+ \text{(low energy)}$$  (3)

where FP is floodplain.

In sand-bed rivers, banks and bed are both more mobile and adjustments are more rapid. Thus in rivers like the Hawkesbury, which is sand dominant, such a change may be similar to the initial case described (equation (1)).

Very few observations have been made on strictly suspended-load streams. Changes in the perimeter may not be great because of perimeter coherence and, where floodplain, cut-off and backswamp storages are great, it seems probable that overbank storage may be the main mechanism coping with larger and more frequent floods. More sand may be introduced to the channel from catchments which could influence depths and local levees but not much research has been attempted on such rivers.

$$Q^+ Q_s^+ \approx w^0 d^- \text{ FP inundation} + \text{Levees}^+$$  (4)

In sub-humid areas of the inland, it is possible that the additional rainfall associated with the FDR may in fact enhance the vegetation cover which in turn could reduce sediment delivery to the channel. In this case:

$$Q^+ Q_S^- \approx w^0 d^+$$  (5)

would prevail, but again no supporting studies are as yet available.

## Changes from FDR to DDR

The last change of this kind took place at about the turn of the century. The amount of information available for that period is much more limited than that for the more recent change from DDR to FDR, but hydrographic surveys between Sackville and Richmond in 1890 (towards the end of the late 19th century FDR) and 46 cross-sections from 1900 for the Penrith reach do allow some insight into the channel at the end of a FDR. Changes occurring over the next 48 years have been documented in some cases. In others there are air photographs taken close to the end of the DDR. These allow an appreciation of the change in a DDR.

In a DDR, discharge is decreased and in a humid vegetation environment sediment discharge should be reduced also, giving:

$$Q^- Q_s^- \approx w^- d^+ \qquad (6)$$

this means that widths should decrease if material is available to be deposited within the channel and depths should increase. The width reduction includes various forms of sedimentation within the wide, shallow FDR channel. Mean annual floods were reduced to a quarter or a half of their FDR magnitude and most flows were contained within the channel (see also Simon, 1992). Surcharging of 10 m banks at Windsor only occurred four times between 1901 and 1948.

Sandy alluvium has been draped between the top of the higher floodplain surface and benches inset within the inherited channel. This is an unusual form of sedimentation in that bedding is inclined to parallel the slope of the banks. Sandy layers alternate with finer deposits, which often contain leaves and charcoal. So a layer of sand in grass may be capped by the silts, leaves and charcoal being deposited on the recession of a flood hydrograph.

On the shallow margins of the channel, further deposition has allowed benches to appear. Again these are sandy with bedding planes more horizontal in this case. These are soon colonised by trees and flood-tolerant grasses and the build up of such benches appears to be fairly rapid. In the Hawkesbury, they have reached the elevation of mean annual floods for DDRs (6 m at Windsor) and can therefore be considered to be incipient floodplains. Similar forms have been described in the Bellingen (Warner, 1992).

Thus in bedload (Bellingen) and mobile, sand-bed (Hawkesbury) streams, low floodplains and in-channel benches have been part of the channel width reduction processes in DDRs. The confined energy of major in-channel floods, plus the anchoring role of bench vegetation, no doubt assisted channel deepening where energy was greater than thresholds of motion in bedload materials. In sand-bed streams, deeper channels prevailed and, in bedload streams, it seems that pools were excavated.

There is even less information about FDR–DDR changes in suspended-load streams.

$$Q^- Q_s^{\pm} \approx w^{\pm} d^{\pm} \qquad (7)$$

According to Schumm's (1977) equations both depth and width can change in either direction. No data are available to corroborate this.

Although levees and higher parts of the floodplains were seldom surcharged (four times in 48 years), tributary back-up flooding occurred more frequently. Fine sediments in such flows and coarse materials moving through levee breaches obviously helped to fill in previously cut chutes. Thus although floodplain heights were not increased by much, back-up flooding did help infill chutes and other low points on floodplains.

In instances where sediments were not available, FDR channels did not change very much in the DDR. This seems to have been the case in some parts of the Clarence, where recent work has revealed no change between DDR and FDR. Little seems to have happened in these channels since the late 19th century (Warner & Martens, 1992) and the change, FDR to DDR, has resulted in:

$$Q^- Q_S^- \approx w^0 d^0 \qquad (8)$$

## Consequences of DDRs and FDRs

The morphological consequences of alternating regimes involve considerable instability or disequilibrium (Phillips, 1992), lagged responses, and complications to alluvial stores or floodplains. In FDRs the utilised channel capacity involves a wide, large channel and the operative floodplain is both high and extensive where valley-floor troughs are wide. The frequency of surcharging of what is the main floodplain can be from less than $Q_2$ in tidal areas where present capacities are less than the mean annual flood ($Q_{2.33}$) to $Q_{10}$ or $Q_{12}$ in obviously enlarged channels (Warner, 1992). In DDRs, as indicated earlier, an incipient floodplain may form which can relate to the mean annual floods of the drier regime. This floodplain is lower, narrower and inset within the FDR channel. In some FDRs it appears that not much of this form survives but, in others, it may be totally removed.

In reality there are often several benches or alluvial shoulders, some of which may be low, lateral bars which are surcharged frequently, some are more established and colonised by vegetation, whilst others may conform with what might be thought of as a floodplain in the DDR. However, contemporary surcharging (FDR) is well above this. Such alluvial forms were investigated initially by Woodyer (1968), and in other studies (Warner et al., 1975), but without knowledge of their probable meaning in terms of alternating regimes.

Thus towards the end of the present FDR, there are channels and floodplains which are complex. The wide enlarged channel still contains in some cases the vestiges of earlier DDR floodplains, as well as bars and partially filled-in pools. On the floodplains, levees have been elevated by more frequent overbank flows and chutes have been cut in the less resistant alluvium to form high and low level floodways. In contrast, at the end of a DDR, there are two distinctive floodplains, and a narrow channel, probably more sinuous and stabilised by woody riparian vegetation. On the high floodplain, old chutes have in many cases 'healed' and there is little indication of high level inundation.

## Sequences of FDRs and DDRs

There are alternating sequences of FDRs and DDRs, where regimes alternate to give a form of cyclic disequilibrium. From the longest flood-stage record at Windsor, it has been possible to show the following sequence:

| | |
|---|---|
| FDR | 1799–1820 |
| DDR | 1821–1856 |
| FDR | 1857–1900 |
| DDR | 1901–1948 |
| FDR | 1949–1988 + |

Frequencies of 6 m and 10 m floods for Windsor are given in Table 9.1, whilst stages for floods of various return periods are shown in Table 9.2. These reveal a 3 m vertical variation in mean annual floods ($Q_{2.33}$) between regimes. Thus the impacts of succeeding regimes have been superimposed one upon another, whether or not a new equilibrium stage had been reached. It seems that floodplain modifications might have been less in the first regime when dense vegetation still prevailed (Figure 9.3a).

**Table 9.1** Windsor: floods over 6 m and 10 m and frequencies

|        |        |       | Number of floods | | | |
|--------|--------|-------|------|------|---------------------|----------------------|
| Period | Regime | Years | 6 m  | 10 m | 6 m yr$^{-1}$ | 10 m yr$^{-1}$ |
| 1799–1820 | FDR | 22 | 14? | 6  | 0.7? | 0.3 |
| 1821–1856 | DDR | 36 | 7?  | 0  | 0.2? | 0.0 |
| 1857–1900 | FDR | 44 | 51+ | 17 | 1.2  | 0.4 |
| 1901–1948 | DDR | 48 | 24  | 4  | 0.5  | 0.1 |
| 1949–1988 | FDR | 40 | 50  | 11 | 1.3  | 0.3 |

*Sources*: 10 m data based on Webb McKeown (1988): the remainder based on Josephson (1885), Tebbutt (Private Observatory values published in 1877, 1882, 1891 and 1916) and Riley (1980). ? = doubtful values.

**Table 9.2** Windsor: stages for $Q_{1.58}$, $Q_{2.33}$, $Q_5$ and $Q_{10}$

|        |        |       | Annual Series | | | |
|--------|--------|-------|---------------|---|---|---|
| Period | Regime | Years | $Q_{1.58}$ | $Q_{2.33}$ | $Q_5$ | $Q_{10}$ |
| 1799–1820 | FDR | 22 | NA   | c.9.0 | 12.9 | 14.4 |
| 1821–1856 | DDR | 36 | NA   | >6.0  | 6.0  | 9.0  |
| 1857–1900 | FDR | 44 | 8.1  | 9.0   | 11.9 | 13.7 |
| 1901–1948 | DDR | 48 | >6.0 | 6.3   | 8.1  | 9.7  |
| 1949–1988 | FDR | 40 | >6.0 | 8.9   | 11.3 | 13.8 |

*Sources*: as in Table 9.1. Bankfull 9.8 m.
*Note*: In Webb McKeown's model 5 year stage = 10.1 and 10 year stage = 12.0 based on long-term means.
NA = not available in the record.

Table 9.3 shows the sequence with the theoretical directions of changes expected for the high floodplains (surface and margins), low floodplains (again surface and margins), the channel, sinuosity, channel planform, and slope. Changes in the independent variables, discharge of water and sediment are shown, together with the ratio of bankfull discharge ($Q_{bf}$) to DDR (given as unity).

In the first FDR (Figure 9.3a) the high floodplain surface was probably raised on both levees and in floodbasins, but not lowered, whereas margins were being eroded. If a low floodplain surface survived, it would be truncated by channel widening but deposition could still occur on the banks, as could stripping. General channel changes would include increases in width, decreases in depth and, as the channel was straightened by $w^+$, the sinuosity would decrease and slope would increase.

Similar changes would have occurred in the second (Figure 9.3c) and third (Figure 9.3e) FDRs, with evidence for these changes from old surveys (1890 and 1900). In FDRs (Table 9.3) $Q$ would certainly have increased; $Q_S$ is shown as likely to go both ways, but increasing in the main. Decreases would reflect more protective vegetation and/or conservation practices. From discharge records the ratios of bankfull discharge for the last FDR to the last DDR was 2:1. There is some suggestion that in the previous FDR (1857–1900) it might have been higher (4:1), but there is no real evidence for the first FDR and DDR.

In Figures 9.3b and 9.3d, upstream and backwater cross-section details are shown for the DDRs. It is evident that $Q$, $Q_S$ and $Q_{bf}$ were all reduced and these caused the

**Table 9.3** Theoretical changes associated with the sequence of *FDR*s and *DDR*s at Windsor

| Regime | Duration | Years | High floodplain | | Low floodplain | | Channel changes | Sinuosity | Channel planform | Slope | $Q_S$ | $Q_{bf}$ | $Q$ |
|---|---|---|---|---|---|---|---|---|---|---|---|---|---|
| | | | Surface | Margins | Surface | Margins | | | | | | | |
| FDR | 1799–1820 | 22 | ±[1] | − | ± | − | $w^+d^-$ | − | Straighter | + | ± | 2? | + |
| DDR | 1821–1856 | 36 | + | +[2] | +[3] | + | $w^-d^+$ | + | Meandering | − | − | 1? | − |
| FDR | 1857–1900 | 44 | ±[1] | − | ±[3] | − | $w^+d^-$ | − | Straighter | + | ± | 2–4? | + |
| DDR | 1901–1948 | 48 | + | +[2] | +[3] | + | $w^-d^+$ | + | Meandering | − | − | 1 | − |
| FDR | 1949–1991 | 43 | ±[1] | − | ± | − | $w^+d^-$ | − | Straighter | + | ± | 2 | + |

1 + levee and floodbasin accretion, crevasse splay accumulation. − stripping and chute cutting; 2 draped alluvium; 3 incipient floodplain. $w$ = width; $d$ = depth; $Q$ = discharge; $Q_S$ = sediment discharge; $Q_{bf}$ = bankfull discharge as ratio of *DDR*.

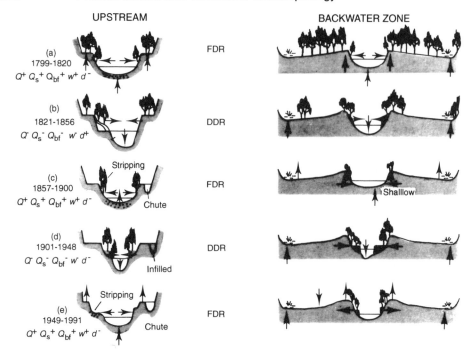

**Figure 9.3**   Sequence of changes advocated by the theory in FDRs and DDRs

channel dimensions to reduce in width and depth to increase. Very limited tributary flood sedimentation might slowly raise floodbasin levels but there would be only very limited flooding of the levees and higher floodplain surface. Lower floodplains would have been building both vertically and laterally, with channel width reductions. A narrow, more sinuous channel would lower the gradient or slope.

The directions of the arrows in Figure 9.3 show the likely direction of change in each landform type. Following a description of the study area and the evidence for channel adjustments as observed, the discussion considers this theoretical sequence and effects on it in light of human-induced changes to the catchment and channel.

## BACKGROUND TO THE STUDY AREA

The Hawkesbury–Nepean River drains some 22 000 km² of country north, west and south of Sydney. Its Mesozoic sandstones form its greenbelt and occupy about 64% of the catchment. Older granites and Palaeozoic metamorphic rocks cover about 26% of the catchment on the western and southwestern fringes. Younger shales cap the sandstones in the depressed centre of the Sydney Basin. They occupy about 10% of the catchment in what is called the Cumberland Plain, just west of Sydney (Figure 9.1).

Precipitation varies from less than 800 mm in the rainshadow area of the plain to over 1500 mm on the Illawarra escarpment south of Sydney and on the eastern parts of the Blue Mountains to the west. Only about 13% of this becomes runoff, but both runoff and rainfall are highly variable on a yearly basis.

The Hawkesbury River forms the lowest part of this system. Its name applies to the channel downstream of the Grose junction with the Nepean River and most of its 140 km length is now tidal. Above the junction with the Warragamba River (Figure 9.1), the upper Nepean has a catchment area of 1700 km². The catchment area of the Warragamba exceeds 8000 km² (all dammed now). At Windsor, just above the study reach, the catchment area is 13 000 km², of which nearly 10 000 km² is affected by dams.

In the study reach (Figure 9.2), the channel leaves the shales where the high floodplain at Windsor is up to 6 km wide and moves back into the sandstones of the Hornsby Plateau. Here the floodplain narrows to between 100 and 200 m (Figure 9.2). This shrinkage of the floodplain causes interesting flood gradients and hydrographs which may have some bearing on the patterns of sedimentation. At Windsor, for instance, hydrograph rising-limb velocities are much higher than those of the recession limb, whilst at Sackville the reverse occurs (Warner, 1990: Figure 9.4).

There has been interest in the Hawkesbury–Nepean ever since the very early days of European settlement. It provided a routeway into the grain areas of the colony and the so-called Macquarie towns (after the governor of that name) are located on this waterway. Flood observations began at Windsor in 1795 and flood heights have been recorded since 1799. Sedimentation in the river always caused concern because it was such an important routeway. Thus hydrographic surveys were perhaps more common than in other less important rivers.

The long-term flood observations have been used not only to demonstrate irregularities in the short-term regime but also the more systematic variations through longer periods of time. These have revealed the alternating nature of the regime between flood and drought domination.

## Flood Records

Stage heights for floods at Windsor have been observed for nearly 200 years. They have been analysed in several places (Josephson, 1885; Riley, 1980; PWD 1983;

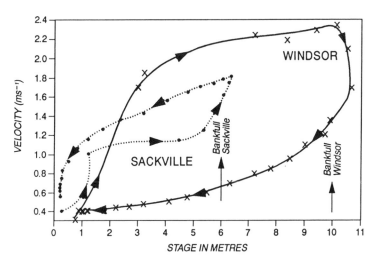

**Figure 9.4**  Relations between velocity and stage for the 5 year floods at Windsor and Sackville (after Webb McKeown, 1988, model)

Webb McKeown, 1988; Warner, 1990). In flood data collection, two stages were considered to be important: 6 m and 10 m. At 6 m these coincide with low inset floodplain elevations and at 10 m with present levee tops. Any flow above 6 m will inundate the low benches and flood basins are flooded by back-up flows. At 10 m and above, levees and all the floodplains are inundated.

As already indicated, periods of frequent large floods have been designated FDRs and those of lower, less frequent floods, DDRs (Table 9.1). The impacts of these alternating regimes are shown at the 6 m and 10 m levels in Table 9.2. In FDRs, 6 m floods occur more frequently than once a year and 10 m floods occur every 2½ to 3 years on average. In DDRs, 10 m floods either do not occur (19th century) or have a recurrence interval of about 12 years. Even 6 m flood recurrences drop to about once every 2 years.

From these data, inundation frequencies can be established for the whole period of European settlement. The main levees have only been topped 34 times in 106 years of FDRs and four times in 84 years of DDRs or 38 times in 190 years. In some places, through buried fence posts, there is a suggestion that the levee might have been raised by up to 1 metre by post-settlement alluviation (PSA). 9 m inundation frequences are 59 for the FDR years and 10 for DDR years, or 69 in all. Floods of 6 m exceed 115 for FDRs (106 years) and 31 for DDRs (84 years).

Actual discharge values are not available below Penrith. Downstream of Richmond, the channel is tidal and subject to backwater effects of floods in major downstream tributaries, notably the Colo River (Figure 9.1). The impacts of tides on floods are minimal in these upper reaches of the Hawkesbury. This is clearly indicated by the Department of Public Works hydrographs for the 1978 flood (PWD, 1979). Modelling by Webb McKeown (1988), however, has provided some information on likely discharges, velocities and stages for 5, 10, 100 and 250 year floods. Marked hysteresis loops exist due to the present channel dynamics and to the effects of the downstream narrowing of the floodplain (Figure 9.2). The drainage of overbank flooding of the wide floodplain at Windsor, as well as likely backwater effects, considerably reduce recession-limb velocities at Windsor, thereby helping to induce in-channel deposition of sands and silts. In contrast the loop is reversed at Sackville, where there is a very narrow floodplain. Rising limbs of the hydrograph are marked by lower velocities than those of the recession. This probably restricts sedimentation (see later). These phenomena are illustrated for the modelled 5 year flood at Windsor and Sackville (Figure 9.4).

Tables 9.4 and 9.5 have been produced from the 5 year flood simulation by the Webb McKeown (1988) model. One metre stage heights, widths ($w$), cross-section areas ($A$) and mean depths ($d$) were obtained from surveyed data. Velocities ($v$), discharges ($Q$), slopes ($s$) and times were derived from the modelled data. Roughness ($n$) was estimated from experience in the channel. Total stream power per unit length of channel ($\Omega$) was obtained from the equation: $\Omega = \gamma Q s$ (Bull, 1979). Unit stream power ($\omega$) was obtained by dividing $\Omega$ by channel width, and the shear stress ($\tau$) was calculated from the equation: $\tau = \gamma R s$ (Magilligan, 1992) where $R$ is the hydraulic radius and $s$ is the energy slope. The median size ($d_{50}$) of particles moved by the tractive force is based on the Graf (1983) equation: $d_{50} = 2709.85 \ Q^{0.32} \ w^{-0.32} \ n^{0.32} \ s^{0.38}$, where $Q$ is discharge, $w$ width, $n$ roughness and $s$ energy slope.

Table 9.4 shows how velocities, discharges, total and unit stream powers, shear stress and size of sediment moved all drop dramatically on the recession limb of the hydrograph at Windsor. Even at bankfull (9.8 m) the discharge drops from 3600 to 2100 m$^3$ s$^{-1}$

**Table 9.4**  A 5 year flood hydrograph for Windsor

| Stage (m) | $w$ (m) | $A$ (m²) | $v$ (m³s⁻¹) | $Q$ (m³s⁻¹) | $d^{2/3}$ (m) | $n$ | $s$ (mm⁻¹) | $\Omega$ (Wm⁻¹) | $\omega$ (Wm⁻²) | $d_{50}$ (mm) | $\tau$ (Nm⁻²) |
|---|---|---|---|---|---|---|---|---|---|---|---|
| 0 | 102 | 240 | 0 | 0 | 1.77 | 0.03 | 0+ | | | | |
| 1 | 112 | 347 | 0.4 | 140 | 2.13 | 0.03 | 0.000032 | 45 | 0.4 | 19 | 1.0 |
| 2 | 118 | 463 | 1.1 | 510 | 2.49 | 0.029 | 0.00016 | 816 | 6.9 | 50 | 6.3 |
| 3 | 124 | 584 | 1.7 | 990 | 2.83 | 0.028 | 0.000285 | 2822 | 22.8 | 64 | 13.5 |
| 4 | 127 | 710 | 2.0 | 1420 | 3.16 | 0.026 | 0.000271 | 3848 | 30.3 | 80 | 15.2 |
| 5 | 133 | 840 | 2.1 | 1760 | 3.44 | 0.025 | 0.000233 | 4101 | 30.8 | 80 | 14.9 |
| 6 | 138 | 975 | 2.18 | 2120 | 3.71 | 0.025 | 0.000216 | 4579 | 33.2 | 81 | 15.4 |
| 7 | 143 | 1115 | 2.22 | 2480 | 3.95 | 0.024 | 0.000182 | 4514 | 31.6 | 75 | 14.3 |
| 8 | 148 | 1260 | 2.24 | 2820 | 4.20 | 0.023 | 0.000150 | 4230 | 28.6 | 73 | 13.4 |
| 9 | 158 | 1413 | 2.28 | 3220 | 4.34 | 0.021 | 0.000122 | 3928 | 24.9 | 68 | 11.0 |
| 9.8 | 175 | 1546 | 2.33 | 3600 | 4.30 | 0.02 | 0.000117 | 4212 | 24.1 | 66 | 10.4 |
| 9.8 | 175 | 1546 | 1.36 | 2100 | 4.30 | 0.02 | 0.00004 | 840 | 4.8 | 35 | 3.6 |
| 9 | 158 | 1413 | 1.1 | 1550 | 4.34 | 0.021 | 0.00003 | 465 | 2.9 | 31 | 2.7 |
| 8 | 148 | 1260 | 0.88 | 1110 | 4.20 | 0.023 | 0.00002 | 222 | 1.5 | 25 | 1.7 |
| 7 | 143 | 1115 | 0.76 | 850 | 3.96 | 0.024 | 0.00002 | 170 | 1.2 | 23 | 1.6 |
| 6 | 138 | 975 | 0.68 | 660 | 3.71 | 0.025 | 0.00002 | 132 | 1.0 | 22 | 1.4 |
| 5 | 133 | 840 | 0.58 | 490 | 3.44 | 0.025 | 0.000018 | 88 | 0.7 | 20 | 1.1 |
| 4 | 127 | 710 | 0.50 | 360 | 3.17 | 0.026 | 0.000017 | 61 | 0.5 | 18 | 1.0 |
| 3 | 124 | 584 | 0.46 | 270 | 2.82 | 0.028 | 0.000021 | 57 | 0.5 | 19 | 0.6 |
| 2 | 118 | 463 | 0.44 | 150 | 2.50 | 0.029 | 0.000026 | 39 | 0.3 | 11 | 1.0 |
| 1 | 112 | 347 | 0.42 | 145 | 2.13 | 0.03 | 0.000035 | 51 | 0.5 | 19 | 1.0 |
| 0 | 102 | 240 | 0⁺ | 0 | 1 | | 0⁺ | | | | |

Durations: 1–6 m 7 hr; l-peak 33 hr; recession 87 hr; total 120 hr.

from rising limb to that of the recession. This is mainly due to the topographic and overbank drainage effects (narrowing of floodplain–Figure 9.2) where water leaving the floodplain causes backwater effects at Windsor. At 9 m even velocities drop from 2.28 m s⁻¹ to 1.1 m s⁻¹. Total power is reduced from over 4000 w m⁻¹ to a range of 40–840 w m⁻¹ on the falling stages. Unit power, size of material and shear stresses are all reduced in similar fashion (Table 9.4).

These values certainly suggest reasons for the heavy bed sedimentation formed in theoretical FDRs (1890 bed depths were very shallow and at low water, shoals were very common). It is interesting to note that maximum total and unit powers are reached at 6 m on the rising stage, that these powers conform to the steepest part of the hydrograph (4–6 m), and that such injections of power are confined to less than 3 hours of the total hydrograph.

In contrast, Table 9.5 shows the 'bathplug' effect at Sackville. Even though the channel is wider, the floodplain is much narrower, being set in a sandstone gorge. The hydrograph is shorter for the 5 year flood (85 hr versus 120 hr). In each case the peak is reached in about 33 hours and this means that the slow outflow recession at Windsor (87 hr) is 25 hours longer than at Sackville.

The peak discharge has been attenuated from 3600 to 3500 m³ s⁻¹ and discharges are lower at this site on the rising stages than on the recession. They vary between 0.5 and 0.9 of the falling stage at the same elevation. In contrast at Windsor, rising stage discharges are 1.5 to 3.9 times those of the same falling stages. This means that total and unit powers are lower, as too are shear stresses and size of materials moved during the hydrograph rise. This would appear to indicate that little deposition is occurring in this gorge on the falling stage.

**Table 9.5** A 5 year flood hydrograph for Sackville

| Stage (m) | $w$ (m) | $A$ (m$^2$) | $v$ (m$^3$s$^{-1}$) | $Q$ (m$^3$s$^{-1}$) | $d^{2/3}$ (m) | $n$ | $s$ (mm$^{-1}$) | $\Omega$ (Wm$^{-1}$) | $\omega$ (Wm$^{-2}$) | $d_{50}$ (mm) | $\tau$ (Nm$^{-2}$) |
|---|---|---|---|---|---|---|---|---|---|---|---|
| 0 | 202 | 558 | 0 | 0 | 1.98 | 0.028 | 0$^+$ | | | | |
| 1 | 205 | 762 | 0.56 | 430 | 2.41 | 0.027 | 0.000039 | 168 | 0.8 | 22 | 1.5 |
| 2 | 206 | 967 | 1.1 | 106 | 2.82 | 0.026 | 0.000103 | 1092 | 5.3 | 44 | 4.9 |
| 3 | 210 | 1175 | 1.14 | 1340 | 3.17 | 0.025 | 0.000081 | 1085 | 5.2 | 43 | 4.6 |
| 4 | 220 | 1390 | 1.15 | 1600 | 3.44 | 0.024 | 0.000064 | 1024 | 4.7 | 38 | 4.1 |
| 5 | 224 | 1612 | 1.19 | 1920 | 3.75 | 0.023 | 0.000053 | 1018 | 4.5 | 39 | 3.8 |
| 6 | 235 | 1842 | 1.54 | 2840 | 3.97 | 0.022 | 0.000073 | 2073 | 8.8 | 47 | 5.8 |
| 6.4 | 240 | 1940 | 1.8 | 3500 | 4.06 | 0.02 | 0.000079 | 2765 | 11.5 | 50 | 6.5 |
| 6 | 235 | 1842 | 1.8 | 3320 | 3.97 | 0.022 | 0.000099 | 3287 | 14.0 | 55 | 7.8 |
| 5 | 224 | 1612 | 1.7 | 2740 | 3.75 | 0.023 | 0.000109 | 2987 | 13.3 | 56 | 7.9 |
| 4 | 220 | 1390 | 1.57 | 2180 | 3.44 | 0.024 | 0.000120 | 2616 | 11.9 | 55 | 7.7 |
| 3 | 210 | 1175 | 1.42 | 1670 | 3.17 | 0.025 | 0.000125 | 2088 | 9.9 | 54 | 7.1 |
| 2 | 206 | 967 | 1.32 | 1280 | 2.82 | 0.026 | 0.000148 | 1894 | 9.2 | 53 | 7.0 |
| 1 | 205 | 762 | 1.13 | 860 | 2.41 | 0.027 | 0.000160 | 1376 | 6.7 | 28 | 6.0 |
| 0 | 202 | 558 | 0 | 0 | 1.98 | 0.028 | 0$^+$ | | | | |

Durations: 1–4 m 4 hr; l-peak 33 hr; recession 52 hr; total 85 hr.

Such power values and shear stresses, as presented in Tables 9.4 and 9.5, are very low compared with those suggested by Magilligan (1992). His work indicated that, for streams with gentle gradients in humid and sub-humid environments, minimum thresholds appear to correspond to shear stresses of about $100\,\mathrm{N\,m^{-2}}$ and unit stream powers of about $300\,\mathrm{W\,m^{-2}}$. In the 5 year flood at Windsor shear stresses only reached just over 15% and at Sackville nearly 8% of the catastrophic thresholds. Unit power is even less: about 11% at Windsor and 5% at Sackville. However, even with such low values in this sand-flanked perimeter, there is still enough power to effect considerable change in FDRs and channel deepening in DDRs.

It seems apparent that the threshold of motion $d_{50}$ values derived from the Graf (1983) equation are too large. However, much of the bed material in these two locations is finer than very coarse sand and this would certainly be moving through most of the 5 year hydrograph at Sackville, but it might well be limited near channel margins at Windsor, where recession velocities range from $1.1\,\mathrm{m\,s^{-1}}$ at 9 m to $0.42\,\mathrm{m\,s^{-1}}$ at 1 m. It is therefore not surprising to find that sand is still deposited in drapes and on truncated incipient floodplains by modern floods.

## Ground Conditions and Human Occupance

Under pre-European settlement conditions, floodplains were densely forested with large swampy areas in adjacent flood basins. Farming was slow to develop and the 20 blocks which were settled downstream of Windsor in the 1790s were devastated by floods of the first FDR. At one stage, wheat farming was relocated to the nearby tributary of South Creek (Jeans, 1972).

In the DDR beginning 1821, settlement was re-established and expanded to other parts of the floodplain. The return of FDR conditions after 1857 until 1900 again caused severe problems with the highest flood (1867) and the second (1900) being experienced. Land resource development for farming probably peaked in the next DDR (1901–1948) together with problems caused by rabbits, noxious weeds, poor conservation practices, gullying and so on (Pickup, 1974; Erskine, 1986).

Many of these problems have been at least partially solved in the present regime, but urban sprawl into the shale areas has further increased polluted stormwater runoff to add to the higher FDR runoff which has destabilised channels and mobilised more sediment.

### The Channel and Floodplain

The channel and floodplain are shown in Figure 9.2 over the 25 km of the study reach. Channel widths at datum vary between about 100 and 200 m; depth is usually up to 5 m but can exceed 20 m in deep holes in outer bends where high sandstone cliffs prevail. The bank tops range between about 10 m at Windsor down to 6 m or more at Sackville and consist of 1–2 m of poorly consolidated sand in the tops of most levees. The alluvial banks are composite in that a narrow beach may be present at low tide, a bench 4 to 6 m high up to 10 m wide may occur below the banktop. Where rotational slumps, which are semi-circular in plan, have involved the older, highly cohesive alluvia, slump steps present an even more irregular profile. In several places banks have been cleared of vegetation near caravan parks and may be battered back to an angle of less than 30 degrees. Where vegetation prevails, willows and casurinas occupy lower levels with banktop eucalyptus in places. There is also a range of exotic vegetation of scrub and small tree types. Figure 9.5 shows typical alluvial and bedrock sections, taken from two actual surveyed cross-sections.

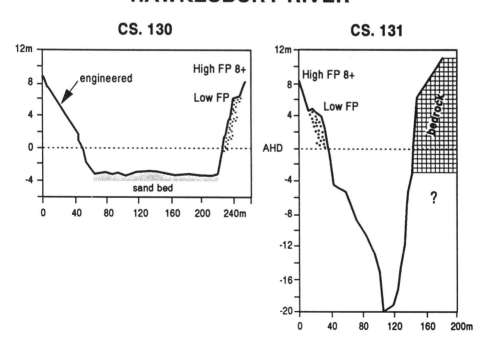

**Figure 9.5**   Typical alluvial (cs. 130) and bedrock (cs. 131) cross-sections for the study area

# EVIDENCE OF SEDIMENTATION AND EROSION

In this section results of observations on sedimentation and erosion are presented to complete the background material necessary prior to a discussion of the theory and how it can be applied in this system.

In this large, steep-banked, tidal channel, there is evidence for both erosion and deposition, much of it very recent. The channel position is more or less fixed and there is very little evidence for lateral migration. Its location is determined by levees which streamline the main channel between alternating sandstone cliffs. In late Pleistocene times when the sea level was up to 120 m lower, the valley was presumably a much deeper and narrower gorge. Although the study reach extends between 98 and 123 km inland, at low tide water level is close to mean sea level. The valley was flooded by the last major transgression which brought sea level to its present elevation some 6000 years ago (Thom & Roy, 1985).

Drilling has indicated that this drowned valley floor is now full of estuarine muds, with coarser deltaic material upstream (Thom *et al.*, in preparation) and that the present alluvia cap these Holocene deposits. Much of the older alluvia is fairly featureless, fine in texture (silt-rich sands) and cohesive. This fines in the direction of the levee toe and flood basins and it forms the bulk of the present alluvial platform. Sand deposits mantle this adjacent to the channel to form levees; they are also draped up the face of the river bank and they form the major part of in-channel benches and the lower floodplain. In contrast to the older alluvium, these deposits are mainly fairly fresh sand, much lighter in colour, low in cohesion and are well stratified, sometimes with interbedded leaf bands, silty layers and artefacts. These deposits have little strength and fail readily when vegetation has been disturbed. In the following sections, sedimentation and erosional evidence is initially considered for the present FDR, and then for the last DDR regime.

## Evidence for FDR Sedimentation and Erosion

The present FDR has prevailed since the late 1940s. In this period, most of the in-channel activity has been erosional. This has occurred because of channel widening in the form of bank erosion, bank failure in semi-circular slumps and as bed degradation, the latter not anticipated by theory. Changes have been established by photogrammetry, by survey and resurvey, and by observation (PWD, 1987; Warner, 1990, 1991a, b).

Overbank sedimentation has occurred on the levees. Fresh sand from floods in 1988 and 1990 has partially buried vegetation over wide areas of the high floodplain levees. At least two sets of fence posts are necessary in some areas to maintain effective field boundaries (Figure 9.6), showing that additions to the levee top have been fairly frequent at least during the last 40–50 years. Below the levee, in the channel, stratified sandy sediments have been progressively draped over the steeper bank in places. This may not always be FDR in origin but there is evidence for very recent sand and silt deposition all the way up to 8–10 m banktops. Truncated (lateral) benches 4–6 m high, and sometimes lower, are formed mainly of stratified sand. Sedimentation is still occurring on these, as indicated by buried vegetation. The rates of this kind of deposition can be checked from coring exotic or introduced trees (Australian native vegetation generally does not exhibit annual growth rings) to determine ages from annual growth rings and checking newly added lateral root systems after each stage of burial (Figure 9.7).

**Figure 9.6**  Partially buried fence posts on levee top at about 112 km (see Figure 9.2)

Preliminary research has revealed annual rates of 20–40 mm y$^{-1}$ in places (D. Martens, pers. comm.), but in other places exposed lateral roots indicate recent erosion.

In FDRs a specific type of landform is formed on convex banks. In the 1890 survey, they were referred to as coves. These embayments indicate flow separation impacts just downstream of tight corners, particularly those where sandstone cliffs form the outer banks. Below these very deep holes exist and these must cause mega-turbulence when high floods pass down this system. Bank erosion can only occur on the opposite bank and flood debris reveals that reverse currents prevail in high flows. The cove is cut into the alluvial bank and often has a fresh sand mound downstream of the embayment (Figure 9.8). These exist where width/curvature radii ratios reach 2 to 3 (Hickin, 1977). In more open corners they are absent or only poorly developed.

Whilst there is much detail in the 1890 (Josephson) survey of banks and channel, there is no real information on sedimentation in the last FDR (1857–1900). However, in places there are wide intertidal bars, sometimes occupying up to half the channel width at low water mark. These represent channel-bed accretion accompanying channel widening in the FDR, as indicated in the theory of expected changes. Such forms are not present in the present FDR except in one reach upstream of Windsor where the channel has been laterally very unstable (McNamara, 1985).

Slackwater deposits, including logs, plastic and rubbish, are sometimes present in large masses downstream of outside bends. Floating debris is caught and held by bank vegetation (Figure 9.9).

**Figure 9.7** New lateral root systems developed following partial burial of a tree trunk on a sandy bench

## Evidence for DDR sedimentation

The last change from FDR to DDR occurred at the beginning of this century (1901) and there were few observations on the channel at that time. It is necessary to rely on the 1890 survey in this area. It is also necessary to use indirect evidence to reconstruct in-channel aggradation associated with fewer and lower floods.

What is known is that only four flows in the 48 year period topped the levees. Consequently, levee sedimentation would have been much less than in FDRs. This means that much of the energy of larger events was kept in-channel and that flood-basin flooding was restricted to tributary back-up flows.

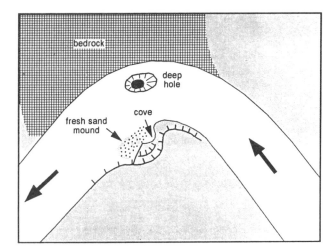

**Figure 9.8**   A diagrammatic representation of a 'cove'

**Figure 9.9**   Slackwater deposits on a concave bank at about 117 km (see Figure 9.2)

In 1890, towards the end of the FDR, the channel was wide and shallow, particularly in the upper reaches of the study area. It has been possible to compare the bank-toe widths for 1890, which may have eroded even more by 1900, with bank-toe widths from the 1949 and 1955 air photographs, at the end of the DDR (PWD, 1987). These 'windows' are as close as it is possible to get to pre-DDR and just post-DDR conditions. The loss of width is a conservative measure of in-channel deposition. In places it exceeds 50 m (Figure 9.10) and averages 25.5 m over the 25.2 km. This amounted to over 64 ha above

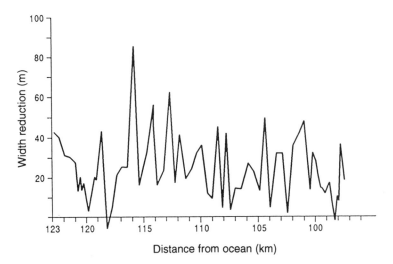

**Figure 9.10** Net accretion between 1890 and 1949/55 for the Hawkesbury River between Wilberforce and Sackville (based on Josephson survey and the PWD 1987 report)

high water mark and, if the banks averaged 4 m high, 2.57 million m³ (3.85 million m³ at 6 m). In the upper 12 km, the average width was 28.3 m. These volume estimates ignore any underwater sedimentation.

Today little remains of this extensive deposit which in its truncated form seems to relate to a DDR floodplain of about 6 m elevation in the upper reaches. This is because since the onset of the recent FDR, the deposit has been laterally truncated at a rate averaging $0.5-0.6 \, \mathrm{m \, y^{-1}}$. Figure 9.11 shows the spatial distribution of this based on photogrammetry for the period 1949/55 to 1980/82 (PWD, 1987) and Figure 9.12 shows the post 1978/80 erosion determined by detailed resurvey of PWD monumented cross-sections for the upper half of the study reach.

Erosion has been estimated two ways: (a) from the length of the eroded section times the average width of erosion, and (b) from cross-section erosion averaged between cross-sections. In the former, erosion amounts to the loss of 29.9 ha or 12 m and in the latter, 35.2 ha or 14 m. In the upper 12 km erosion averaged 17.0 m between 1949/55 and 1978/80 compared with 12–14 m (above) for the whole. In the lower part, the average was 11.2 m. Detailed survey between 1978/80 and 1989/91 in the 12 km revealed a further loss of 6.7 ha or an average width of 5.6 m.

In summary, changes for the upper reach involve:

| | | |
|---|---|---|
| 1890–1949/1955 | Accretion | 28.3 m ⎫ |
| 1949/1955–1979/1980 | Erosion | 17.0 m ⎬ Total erosion 22.6 m |
| 1978/1980–1988/1991 | Erosion | 5.6 m ⎭ |

Thus an average of about 5.7 m of the accretion survives.

For the lower reach, data are incomplete:

| | | |
|---|---|---|
| 1890–1949/1955 | Accretion | 22.9 m |
| 1949/1955–1979/1980 | Erosion | 11.3 m |

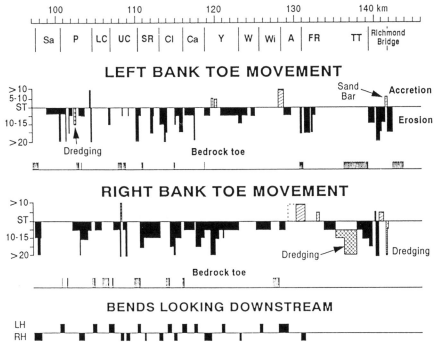

Figure 9.11   Left and right bank accretion and erosion of the Hawkesbury River between Sackville and Richmond Bridge. (Reach abbreviations shown in Figure 9.2: LH and RH refer to left and right hand bends looking downstream, based on photogrammetric analysis carried out by Sinclair Knight and Partners for NSW Department of Public Works)

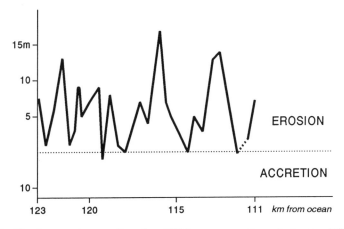

**Figure 9.12**   Erosion and accretion for PWD cross-sections between Wilberforce and Ebenezer–1978/80 to 1989/91

This bench or truncated floodplain survives in its narrowed form in many places. It consists of stratified sands and observations have been made on interbedded artefacts which confirm its recent age. In the upper parts where recession energy is lower and conducive to higher sedimentation rates, artefacts have included: pieces of sawn timber, old iron bolts, bits of glass and bottles and builders' waste (some reworked into a thin lens). Later materials include: abandoned cars, plastic and beer cans.

### Evidence of Sedimentation and Erosion associated with Human Activities

The total catchment 'affected' by human activities is a fairly small proportion of the 22 000 km$^2$. However, changes have been intense in the areas just upstream of the study reach. Almost all the shale plains have been cleared of forest and many parts have been urbanised. Adjacent areas of the lower Blue Mountains have also been urbanised. Thus sediment yields have increased mainly from local sources. They have been reduced from those areas affected by farm storage dams which help irrigate pastures in the rainshadow area.

Introduced materials probably reached a peak in the drought-dominated regime (1901–1948), when droughts, poor farming practices, rabbits, noxious weeds and so on increased sediment delivery to the channel from sheet erosion, rills and gullies (Pickup, 1976). Urban-based sediments have been high in developing stages and much of this area has been urbanised since World War II in the present FDR. However, conservation practices are now being introduced even there, with detention ponds and similar structures.

Changes to channels have taken place to affect erosion and sedimentation in four ways:

1. The introduction of riparian weirs above Penrith in the early 1900s created not only ponds for water retention but also minor sediment traps.
2. The building of the four Nepean dams (1907–1936) and then the Warragamba Dam (1960) has effectively curtailed all bedload movement from over 9000 km$^2$ (yet runoff over the Warragamba Dam in the FDR 1970s was higher than in the DDR 1940s before the dam existed – to give highly sediment-deficient discharges).
3. Sand and gravel extraction from the channel, floodplains and terraces between Penrith and Richmond has involved tens of millions of cubic metres. Channel removals there and below Windsor have had a greater impact in sediment starvation.
4. Flood-mitigation works have restricted tributary back-up flooding by the use of flood gates, and levee protection works have kept all but the highest floods in the channel, where flow energy is not dissipated by overbank events.

Many of these impacts have been considered in earlier works (Warner, 1984, 1987a; Erskine & Warner, 1988; Warner & Bird, 1988).

## DISCUSSION

In reality the theoretical sequence of adjustments imposed by alternating regimes has been modified by the impacts of human actions in the channel and in the catchment. Only 200 years ago this was a 'pristine' landscape with the only significant human impact being aboriginal firing of parts of the landscape. It was after 1788 that European settlers came to this area. There was a slow initial colonization of the Hawkesbury valley which suffered setbacks as a result of flooding in the first FDR. However, human impacts have increased progressively with farming and mainly post-1945 urbanisation, with small-scale water developments (dams in 1901–48 DDR) and then large-scale impacts in the present FDR (post-1949) like the Warragamba Dam (1960), flood-mitigation works (1950 onwards) and widespread aggregate extraction (1950 onwards).

Some of these changes, particularly in the latter regimes, have been incorporated in Figure 9.13, but it is difficult to show the complete range. What is illustrated there is an upstream, unleveed cross-section with narrow floodplains and a wide leveed floodplain sequence of a backwater-tidal river system. Lower gorge-like sections, highly resistant perimeters, bedload and wholly suspended-load and other types of perimeters have not been incorporated. Shown are 'mobile' (Pickup, 1984) cross-sections which respond fairly readily to regime shifts and to large-scale human activities.

Thus the cyclic disequilibria (Nanson & Erskine, 1988) associated with regime alternation are forced to accommodate 'progressive' impacts of human activities, not all of which lead to degradation. Water conservation on the Cumberland Plain for irrigation of pastures, soil conservation and removal of vermin, and urban stormwater retention basins have all tended to reduce water and sediment discharges in the present regime, but these impacts have been outweighed by the high runoff in the FDR and sediment starvation imposed by the dams and aggregate removal. Flood-mitigation structures tend to keep moderate floods in the channel. These in turn increase power there which places additional stress on banks and bed. Bedload is moved further into the estuary (Warner & Paterson, 1987) and the flooding of floodbasins and backswamp is decreased by floodgates across tributary entrances (Figure 9.13e).

Under alternating regimes two floodplains can exist with the high floodplain becoming virtually a terrace in DDRs, whereas in FDRs it is surcharged every 2 to 3 years. Low floodplains build in DDRs, as has been shown in Figure 9.10 and by the observations of artefacts in the stratified sands. These are progressively destroyed in FDRs. The availability of sand to create these is to some extent dependent on catchment instability. Obviously in the first DDR there was much less activity in the catchments, lower erosion rates and less material delivered to the channel (Figure 9.13b). The second DDR (1901–1948), was the greatest period of land degradation (Pickup, 1974, 1976; Erskine, 1986) and much more material was available in the channel for reworking (Figures 9.10 and 9.13d). Higher energy floods in the present FDR have less sand to convey, through starvation from upstream, but they deposit sand on the levees frequently to bury fence posts (Figure 9.6) and they are generally deepening the channel in the study reach. This is probably the greatest deviation from the theoretical sequence proposed (Figure 9.13e). Such changes which predominate in the upper part of the study reach, where the main floodplain is much wider, can be explained in the rapid power reductions in the recession limbs of overbank hydrographs (Table 9.4). In the gorge at Sackville, power increases in the hydrograph recession and there is less contemporary sedimentation.

There have been no simulations of DDR hydrographs but it must be presumed that a 2 year flood would be of much shorter duration and would peak at 6 m at Windsor and 4 m or so at Sackville. Rising limbs would be similar but recessions with no large floodplain drainage would be shorter without necessarily decreasing power very much at Windsor or increasing it at Sackville. In the former case this would help degrade the bed.

In areas where perimeter conditions are more resistant, changes are slower to occur. Even on the floodplains erosion would be slower under forested conditions (Figure 9.13a). However, after clearance, high energy overbank flows could cut chutes or involve some floodplain stripping (Nanson, 1986) (Figures 9.13c and e). There is evidence for chute

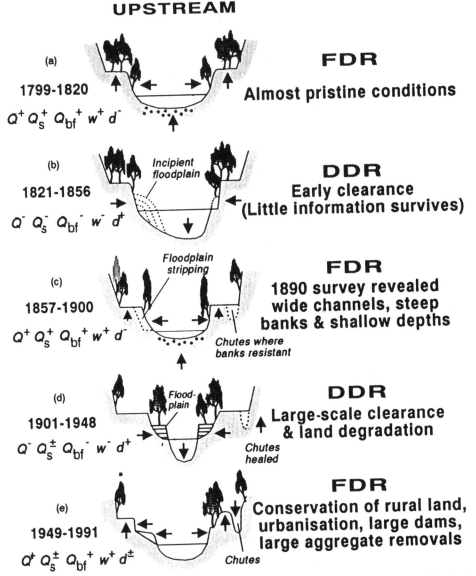

**Figure 9.13** Modifications to the theoretical sequence imposed by human activities in the catchment and channel

cutting at Charity Creek on the Manning River (see Nanson, 1986) in the later 19th century FDR on the original estate maps. Air photographs taken in the 1940s reveal that such overflow channels had largely been filled in by flood deposits (Figure 9.13d). Thus floodplain stores, both on the surface and at the margins, were affected in this alternating regime sequence.

Although adjustments feature prominently in this discussion, it is by no means certain that they are ever 'completed'. Equilibrium is perhaps seldom attained (Phillips, 1992) because of high perimeter resistance in high energy areas (source and armoured reaches – Pickup (1984)) and relatively high perimeter resistance in suspended-load channels where

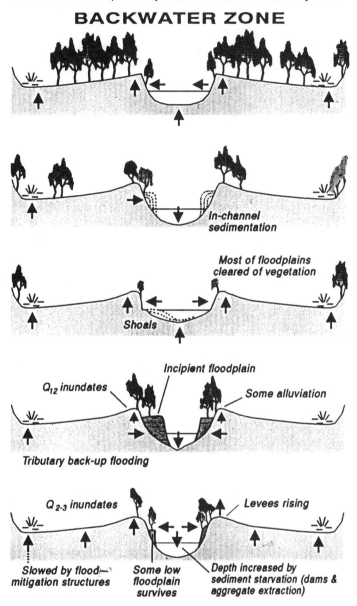

**Figure 9.13**(continued)

power may be very low. The latter is particularly the case in the lower coastal deltas where flood velocities are low (1 m s$^{-1}$), where banks are highly coherent and where there are large overbank water storages. This means that lagged adjustments are carried into subsequent regimes. This is particularly true of the last DDR incipient floodplain. It has survived over 40 years into the present regime, perhaps because of attenuated, dam-affected flood flows and other bank-protection works.

The highest rate of adjustment is in the mobile zone where banks and bed change readily with variations in water and sediment discharge. This creates spatial variability in adjustment and further long-profile disequilibrium (Warner, 1987b), thereby helping to explain why downstream hydraulic geometries are so 'noisy' in New South Wales.

In-channel benches have attracted attention in the past (Kilpatrick & Barnes, 1964; Woodyer, 1968; Warner et al., 1975). They were related to certain flow return periods, using the entire hydrologic record as a single series. Now the dating of the incipient floodplain (1901–1948) in the Hawkesbury indicates that they may in part be related to alternating natural regimes or indeed alternating induced regimes (Sherrard & Erskine, 1991). The main floodplain has a surcharging recurrence greater than the mean annual flood in the entire flood record because this includes DDRs, when it is virtually a terrace. Its roots are undoubtedly Holocene, with 3000–5000 years BP being a common date (Warner, 1992), but the upper part of the surface includes post-settlement alluvium (Figure 9.6).

So, even though regime shifts are major influences on channel instability and contemporary adjustments in channels and on floodplains, progressive as well as single-action human activities do influence the balance of $Q$ and $Q_S$ moving into and through the channels.

Consequently, spatial refinements to the theory must include the nature and degree of human impacts in the catchment and in the channel, as well as natural elements like the condition, width and gradients of floodplains; the relative resistance of the channel perimeter to different flow rates and frequencies of event-forcing energies and the impacts of catastrophic events. Knowledge of all these helps to 'explain' why the relatively simple theory does not hold in many cases.


## CONCLUSIONS

It is appropriate to offer a theory of potential channel changes and floodplain modifications related to alternating FDRs and DDRs. The theory suggests that in relatively mobile areas adjustments generally include: removal of floodplain through channel width increases and depth decreases in FDRs, and an increase in the alluvial store with width decreases but depth increases in DDRs, as the channel adjusts to smaller floods. This sequence is lagged where perimeters are slower to react. In high energy upstream environments, floodplains are modified by chute cutting or stripping to increase floodway areas in FDRs. In low energy downstream areas, flood incidence increases but erosional impacts are not great on wider floodplains. The availability of sediment from upstream, from channel erosion, and from tributaries is important in DDRs for construction of incipient floodplains and other in-channel sedimentation. Where this is not available, lower rates of DDR modifications may be experienced. Sediment supplies from catchment and channel are intimately related to the level of human activity in developing and densely settled environments. These then have been important in the Hawkesbury, particularly in the 1901–1948 DDR in-channel deposition and in the present FDR with increases in channel depth through sediment starvation. This is contrary to the theory.

The theory enables some distinctions to be made between 'natural' and human-induced changes. This is relevant to the understanding of channel and floodplain changes and, more importantly, to the management of unstable systems. Attempts to stabilise, rehabilitate and otherwise improve channels made without appreciation of instability would probably be wasteful in terms of resources. The understanding of the causes of erosion and sedimentation and their reversals has to be of some value. There is a good case in such alternating unstable systems to have passive buffer zones as an active part of management strategy (Warner, 1988). There is also a need to understand the system if we are to resolve the irreconcilable concepts of regime theory and environmental change (Lewin *et al.*, 1988).

## REFERENCES

Bull, W. B. (1979). Threshold of critical power in streams. *Geological Society of America Bulletin*, **90**, 453–464.

Erskine, W. D. (1986). River metamorphosis and environmental change in the Hunter Valley, New South Wales, Unpublished Ph.D. Thesis, University of New South Wales.

Erskine, W. D. and Warner, R. F. (1988). Geomorphic effects of alternating flood- and drought-dominated regimes on NSW coastal rivers. In Warner, R. F. (ed.), *Fluvial Geomorphology of Australia*, Academic Press, Sydney, pp. 223–244.

Graf, W. L. (1983). Flood-related channel change in an arid-region river. *Earth Surface Processes and Landforms*, **8**, 125–139.

Hickin, E. J. (1977). Hydraulic factors controlling drainage migration. In Davidson-Arnott, R. and Nickling, K. (eds), *Proceedings 5th Guelph Symposium*, Geo-Abstracts, Norwich, pp. 59–66.

Jeans, D. N. (1972). *An Historical Geography of NSW to 1901*, Longman Cheshire, Melbourne.

Josephson, J. P. (1885). History of floods in the Hawkesbury River. *Journal of the Royal Society of NSW*, **19**, 97–105.

Kilpatrick, F. A. and Barnes, H. H. (1964). Channel geometry of the Piedmont streams, as related to frequency of floods. *United States Geological Survey Professional Paper*, **422–E**.

Lewin, J., Macklin, M. G. and Newson, M. D. (1988). Regime theory and environmental change – irreconcilable concepts?. In White. W. R. (ed.), *International Conference on River Regimes*, John Wiley and Sons, Chichester, pp. 431–445.

Magilligan, F. J. (1992). Thresholds and the spatial variability of floodpower during extreme floods. In Phillips, J. D. and Renwick, W. H. (eds), *Geomorphic Systems*, a special issue of *Geomorphology*, **5**, 373–390.

McNamara, R. L. (1985). Channel changes in the Upper Hawkesbury River, New South Wales, since 1890, Unpublished BA Thesis, University of Sydney.

Nanson, G. C. (1986). Episodes of vertical accretion and catastrophic stripping: a model of disequilibrium floodplain development. *Geological Society of America Bulletin*, **97**, 1467–1475.

Nanson, G. C. and Erskine, W. D. (1988). Episodic changes of channels and floodplains on coastal rivers of NSW. In Warner, R. F. (ed.), *Fluvial Geomorphology of Australia*, Academic Press, Sydney, pp. 201–221.

Phillips, J. D. (1992). The end of equilibrium? In Phillips, J. D. and Renwick, W. H. (eds), *Geomorphic Systems*, a special issue of *Geomorphology*, **5**, 195–202.

Pickup, G. (1974). Channel adjustment to changed hydraulic regime, in the Cumberland Basin, NSW, Unpublished Ph.D. Thesis, University of Sydney.

Pickup, G. (1976). Geomorphic effects of changes in runoff, Cumberland Basin, NSW, *Australian Geographer*, **13**, 188–195.

Pickup, G. (1984). Geomorphology of tropical rivers. I. Landforms, hydrology and sedimentation in the Fly and Lower Purari, Papua New Guinea. *Catena Supplement*, **5**, 1–17.

PWD (Department of Public Works, NSW) (1979). Hawkesbury River: March 1978 Flood Report. Report by Metropolitan District Engineer.

PWD (Department of Public Works, NSW) (1983). Hawkesbury River Flood Stage Heights – Yarramundi to Ebeneezer.

PWD (Department of Public Works, NSW) (1987). Channel geometry, morphological changes and bank erosion. Hawkesbury River Hydraulic and Sediment Transport Processes, Report No. 10.

Riley, S. J. (1980). Aspects of the flood record at Windsor, *Proceedings of the 16th Conference of the Institute of Australian Geographers*, Newcastle, pp. 325–340.

Schumm, S. A. (1977). *The Fluvial System*, Wiley, New York.

Sherrard, J. J and Erskine, W. D. (1991). Complex response of a sand-bed stream to upstream impoundment. *Regulated Rivers: Research and Management*, **6**, 53–70.

Simon, A. (1992). Energy, time, and channel evolution in catastrophically disturbed fluvial systems. In Phillips, J. D. and Renwick, W. H. (eds), *Geomorphologic Systems*, a special issue of *Geomorphology*, **5**, 345–372.

Thom, B. G. and Roy, P. S. (1981). Relative sea levels and coastal sedimentation in Southeastern Australia in the Holocene. *Journal of Sedimentary Petrology*, **55**, 257–264.

Warner, R. F. (1971). Alluvial and terrace geomorphology in a coastal valley of northern New South Wales. *Photo Interpretation*, **71/1**, 17–21.

Warner, R. F. (1983). Channel changes in the sandstone and shale reaches of the Nepean River, New South Wales. In Young, R. W. and Nanson, G. C. (eds), *Aspects of Australian Sandstone Landscape*, Australia New Zealand Geomorphology Group Special Publication No. 1, pp. 106–119.

Warner, R. F. (1984). Impacts of dams, weirs, dredging and climatic changes: An example from the Nepean River, New South Wales, Australia, Paper for 25th International Geographical Union Congress, Paris.

Warner, R. F. (1987a). The impacts of alternating flood-and drought-dominated regimes on channel morphology at Penrith, NSW, Australia, *International Association for Scientific Hydrology*, Publication No. 168, 327–338.

Warner, R. F. (1987b). Spatial adjustments to temporal variations in flood regime in some Australian rivers. In Richards, K. S. (ed.), *Rivers: Environmental Process and Morphology*, Blackwell, Oxford, pp. 14–40.

Warner, R. F. (1988). Environmental management of coastal rivers. *Planner*, **3**, 32–35.

Warner, R. F. (1990). Channel changes, regime shifts and their implications in the upper Hawkesbury River, NSW. Report for Mitchell McCotter and NSW Department of Public Works.

Warner, R. F. (1991a). Impacts of environmental degradation of rivers, with some examples from the Hawkesbury–Nepean system. *Australian Geographer*, **22**, 1–13.

Warner, R. F. (1991b). Water and sediment problems in the Hawkesbury–Nepean River, NSW, Australia. *Mitteilungsblatt des Hydrographischen Dienstes in Oesterreich*, No. 65/66, 45–46.

Warner, R. F. (1992). Floodplain evolution in a New South Wales coastal valley, Australia: spatial process variations. *Geomorphology*, **4**, 447–458.

Warner, R. F. and Bird, J. F. (1988). Human impacts in river channels in New South Wales and Victoria. In Warner, R. F. (ed.) *Fluvial Geomorphology of Australia*, Academic Press, Sydney, pp. 343–363.

Warner, R. F. and Martens, D. (1992). Regime shifts and their impacts on northern New South Wales rivers Australia, Abstracts *27th International Geographical Congress*, Washington, p. 664.

Warner, R. F. and Paterson, K. W. (1987). Bank erosion in the Bellinger Valley, NSW: definition and management. *Australian Geographical Studies*, **25**, 3–14.

Warner, R. F., Sinclair, D. and Ewing, J. (1975). A comparative study of relations between bedrock meander wavelengths, benchfull discharges and channel perimeter sediments for the Sydney Basin and the northeast of New South Wales. In Douglas, I., Hobbs, J. E. and Pigram, J. J. *Geographical Essays in Honour of Gilbert J. Butland*, University of New England, Armidale, pp. 159–212.

Webb McKeown and Associates Pty Ltd. (1988). Warragamba dam – major flood protection works: Windsor flood levels, Report for Sydney Water Board.

Woodyer, K. D. (1968). Bankfull frequency in rivers. *Journal of Hydrology*, **6**, 114–142.

# 10 Analytical Approach to Flow Resistance in Gravel-bed Channels with Vegetated Banks

RICHARD MASTERMAN and COLIN R. THORNE
*Department of Geography, University of Nottingham, UK*

## ABSTRACT

Many natural rivers have a gravel-bed and finer-grained, vegetated banks. This paper introduces a theoretical, predictive method to take into account the influence of bank vegetation on flow resistance in such rivers. The channel cross-section is divided into bank areas, where resistance is determined by vegetation roughness, and a bed area, where resistance is determined by sediment size. Rational flow resistance equations are developed for each area, based on the distribution of boundary shear stress. A roughness height based on stiffness is used for flexible vegetation and a wake velocity correction factor is used for tall vegetation that does not bend in the flow. The model is used to assess the contribution of the vegetated bank areas to the overall flow resistance of the channel. The effects of changes in bank vegetation and of different types of vegetation on channel capacity are modelled. The results of numerical simulations indicate that the effects of bank vegetation decline rapidly as width–depth ratio increases.

## INTRODUCTION

Vegatation influences river channel form and processes at a range of scales (Gregory & Gurnell, 1988). Its effect on flow resistance depends on the type of vegetation, its location in the channel or on the banks, and factors such as spacing, density, extent, height and stiffness (Thorne, 1990). Studies of vegetation in natural channels have emphasised the reductions in channel capacity that result from seasonal growth (Powell, 1978; Watts & Watts, 1990), or from ecological succession over a period of years (Gwinn & Ree, 1980). These studies are largely concerned with the encroachment of bank vegetation into the channel and the growth of in-channel vegetation. There is also a substantial literature on roughness and resistance aspects of grass-lined channels, of flow across vegetated floodplains and of the ecological effects of water flow. A comprehensive bibliography has been prepared by Dawson & Charlton (1988). However, many of these studies do not reflect the situation in natural rivers and flood control channels where there is a sediment bed and vegetated banks. This paper considers a method that allows a detailed analysis of flow resistance, based on a theoretical knowledge of flow in open channels.

*Process Models and Theoretical Geomorphology*. Edited by M. J. Kirkby
©1994 John Wiley & Sons Ltd

The contribution of vegetation to flow resistance in open channels has traditionally been estimated using one of a number of empirical methods that involve determining the Manning's 'n' resistance coefficient (Cowan, 1956; Chow, 1959; Barnes, 1967). But for many design and analysis applications it is preferable to determine flow resistance in channels using the Darcy–Weisbach friction factor:

$$f = \frac{8gRS}{v^2} \tag{1}$$

where $f$ is the Darcy–Weisbach friction factor; $g$ is the acceleration due to gravity; $R$ is the hydraulic radius; $S$ is the energy slope; and $v$ is the average downstream velocity in the cross-section. This is a dimensionless term that has been recommended for use in studies of open channel flow (ASCE, 1963) and has a theoretical basis derived from studies of flow in pipes. However, because of the complex nature of flow resistance, empirical elements and simplifications are still introduced in calculations of a resistance parameter (Rouse, 1965; Bathurst, 1982).

Resistance can be separated into skin, free surface and form resistance in straight channels with uniform and fully turbulent flow. While form resistance can be important in sand-bed streams (Simons et al., 1965), it is considered to be negligible in gravel-bed rivers (Hey, 1979). The contribution of the free surface to flow resistance is dependent on the Froude number (Bathurst et al., 1981). Where the flow is subcritical, the contribution of the free surface to flow resistance is minimal. Therefore, in many gravel-bed streams flow resistance can be approximately determined from a measure of the skin friction.

A theoretically based or 'rational' analysis of flow resistance in open channels was developed by Keulegan (1938). This approach assumes that the apparent shear at any point on the boundary of a cross-section depends on the velocity at that point and that turbulence can be characterised by a universal constant (Richards, 1982). Thus, the shear stress at a point can be represented by:

$$\tau_0 = kv^2 \tag{2}$$

where $k$ is a roughness coefficient and $v$ is velocity.

The physical implications of this approach are that velocity profiles perpendicular to the solid boundary are logarithmic and the hydraulic effect of friction can be determined from a measure of the relative roughness in a channel. The average velocity in a cross-section can be determined by integration. The effect of channel shape on shear stress distribution is an additional factor evaluated by Keulegan (1938).

Where channels satisfy these conditions flow resistance can be predicted by determining a parameter of relative roughness that is the ratio of the roughness height of the surface to the hydraulic radius (Hey, 1979; Bathurst, 1982). Assuming that flow resistance is entirely dependent on surface roughness and that the roughness height of the surface is greater than a critical roughness below which flow is said to be hydraulically smooth (Chow, 1959), a practical method for calculating flow resistance in gravel-bed rivers can be derived based on Keulegan's approach using the Colebrook–White equation (Hey, 1979):

$$\frac{1}{\sqrt{f}} = a + c \log\left(\frac{R}{k}\right) \tag{3}$$

where $k$ is a roughness height, $c$ is a coefficient derived from the von Karman constant, usually taken to be 0.4, and $a$ is a coefficient determined by the cross-sectional shape of the channel. From the hydraulic geometry of a cross-section both mean velocity and the discharge capacity of a channel can be predicted.

Equations of this form have been calibrated and applied by, among others, Hey (1979). His equation for gravel-bed rivers has the form:

$$\frac{1}{\sqrt{f}} = 2.03 \log\left(\frac{a_s R}{3.5 D_{84}}\right) \tag{4}$$

where $a_s$ is a shape correction factor defined by:

$$a_s = 11.1\left(\frac{R}{d_{max}}\right)^{-0.314} \tag{5}$$

$D_{84}$ is the intermediate diameter of gravel of which 84% of the bed material is finer than or equal to (Bathurst, 1982) and $d_{max}$ is the maximum depth of the channel. Applications indicated that predictions of bankfull discharge were within 15% of the actual values for riffle cross-sections although there were larger discrepancies for pool cross-sections (Hey, 1979). In cobble and boulder-bed rivers, the equation was found to give reasonable accuracy ($\pm 20\%$) so long as the relative roughness of the bed, defined as the ratio of the roughness height of the bed material with the depth of flow, was less than about 0.5 (Thorne & Zevenbergen, 1985).

## CATEGORISING VEGETATION

A number of methods exist to estimate flow resistance in vegetated channels. These methods are reviewed in a Hydraulics Research Station Report (1985). The report attempts a rationalisation of the various approaches by categorising them in terms of which phases of flow they are concerned with, whether the relative depth of the flow is shallow (less than the deflected height of the vegetation) or deep (greater than the deflected height of the vegetation), and whether the spacing of the vegetation represents dense or open cover. The Hydraulics Research Station Report (1985) also classifies vegetation in channels to take account of ecological, hydraulic and the relative height of vegetation to depth of flow factors (Table 10.1). No single approach to estimating flow resistance in vegetated channels can hope to take account of all the aspects of the problem.

The division between dense and open vegetation has been established from experimental data in laboratory flumes. Open vegetation occurs where significant flow within the vegetation stand occurs. Li & Shen (1973) have set theoretical limits for open vegetation of greater than six stem diameters in the downstream direction and three stem diameters in a transverse direction.

The terms erect, inclined and prone represent the hydraulic states of vegetation. The differences in these states have been explored and analysed by Kouwen & Unny (1973).

**Table 10.1**   Categorising dense open vegetation

|          | $d < k_w$ | $d > k_w$ |
|----------|-----------|-----------|
| DENSE    | erect     | (erect)   |
|          | inclined  | inclined  |
|          |           | prone     |
| OPEN     | erect     | erect     |
|          | inclined  | inclined  |
|          |           | prone     |

*Note*: In this classification $d$ refers to the depth of the channel and $k_w$ refers to the effective roughness height of the vegetation taking into account any deflection by the flow of water.

Velocity profiles within prone vegetation appear to behave differently from those within erect vegetation (Saowapon & Kouwen, 1989). The reason for this may be due to the increase in influence of the substrate on which the vegetation grows on flow resistance. The degree of deflection of flexible vegetation has been used as a stability criterion in the design of channels (Kouwen & Li, 1980a). As vegetation is deflected by stronger flows, shear stresses generated by the flow of water begin to act on the substrate material, increasing the potential for erosion.

Analyses of flow resistance in natural rivers and flood control channels have usually ignored the effects of bank vegetation. However, studies of the hydraulic geometry of channels indicate that variations in width are correlated with differences in vegetation (Charlton *et al.*, 1978; Andrews, 1984; Hey & Thorne, 1986). While these effects may partially be due to changes in the stability of the bank material brought about by different types of vegetation, the influence on flow resistance on the form of stable channels may also be significant. In any natural channel, there is a balance to be struck between stability of the banks and the conveyance that results from the channel form.

A rational approach to predicting flow resistance in grass-lined channels has been developed by Kouwen *et al.* (1969). The effective roughness height of flexible vegetation is dependent on vegetation stiffness and the strength of flow over the vegetation (Kouwen & Li, 1980b). Stiffness (*mei*) is a composite parameter that takes into account the density, elasticity and bendiness of vegetation. Kouwen has indicated that for grasses stiffness can be determined in the field either from the height of the grass or by dropping a board to test stiffness directly (Kouwen, 1988). Figure 10.1 illustrates this method. The distance from the edge of the board to the ground has been related to vegetation stiffness (Eastgate, 1966). This approach simulates the action of water flowing over vegetation. The effective roughness height ($k_w$) of the vegetation can then be predicted from:

$$k_w = 0.14\,h \left( \frac{\left( \frac{mei}{\tau_0} \right)^{0.25}}{h} \right)^{1.59} \tag{6}$$

where $h$ is the height of the vegetation, and $\tau_0$ is the mean boundary shear stress, given by $\tau_0 = \gamma dS$, where $d$ is the depth of the channel and $S$ is the channel-bed slope. The effective roughness height of vegetation ($k_w$) thus replaces $k$ in equation (3).

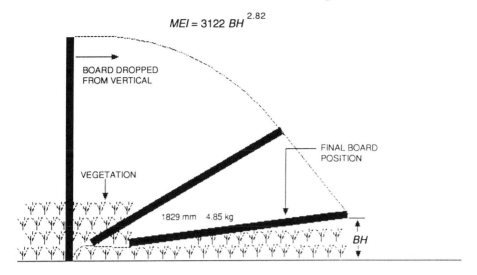

**Figure 10.1**   The board drop test (after Kouwen, 1988. Reproduced by permission of the International Association for Hydraulic Research)

Vegetation behaves differently depending on whether it is erect or prone in the flow, so that the coefficients $a$ and $c$ in equation (3) also vary with the state of proneness of the vegetation (Kouwen and Unny, 1973).

For grasses, the *mei* parameter correlates reasonably well with the height ($h$) of the grass (Kouwen, 1988). For green, growing grass the relationship is:

$$mei = 319\,h^{3.3} \tag{7}$$

and for dormant or dead grass it is:

$$mei = 25.4\,h^{2.26} \tag{8}$$

Application of the *mei* values so produced is restricted at present to only those grasses which have been tested. Although the above equations apply to bed vegetation, observations indicate that there is no difference in the way that flexible bank vegetation responds to flowing water.

Kouwen *et al.*'s (1969) laboratory analysis used articial flexible elements to simulate vegetation. A logarithmic velocity profile was present above the deflected strips. Gourlay (1970) confirmed that a similar profile was present above real vegetation. In Kouwen *et al.*'s (1969) analysis the shear force on the soil/water interface is displaced to the tips of the vegetation or some level within the vegetation, so that a small residual velocity is all that occurs in the vegetation. Saowapon & Kouwen (1989) have developed a physically based theoretical analysis for flow within vegetation that appears to agree with this analysis, at least for erect and waving vegetation.

Pethick *et al.* (1990) demonstrate that velocity profiles within open salt marsh vegetation covers can also be fitted to a logarithmic curve. Generally, where flow is through open cover, allowing water to flow between stems, a different approach is required taking account of the effects of vegetation on wake velocities (Li & Shen, 1973). This approach has been used by researchers who have investigated floodplain roughness due to non-submerged vegetation (Klassen & Zwaard, 1974; Thompson & Roberson, 1975; Petryk & Bosmajian III, 1975; Pasche & Rouve, 1985). These analyses assume that the shear at the soil/water interface is negligible and that the maximum shear is determined by a drag force on the vegetation. Calculations of the resistance factor can be made by estimating the density of vegetation and taking account of its patterns of growth (Li & Shen, 1973).

Where the height of the vegetation, or the deflected height of flexible vegetation, is greater than the depth of the channel, then an analysis based on the drag coefficients of the individual vegetation elements can be used (Li & Shen, 1973; Thompson & Roberson, 1976). In this procedure an underlying roughness and resistance parameter can be first established and the turbulence effects of the protruding vegetation are then taken account of separately by calculating wake velocities from empirical equations.

## SHEAR STRESS DISTRIBUTIONS

Shear stress distributions around the boundary of complex cross-sections are virtually impossible to predict from theoretical considerations without making considerable simplifying assumptions. Bathurst (1979) showed that shear stress distributions become more uniform around the boundary of channel cross-sections as the Reynolds number of the flow increases. Knight (1981) carried out experiments that indicated that shear stress distributions around the boundary of cross-sections in open channels were dependent on the width/depth ratio of the channel and the roughness of the bed. Flintham & Carling (1988) have developed a tentative empirical model to predict shear stress distributions from the relative roughness of the bed and bank perimeters in a cross-section and the relative lengths of these perimeters.

Using the method developed by Flintham & Carling (1988), shear stress distributions around the boundary of a cross-section can be predicted for a channel divided into three sub-areas (Figure 10.2) from the the following series of equations:

$$\frac{\tau_w}{\gamma dS} = 0.01 \% SF_w \left( \frac{w + P_b}{2P_w} \right) \tag{9}$$

where $\tau_w$ is the wall shear stress, $\% SF_w$ is the percentage shear force carried by the side walls, $w$ is the channel flow width, $P_b$ is the wetted perimeter of the bed and $P_w$ is the wetted perimeter of the vegetated banks. $\% SF_w$ is calculated from

$$\log \% SF_w = -1.4026 \log \left( \frac{P_b}{P_w tr} + 1 \right) + 2 \tag{10}$$

where $tr$ is a threshold parameter calculated from

$$\log tr = -0.203 \log \left( \frac{D}{k_w} \right) + 0.176 \tag{11}$$

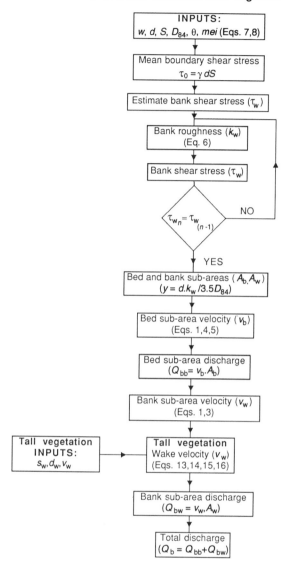

**Figure 10.2**  Flow chart showing the sequence of algorithms to calculate discharge in gravel-bed channels with vegetated banks

where $D$ is the bed roughness parameter and $k_w$ is the bank roughness parameter. Thus, the shear stress distribution between the bed and bank areas is function of the ratio of the effective roughness of the bed area and the effective roughness of the bank areas.

## FLOW RESISTANCE OF CHANNELS WITH COMPOUND ROUGHNESS

Channel discharge capacity in channels with compound roughness can be predicted using a procedure that combines the methods for predicting flow resistance in sediment-bed

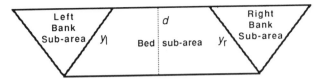

**Figure 10.3**   Bed and bank sub-areas

rivers and grass-lined channels (Figure 10.2). This procedure is based on division of the cross-section of the channel into three sub-areas (Figure 10.3). The divisions are made on the basis of the relative influence of the effective roughness parameters in the bed sub-area and each bank sub-area (Hey, 1979). The boundaries between bed and bank flow are constructed from determining $y_l$ and $y_r$ in Figure 10.3, using the method described in Hey (1979) for determining the line where the velocity dependent on the bed roughness is equal to that dependent on bank roughness. With this information it is then possible to calculate mean velocities and a predicted discharge for each sub-area, since

$$Q_a = v_a A_a \tag{12}$$

where $Q_a$ is the discharge through a sub-area, $v_a$ is the mean velocity in that sub-area, and $A_a$ is its cross-sectional area. The predicted channel discharge capacity is determined by adding together the predicted discharges for the sub-areas.

The effective roughness parameters for the bank sub-areas are determined from equation (6). This equation includes a term for the boundary shear stress and the method developed by Flintham & Carling (1988) can be used to predict average boundary shear stress distributions separately for the bank sub-areas and the bed sub-area. Since boundary shear stress is itself dependent on the ratio of the effective roughness of each bank sub-area with the effective roughness of the bed sub-area, an iterative procedure must be used to determine the predicted boundary shear stress distribution.

In this procedure initial effective roughness heights for the two bank sub-areas are calculated on the basis of the average channel boundary shear stress ($\gamma dS$) and then the shear stress distribution is predicted using the procedure presented by Flintham & Carling (1988). This new value is then used to calculate new effective roughness heights for the bank sub-areas and the associated shear stress distribution between the bank sub-areas and the bed area. The iterative procedure continues until differences in the parameters between iterations are insignificant.

Emergent, non-flexible bank vegetation may be incorporated into the analysis through an approach proposed by Thompson & Roberson (1976) developed from a theoretical approach outlined by Li & Shen (1973). In this approach the wake velocities downstream of emergent, non-flexible vegetation are calculated using one of four equations according to the patter of the vegetation and the ratio of spacing of vegetation to its diameter ($S_w/d_w$). For staggered spacing the equations are:

$$\frac{u_w}{v} = 1.06 \left(\frac{S_w}{d_w}\right)^{0.04} \qquad 4 \leqslant \frac{S_w}{d_w} \leqslant 10 \tag{13}$$

and

$$\frac{u_w}{v} = 0.93 \left(\frac{s_w}{d_w}\right)^{0.015} \qquad 10 \leqslant \frac{s_w}{d_w} \leqslant 100 \qquad (14)$$

For parallel spacing the equations are:

$$\frac{u_w}{v} = 0.48 \left(\frac{s_w}{d_w}\right)^{0.14} \qquad 4 \leqslant \frac{s_w}{d_w} \leqslant 20 \qquad (15)$$

and

$$\frac{u_w}{v} = 0.70 \left(\frac{s_w}{d_w}\right)^{0.08} \qquad 20 \leqslant \frac{s_w}{d_w} \leqslant 100 \qquad (16)$$

where $u_w$ is the flow velocity in the wake area, $v$ is the approach flow velocity, $s_w$ is the spacing of the elements in a longstream direction and $d_w$ is the diameter of the elements. Spacing can be assessed qualitatively according to the arrangement of plants on the bank. These equations are the results of empirical relationships determined from a mathematical model and so may need to be amended in the light of field experience.

These corrected velocities take into account the longstream effect of emergent, non-flexible vegetation. The corrected velocities can be used in calculations of the discharge capacity at the cross-section.

The method outlined does not take account of all the factors that are involved in flow resistance (Bathurst, 1982). Shear stresses are assumed to be uniform within each segment and the effects of secondary currents and momentum exchange between segments (Myers, 1978) are ignored. Bedforms in gravel-bed rivers may also play a greater role in flow resistance than is generally accepted (Hey, 1988; Robert, 1990). Shape effects are more complex than has been considered (Kazemipour & Apelt, 1979). Nevertheless, the simplifying steps that have been taken are essential to allow the development of a first approximation, rational method for calculating flow resistance and velocities, and, thus, predicting channel capacities and flow-dependent bank stability in gravel-bed channels with vegetated banks.

## MODEL EVALUATION

The inputs to the model are flow width, depth, bed material $D_{84}$, right and left bank vegetated wetted perimeters, bank angles, vegetation stiffness (which for grasses can be derived from height with reasonable accuracy (Kouwen, 1988); and for emergent, non-flexible vegetation, the diameter and spacing in the longstream direction. Using hypothetical data, differences in channel discharge capacity and velocities in the bank region that result from different vegetation types or changes in bank vegetation can be simulated.

**Table 10.2**   Input data for cases 1 and 2

| Parameter | Case 1 | Case 2 |
|---|---|---|
| Width (m) | 4–33 | 5, 10, 20, 30 |
| Maximum depth (m) | 1 | 1 |
| Water surface slope | 0.005 | 0.005 |
| $D_{84}$ (m) | 0.118 | 0.118 |
| Channel side slope (°) | 45 | 45 |
| Vegetation stiffness (N m$^{-2}$) | 153 | 1.6–1504 |

## Predicting Channel Capacity

Two examples indicate how the model can be used to predict the effects of bank vegetation on overall discharge capacity. In the first case a comparison is made between channels of different width–depth ratios. In the second case the effects of seasonal vegetation growth on the discharge capacity of channels with different width–depth ratios but similar geometries are simulated. The input data used for both simulations are present in Table 10.2.

### Case 1: Variation with Width–Depth Ratio

In Case 1 channel width is varied to allow a comparison between channels with different width–depth ratios. The stiffness valued used, 153 N m$^{-2}$, corresponds to densely packed grasses with a height of about 0.8 m (Kouwen, 1988) that is reduced to 0.3 m by deflection during high flows. Figure 10.4 shows a plot of the discharge through the bank sub-areas as a percentage of the total discharge against the width–depth ratio for the simulated channels. This plot illustrates that the discharge capacity in the bank areas only becomes significant (>5%) in channels of width–depth ratio less than 9 (Figure 10.4).

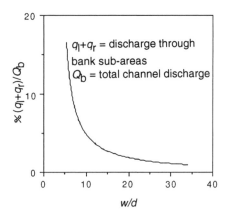

**Figure 10.4**   Change in bank sub-area discharge as a percentage of total discharge with width to depth ratio. The discharge capacity in the bank area is significant (>5%) in channels of width–depth ratio less than 9

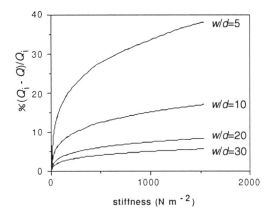

**Figure 10.5** Percentage reductions in discharge with seasonal bank vegetation growth. There is a 38% reduction in discharge capacity for fully grown vegetation when the width–depth ratio is 5; with a width–depth ratio of 20 the discharge reduction for fully grown vegetation is 8%. With a width–depth ratio of 30 the discharge reduction is less than 6%

### Case 2: Variation with Seasonal Growth

In Case 2 the effect of seasonal vegetation growth (represented by increasing stiffness) is simulated for trapezoidal channels with constant side slope angles of 45°. It is assumed that winter vegetation has initial stiffness value of 1.6 N m$^{-2}$ and when fully grown a stiffness value of 1504 N m$^{-2}$. The initial stiffness value is within the range found for grasses (Kouwen & Li, 1980), and the value for fully grown vegetation is an estimate for dense stands of typical riparian species. This estimate is derived by extending the correlation of stiffness with stem height obtained for grasses (Kouwen, 1988). The value may therefore either represent flexible grass species growing to a height of 1.6 m, or shorter flexible riparian species with greater stiffness. In either case the deflected height in high flows would be reduced to 0.5 m. The reduction in discharge capacity due to the growth of vegetation is modelled for channels with width–depth ratios of 5, 10, 20 and 30 in Figure 10.5, where $Q$ is the changing discharge capacity of the channel and $Q_i$ is the initial discharge capacity, when bank vegetation is dormant.

The plots indicate that reductions in discharge capacity for channels with the configurations given in Table 10.2 become less significant as the width–depth ratio increases. Whereas there is a 38% reduction in discharge capacity for the fully grown vegetation when the width–depth ratio is 5, with a width–depth ratio of 20 the discharge reduction for similar vegetation is only 8%. With a width–depth ratio of 30 the discharge reduction is less than 6%.

### Stability Criteria

Kouwen & Li (1980) developed a physically based stability criteria for grass-lined channels based on the deflection ratio of the grass used as a channel lining, determined by $k/h$. Their data suggest that once the deflection ratio of the grass reaches about 0.6 for short grass and 0.35 for long grass then significant erosion of the substrate material occurs.

**Table 10.3**   Green and dormant bank vegetation deflection ratios ($k_w/h$)

| Width (m) | Depth (m) | Green vegetation | | | Dormant vegetation | | |
|---|---|---|---|---|---|---|---|
| | | Discharge ($m^3 s^{-1}$) | $V_w$ ($m s^{-1}$) | $k_w/h$ | Discharge ($m^3 s^{-1}$) | $V_w$ ($m s^{-1}$) | $k_w/h$ |
| 20 | 0.5 | 4.2 | 0.02 | 1 | 4.3 | 0.08 | 0.59 |
| 21 | 1.0 | 15.0 | 0.196 | 0.82 | 15.7 | 0.29 | 0.42 |
| 22 | 1.5 | 31.2 | 0.376 | 0.68 | 32.9 | 0.49 | 0.35 |
| 23 | 2.0 | 52.2 | 0.539 | 0.61 | 55.3 | 0.668 | 0.31 |
| 24 | 2.5 | 77.8 | 0.692 | 0.56 | 82.8 | 0.830 | 0.29 |
| 25 | 3.0 | 108.1 | 0.838 | 0.52 | 115.3 | 0.985 | 0.27 |
| 26 | 3.5 | 142.4 | 0.972 | 0.48 | 152.76 | 1.13 | 0.25 |

By applying these criteria to bank vegetation it is possible to gain some insight into the influence of bank vegetation on the stability of alluvial river banks, from the perspective of its impact on energy dissipation in the near-bank zone. This ignores any effect vegetation may have in increasing root reinforcement of the bank material.

Table 10.3 shows a simulation of the relative effectiveness of green and dormant vegetation in dissipating energy in the near-bank zone. The vegetation is 0.5 m high and equations (7) and (8) are used to determine vegetation stiffness for green and dormant vegetation, respectively. The table records the near-bank velocities and deflection ratios as discharge is increased for a trapezoidal channel with bank slopes of 45° and a water surface slope of 0.001.

As would be expected from the analysis in the previous section, the effect of the difference between green and dormant vegetation on discharge capacity is small. However, the effect on mean velocities in the bank sub-areas and on the deflection ratio of the bank vegetation is more marked. Assuming a critical deflection ratio of 0.5, for medium length grass, before the substrate material is exposed to shear stresses, for this type of vegetation, then green vegetation continues to be effective in dissipating energy at discharges greater than $120 \, m^3 \, s^{-1}$. Dormant vegetation, by contrast, becomes ineffective at dissipating energy at discharges as low as $10 \, m^3 \, s^{-1}$. At this discharge the bank material is exposed to the potential for erosion although the predicted mean velocities in the bank sub-areas are unlikely to cause significant erosion until higher discharges are achieved. Figure 10.6 shows these respective thresholds graphically.

## Emergent Vegetation

An attempt to include emergent, non-flexible bank vegetation by taking account of how this type of vegetation affects velocities in the bank sub-areas has been made, using equations (13) to (16). Table 10.4 summarises the influence this type of vegetation has on mean velocities in the bank sub-areas for different arrangements of emergent, non-flexible vegetation. Calculations have been carried out for a trapezoidal channel with a width of 20 m, a depth of 1 m and side slope of 45° and with underlying flexible vegetation of $59 \, N \, m^{-2}$ stiffness corresponding to a 0.6 m high grass cover on the banks.

The predicted effect on channel discharge in the cases analysed is insignificant. The difference between predicted discharge with and without emergent, non-flexible

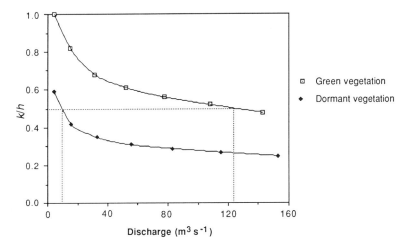

**Figure 10.6** Deflection ratio ($k_w/h$) versus discharge for green and dormant vegetation. Green vegetation is effective in dissipating energy at discharges greater than 120 m³ s⁻¹. Dormant vegetation becomes ineffective at dissipating energy at discharges as low as 10 m³ s⁻¹

vegetation is less than 1%. Table 10.4 indicates that there is a reduction of velocity in the bank sub-area of between 30 and 60%, depending on the arrangement and density of the emergent vegetation. Velocities in the bank sub-areas are a critical factor in determining whether or not the bank substrate material is exposed to the potential for erosion. As velocities are reduced there will be a corresponding increase in the deflection ratio of the flexible vegetation, leading to enhanced stability of the bank using the criteria developed by Kouwen & Li (1980). The implications of this are that banks with emergent, non-flexible vegetation and an underlying layer of flexible vegetation are able to protect the bank substrate material from erosive shear forces at higher discharges than would otherwise be the case. The model predicts that this enhanced protective influence of emergent bank vegetation does not significantly affect overall channel capacities.

**Table 10.4**  Effects of emergent, non-flexible vegetation on near-bank velocities

| Spacing/ diameter of ($s_w/d_w$) | Parallel spacing | | Staggered spacing | |
| --- | --- | --- | --- | --- |
| | Wake velocity of near-bank zone ($V_w$) | Wake velocity/ approach velocity ($V_w/u$) | Wake velocity of near-bank zone ($V_w$) | Wake velocity/ approach velocity ($v_w/u$) |
| 4 | 0.212 | 0.352 | 0.407 | 0.675 |
| 8 | 0.233 | 0.386 | 0.419 | 0.695 |
| 10 | 0.241 | 0.340 | 0.422 | 0.700 |
| 12 | 0.247 | 0.410 | 0.351 | 0.582 |
| 20 | 0.265 | 0.439 | 0.353 | 0.585 |
| 40 | 0.342 | 0.567 | 0.357 | 0.592 |
| 60 | 0.353 | 0.585 | 0.359 | 0.595 |
| 80 | 0.361 | 0.599 | 0.361 | 0.599 |
| 100 | 0.368 | 0.610 | 0.362 | 0.600 |

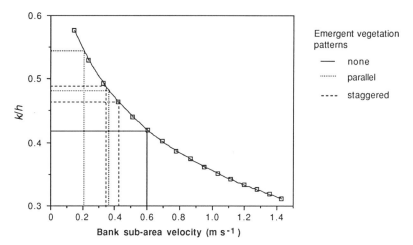

**Figure 10.7**  Deflection ratio ($k/h$) versus bank sub-area velocities for a range of 20 m wide channels

Figure 10.7 illustrates a plot of the deflection ratio against predicted velocities in the bank sub-areas for channels with identical width and bank vegetation characteristics to the channel used to generate data in Table 10.4. Using this graph it is possible to estimate the size of the increase in the deflection ratio of the bank vegetation that results from the reduced near-bank velocities predicted when different arrangements of emergent, non-flexible vegetation are present.

In the absence of emergent, non-flexible vegetation the predicted near-bank velocity is $0.608 \text{ m s}^{-1}$ and the deflection ratio is 0.42. Where emergent bank vegetation is present reduced near-bank velocities are predicted, ranging from 0.351 to $0.422 \text{ m s}^{-1}$ for staggered configurations and from 0.212 to $0.368 \text{ m s}^{-1}$ for parallel configurations. The corresponding deflection ratios range from 0.46 to 0.49 for staggered configurations and from 0.48 to 0.54 for parallel configurations.

To reduce the deflection ratio by 10% in the kind of channel being considered with 0.6 m high flexible vegetation requires an increase in discharge of the order of 50%. It is, therefore, apparent that combinations of emergent, non-flexible bank vegetation and flexible bank vegetation may have a significant impact on bank stability through the dissipation of energy and protecting the bank substrate from exposure to shear stresses.

It should, however, be noted that the model predicts only mean values of velocities and shear stresses in the sub-areas. Localised increases in shear stresses due to impinging and redirected flow and separation in wake zones may still cause significant local erosion to occur in places where the model predicts an overall degree of stability.

## DISCUSSION

The procedure outlined in this paper provides a theoretically based method for taking into account the effects of bank vegetation on the channel flow parameters. The method can be used to predict what effect changes in bank vegetation will have on the discharge capacity of a channel and to make comparisons of bank vegetation effects between channels. The analysis carried out shows that it is possible to relate these effects of bank vegetation to the width–depth ratio of channels. This suggests that any reductions in channel capacity that result from typical bank vegetation are only discernible for those channels where the width–depth ratio is less than 16.

The procedure can also be used to analyse the effect that flexible vegetation has on shear stress distributions. This has significant implications for studies on the influence of vegetated banks on channel morphology, as the model provides a theoretically based means of quantifying the extent that flexible bank vegetation interacts with the flow dynamics of the channel to increase the effective stability of the bank. This is just one aspect of the bank's stability. The overall stability of the bank will depend additionally on geotechnical and hydrologic properties of the bank material and any effects vegetation may have on these factors. Hence, the net impact of vegetation of a specific type may be positive or negative.

Emergent, non-flexible bank vegetation is included in the model using a theoretically derived method that predicts the wake velocities produced by different arrangements of this type of vegetation. These predictions can also be used to determine the effective stability of vegetated banks in terms of the interaction of bank vegetation with the flow dynamics.

The geomorphological implications of this analytical approach can be summarised in terms of the influence bank vegetation has on the hydraulic geometry of alluvial channels and in terms of the interactions of bank vegetation with the processes of fluvially controlled erosion and mass-wasting of banks.

The model's predictions support the observation that some aspects of the hydraulic geometry of alluvial rivers are partially dependent on the type of vegetation on the bank. Hey & Thorne (1986) showed that width predictions of alluvial rivers, based on discharge, are improved when the bank vegetation is considered. Model predictions illustrate that vegetation can protect the bank from potentially erosive flows without having a significant effect on discharge capacity. This protection is additional to any effects of root reinforcement the vegetation may have on the bank material properties. Predictions suggest that combinations of emergent, non-flexible and flexible vegetation are better able to protect banks than either one alone. Channels with this type of vegetation are, therefore, likely to be narrower than those with unvegetated banks for the same bankfull discharge. Removal of bank vegetation or a decline in its stiffness due to unsympathetic operational maintenance leaves the bank less well protected and exposes it to the potential for increased rates of erosion and widening.

Predictions of the interactions of bank vegetation with the flow dynamics of a channel, derived from this analytical approach, suggest that the dissipation of energy by the vegetation and the consequent reduction in near-bank velocities are key factors in understanding bank erosion processes. The deflection ratio for flexible vegetation may be a useful means of establishing a critical threshold for exposure of the underlying substrate to the potential for erosion. Once vegetation becomes prone its protective ability is markedly reduced as boundary shear stresses become more uniform around the wetted perimeter of a cross-section.

## NOTATION

$a$  = a coefficient
$A_a$ = sub-area cross-sectional area

$a_s$ = shape correction factor
$c$ = a coefficient
$d$ = mean depth
$d_{max}$ = maximum depth
$D$ = grain size parameter
$d_w$ = vegetation diameter
$D_{84}$ = 84th percentile grain size
$f$ = Darcy–Weisbach friction factor
$g$ = acceleration due to gravity
$h$ = vegetation height
$k$ = roughness coefficient
$k_w$ = vegetation roughness
$mei$ = vegetation stiffness
$P_b$ = wetted perimeter of the bed
$P_w$ = wetted perimeter of the banks
$Q_a$ = sub-area discharge
$R$ = hydraulic radius
$S$ = channel-bed slope
$s_w$ = vegetation spacing
$tr$ = threshold parameter
$u_w$ = wake velocity
$v$ = average velocity in cross-section
$v_a$ = sub-area average velocity
$w$ = channel flow width
$\theta$ = bank angle
$\tau_0$ = mean boundary shear stress
$\tau_w$ = bank shear stress
$\%SF_w$ = percentage shear force carried by banks

## REFERENCES

Andrews, E. D. (1984). Bed-material entrainment and hydraulic geometry of gravel-bed rivers. *Geological Society of America Bulletin*, **95**, 371–378.

ASCE (1963). Frictional factors in open channels. *Journal of the Hydraulics Division, ASCE*, **89**, 97–143.

Barnes, H. H. (1967). Roughness characteristics of natural channels. *Geological Survey Water-Supply Paper* 1849, US Government Printing Office, Washington, DC.

Bathurst, J. C. (1979). Distribution of boundary shear stress in rivers. In Rhodes, D. D. and Williams, G. P. (eds), *Adjustments of the Fluvial System*, Proceedings Volume of the 10th Annual Geomorphology Symposium Series held at Binghamton, NY, Sept 21–22 1979. George Allen and Unwin, pp. 95–116.

Bathurst, J. C. (1982). Theoretical aspects of flow resistance. In Hey, R. D., Bathurst, J. C. and Thorne, C. R. (eds), *Gravel-bed Rivers*, John Wiley & Sons, Chichester, pp. 83–108.

Bathurst, J. C., Li, R-M and Simons, D. B. (1981). Resistance equation for large scale roughness. *Journal of the Hydraulics Division, ASCE*, **107**, 1593–1613.

Charlton, F. G., Brown, P. M. and Benson, R. W. (1978). Hydraulic geometry of some gravel rivers in Britain, Report No. IT 180, July 1978, Hydraulics Research Station, Wallingford, UK.

Chow, V. T. (1959). *Open Channel Hydraulics*, McGraw-Hill Book Company, New York.

Cowan, W. L. (1956). Estimating hydraulic roughness coefficients. *Agricultural Engineering*, **37**, 473–475.

Dawson, F. H. and Charlton, F. G. (1988). *Bibliography on the Hydraulic Resistance or Roughness of Vegetated Watercourses*, Freshwater Biological Association, UK. 50 pp.

Eastgate, W. I. (1966). Vegetated stabilisation of grasses, waterways and dam bywashes. MEng.Sc. Thesis, University of Queensland, Australia.

Flintham, T. P. and Carling, P. A. (1988). The prediction of mean bed and wall boundary shear in uniform and compositely rough channels. In White, W. R. (ed.), *International Conference on River Regime*, John Wiley & Sons, Chichester, pp. 267–287.

Gourlay, M. R. (1970). Discussion of flow retardance in vegetated channels. *Journal of the Irrigation and Drainage Division, ASCE*, **96** (IR3), 351–357.

Gregory, K. J. and Gurnell, A. M. (1988). Vegetation and river channel form and process. In Vines, H. A. (ed.), *Biogeomorphology*, Basil Blackwell, Oxford, pp. 11–42.

Gwinn, W. R. and Ree, W. O. (1980). Maintenance effects on the hydraulic properties of a vegetation-lined channel. *Transactions of the American Society of Agricultural Engineers*, **23** (3), 636–642.

Hey, R. D. (1979). Flow resistance in gravel-bed rivers. *Journal of the Hydraulics Division, ASCE*, **105**, 365–379.

Hey, R. D. (1988). Bar form resistance in gravel-bed rivers. *Journal of the Hydraulics Division, ASCE*, **114** (12), 1498–1508.

Hey, R. D. and Thorne, C. R. (1986). Stable channels with mobile gravel beds. *Journal of Hydraulic Engineering, ASCE*, **112** (HY8), 671–689.

Hydraulics Research (1985). The hydraulic roughness of vegetated channels. *Report No. SR 36*, Hydraulics Research, Wallingford, 24 pp.

Kazemipour, A. K. and Apelt, C. J. (1979). Shape effects on resistance to uniform flow in open channels. *Journal of Hydraulic Research*, **17** (2), 129–147.

Keulegan, G. H. (1938). Laws of turbulent flow in open channels. *Journal of Research of the National Bureau of Standards*, **21**, 707–741. Research Paper RP1151.

Klaasen, G. J. and van der Zwaard, J. J. (1974). Roughness coefficients of vegetated flood plains. *Journal of Hydraulic Research*, **12**, 42–63.

Knight, D. W. (1981). Boundary shear in smooth and rough channels. *Journal of the Hydraulics Division, ASCE*, **107** (HY7), 839–851.

Kouwen, N. (1988). Field estimation of the biomechanical properties of grass. *Journal of Hydraulic Research*, **26** (5), 559–568.

Kouwen, N. and Li, R-M. (1980a). A stability criteria for vegetated waterways. In International Symposium on Urban Storm Runoff (University of Kentucky, Lexington, July 28–31, 1980), pp. 203–210.

Kouwen, N. and Li, R-M. (1980b). Biomechanics of vegetative channel linings. *Journal of the Hydraulics Division, ASCE*, **106**, 1085–1103.

Kouwen, N. and Unny, T. E. (1973). Flexible roughness in open channels. *Journal of the Hydraulics Division, ASCE*, **99**, 713–728.

Kouwen, N., Unny, T. E. and Hill, A. M. (1969). Flow retardance in vegetated channels. *Journal of the Irrigation and Drainage Division, ASCE*, **95** (IR2), 329–342.

Li, R-M. and Shen, H. W. (1973). Effect of tall vegetation on flow and sediment. *Journal of the Hydraulics Division, ASCE*, **99** (HY5), 793–814.

Myers, W. R. C. (1978). Momentum transfer in a compound channel. *Journal of Hydraulic Research*, **16** (2), 139–150.

Pasche, E. and Rouvé, M. (1985). Overbank flow with vegetatively roughened floodplains. *Journal of Hydraulic Engineering, ASCE*, **111**, 1262–1278.

Pethick, J., Leggett, D. and Husain, L. (1990). Boundary layers under salt marsh vegetation developed in tidal currents. In Thornes, J. B. (ed.), *Vegetation and Erosion*, John Wiley & Sons, Chichester, pp. 113–124.

Petryk, S. and Bosmajian III, G. (1975). Analysis of flow through vegetation. *Journal of the Hydraulics Division, ASCE*, **101**, 871–884.

Powell, K. E. C. (1978). Weed growth – a factor of channel roughness. In Herschy, R. W. (ed.), *Hydrometry, Principles and Practice*, John Wiley & Sons, Chichester, pp. 327–352.'

Richards, K. S. (1982). *Rivers: Form and Process in Alluvial Channels*. Methuen, London, 361 pp.

Robert, A. (1990). Boundary roughness in coarse-grained channels. *Progress in Physical Georgrapy*, **14** (1), 42–70.

Rouse, H (1965). Critical analysis of open-channel resistance. *Journal of the Hydraulics Division, ASCE*, **91** (HY4), 1–25.

Saowapon, C. and Kouwen, N. (1989). A physically based model for determining flow resistance and velocity profiles in vegetated channels. In Yen, B. C. (ed.), *Proceedings of the International Conference on Channel Flow and Catchment Runoff*: Centennial of Manning's Formula and Kuichling's Rational Formula, 22–26 May 1989 at the University of Virginia, pp. 559–568.

Simons, D. B., Richardson, E. V. and Nordin, C. F. (1965). Sedimentary structures generated by flow in alluvial channels. *Society of Economic Paleontologists and Mineralogists, Special Publication*, **121**, 34–52 (cited in Schumm, S. (ed.) (1972). River Morphology, Dowden, Hutchinson & Ross Inc., Pennsylvania, pp. 314–335).

Thompson, G. T. and Roberson, J. A. (1976). A theory of flow resistance for vegetated channels. *Transactions of the American Society of Agricultural Engineers*, **19** (2), 288–293.

Thorne, C. R. (1990). Effects of vegetation on riverbank erosion and stability, in Thornes, J. B. (ed.), *Vegetation and Erosion*, John Wiley & Sons, Chichester, pp. 125–144.

Thorne, C. R. and Zevenbergen, L. W. (1985). Estimating mean velocity in mountain rivers. *Journal of Hydraulic Engineering, ASCE*, **111** (4), 612–624.

Watts, J. F. and Watts, G. D. (1990). Seasonal change in aquatic vegetation and its effect on river channel flow. In Thornes, J. B. (ed.), *Vegetation and Erosion*, John Wiley & Sons, Chichester, pp. 257–607.

# Part 3

## VALLEY HEADS

# 11 Landscape Dissection and Drainage Area–Slope Thresholds

DAVID R. MONTGOMERY
*Department of Geological Sciences and Quaternary Research Center, University of Washington, USA*

and

WILLIAM E. DIETRICH
*Department of Geology and Geophysics, University of California, Berkeley, USA*

## ABSTRACT

Development of the distinct ridges and valleys that define landscape dissection may be controlled by either a spatial competition between slope-dependent and area-dependent sediment transport processes or by exceedence of a spatially and temporally variable channel initiation threshold. Field observations indicate that channel heads in many humid, soil-mantled environments represent a change in sediment transport processes, rather than a change in process dominance. In environments where sediment transport by sheetwash and rainsplash occur, however, the development of defined incision may reflect the spatial dominance of overland flow over rainsplash that infills incipient channels. Simple models of channel initiation processes predict drainage area–slope thresholds for channel initiation by overland flow and landsliding that are similar to observed relations in regions where these processes control channel initiation. A drainage area–slope threshold for soil saturation also can be used to define the spatial extent of saturation overland flow. Plotted together on a graph of drainage area versus slope these thresholds predict both a region of the landscape dominated by diffusive sediment transport that corresponds to undissected hillslopes and a region of generally non-erosive overland flow corresponding to unchanneled valleys where occasional evacuation of stored sediment occurs, and regions of the landscape where sediment transport occurs by overland flow and landsliding. Over geologic time scales, the short-term variability of these thresholds may result in valley incision upslope of the mean channel head location. Maintenance of the upslope ends of the valley network thus may reflect either the spatial signature of temporal variability of channel head locations or a spatial transition in the dominant sediment transport processes. In either case, landscape dissection is sensitive to both long-term change in the controls on channel initiation and short-term variance in channel head locations.

## INTRODUCTION

The degree to which a landscape is dissected traditionally is described by the drainage density, the total stream length divided by drainage area. Unfortunately, the extent of the

*Process Models and Theoretical Geomorphology.* Edited by M. J. Kirkby
©1994 John Wiley & Sons Ltd

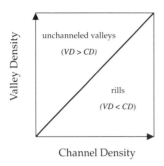

**Figure 11.1** Schematic illustration of potential relations between channel and valley density. Landscapes in which the valley density exceeds the channel density will have unchanneled valleys. Landscapes in which the channel density exceeds the valley density have channels, or rills, developed on convex or planar hillslopes

stream network is variable (Gregory & Walling, 1968) and thus there is not a unique drainage density for a given landscape. Less ambiguous descriptions of landscape dissection are provided by the extent of valleys (areas of topographic convergence) and channels (avenues of sediment transport within definable banks). Valley and channel densities are morphologically and spatially distinct, as the extent of channel and valley networks only rarely coincide (Figure 11.1). Unchanneled valleys often separate channel heads from drainage divides in humid, soil-mantled landscapes, whereas in arid or disturbed landscapes small channels or rills often extend beyond the confines of the valley network and onto the intervening hillslopes. Channel and valley densities thus provide a less ambiguous description of landscape dissection than does the drainage density. Understanding the process of landscape dissection into distinct hillslopes and valleys depends, however, on establishing relations between the controls on valley and channel densities, as erosion by streamflow, dissolution, and debris flows incises valleys into the landscape.

In soil-mantled landscapes the distinction between valleys and ridges is pertinent, as there is a scale beyond which smaller valleys are not present (Dietrich *et al.*, 1987; Montgomery & Dietrich, 1992). However, there may not be such a scale in rocky landscapes in which sediment transport is limited by the rate at which weathering processes generate material available for transport. In such landscapes, diffusive processes may be essentially inactive and incisive transport processes may act over virtually all scales. Consequently, the present discussion is restricted to transport-limited landscapes.

Most landscape evolution models assume either what we term the Gilbert hypothesis or the Horton hypothesis about processes controlling the development of landscape dissection. In a series of papers, Gilbert (1877, 1909) argued that ridge erosion is controlled by slope-dependent processes that give rise to convex slopes, whereas valley development is a product of both runoff and slope (see review in Montgomery, 1991). In the Gilbert hypothesis then, ridge and valley topography arises from the varying spatial dominance of diffusive (slope-dependent) and incisive (slope and discharge-dependent) processes. Smith & Bretherton (1972) subsequently demonstrated mathematically, for the particular case of a landscape undergoing steady state erosion, that lateral perturbations would tend to grow (and presumably form valleys) where incisive processes dominate over diffusive processes. Instead of the dominance of one process over another, Horton (1945) proposed that erosion and subsequent valley development only occurs where a threshold of ground surface resistance to overland flow is exceeded. He formulated a quantitative theory that predicted this threshold distance, $x_c$, downslope from the drainage divide. Subsequent models which assumed that a threshold of erosion must be crossed for channels to develop (e.g., Schaefer, 1979; Dietrich *et al.*, 1986; Montgomery & Dietrich, 1988, 1989; Montgomery, 1991; Willgoose *et al.*, 1991) are similar in spirit to the Horton hypothesis. Montgomery and Dietrich (1992) further argued that the variance in the position of channel head locations may define the headward extent of valleys. This observation suggests a resolution of the Gilbert and Horton hypotheses in which the Gilbert hypothesis is most relevant to valley maintenance over geologic time scales (i.e., $10^4$ to $10^6$ yr) and the Horton hypothesis is most relevant to channel initiation and sediment transport over shorter geomorphic time scales (i.e., $10^2$ to $10^3$ yr).

The extent of both channels and valleys reflect the interaction of erosive processes, the resistance of the ground surface to those processes, and processes acting to infill incipient incision. Channel heads may be defined on the basis of the upslope extent of sediment transport concentrated within clearly demarkated banks. The extent of valleys, on the other hand, may be defined by the upslope extent of convergent topography. Incisive processes act to create topographic depressions, generate relief, and incise hillslopes, while diffusive processes tend to infill topographic depressions, degrade relief, and round hillslopes. Consequently, either the transition from convex to concave slopes or from convergent to divergent topography has been considered to correspond to the transition from diffusive to incisive transport, in accordance with the Gilbert hypothesis.

Several previous workers in effect equated this transition with both channel and valley heads (Smith & Bretherton, 1972; Kirkby, 1980, 1987; Tarboton *et al.*, 1992). This is consistent with the Gilbert hypothesis and involves the testable assumptions that above the channel head diffusional sediment transport prevents incision by infilling incipient channels, while below the channel head incisive sediment transport processes overwhelm diffusional transport and maintain a valley. In general, sediment transport at any point on the landscape may be expressed as the sum of diffusive and incisive components, which may be expressed as

$$q_s = f(S) + f(q,S) \tag{1}$$

where $q_s$ is the sediment flux, $S$ is the slope, and $q$ is discharge. The first term on the right-hand side of equation (1) represents sediment transport by diffusive processes and the second term represents transport by incisional (or advective (Loewenherz, 1991)) processes. This model, first presented by Smith and Bretherton (1972), assumes that channels and valleys correspond to areas where $q_s/q < \delta q_s/\delta q$. In this case, discharge-dependent sediment transport occurs upslope of the convex/concave slope transition, but the tendency to incise will be overwhelmed by diffusive transport (Dunne, 1980).

The Horton hypothesis requires that channel and valley maintenance results from exceedence of an erosional threshold. Although Horton originally maintained that no erosion occurs up slope of $x_c$, the Horton hypothesis can be extended to include the more general case in which sediment transport above the channel head occurs only by slope-dependent processes, whereas below the channel head sediment transport also is a function of discharge, or drainage area. Thus, according to the Hortonian model the channel head is a point where sediment transport processes change, rather than a transition in the dominant transport process, as held under the Gilbert hypothesis. Dietrich & Dunne (1993) expressed this condition as

$$q_s = f(S) \qquad x < x_c \tag{2a}$$

above the channel head, whereas at and downslope of the channel head sediment transport may be described as

$$q_s = f(S) + f(q,S) \qquad x \geqslant x_c \qquad (2b)$$

where $x_c$ is the threshold for sediment transport by incisive processes. A threshold control on channel initiation implies that channel heads need not correspond to the transition from convex and concave slopes. Rather, at any given point in time channeled transport may start some distance downslope from drainage divides regardless of slope curvature. The channel head position may fluctuate up and downslope in response to climatic, land-use, or even seasonal variation, and the extent of valleys (and the consequent transition from convex to concave topography) will reflect a balance between periodic headward extention of the channel head and relatively continuous diffusive processes. These hypotheses for channel initiation and valley maintenance involve testable assumptions about sediment transport and channel initiation processes, which are discussed in the sections below.

## CHANNEL INITIATION

While there are a number of mechanisms for initiating channels and maintaining channel heads, we restrict our discussion to landscapes in which channel initiation is dominated by overland flow (Horton, 1945; Kirkby & Chorley, 1967), seepage erosion (Dunne, 1980; Montgomery, 1991), and landsliding (Dietrich et al., 1986; Montgomery & Dietrich, 1989). Channels initiated and maintained by these processes tend to have characteristic forms, which can be grouped into gradual and abrupt morphologies (Montgomery & Dietrich, 1988, 1989), although a more elaborate classification scheme has been offered by Dietrich & Dunne (1993). Within any given area, several channel initiation processes may be active, reflecting variations in slope, soil type and thickness, and vegetation type and density.

Processes controlling each of these channel initiation mechanisms can be described as threshold phenomena. Channel initiation by overland flow where diffusional dampening is effective (see Dunne (1980) and Loewenherz (1991)) may be assumed to occur wherever the flow chronically exerts a boundary shear stress in excess of critical for initial motion. This is the assumption explicitly proposed by Horton (1945) and subsequently adopted by others (i.e., Schaefer, 1979; Montgomery, 1991; Willgoose et al., 1991; Dietrich et al., 1992). Channel initiation by seepage erosion may be considered to be controlled by either seepage gradients capable of entraining surficial materials (Dunne, 1990) or, in cohesive material, by sufficient drainage area to support channel head spalling and overland flow to remove failed material (Montgomery, 1991). Channel initiation by landsliding is controlled by critical pore pressures in the colluvium and thus can be modelled as occurring when sufficient throughflow converges at a location to generate pore pressures necessary to trigger slope instability (Dietrich et al., 1986; Montgomery & Dietrich, 1988, 1989). Consequently, the channel initiation threshold in equation (2) may be formulated as a function of either the overland flow discharge necessary to overcome the erosional resistance of the ground surface, the head gradient necessary for seepage to entrain surface materials, subsurface discharge sufficient to

**Figure 11.2**   Small channels, or rills, developed on a convex hillslope in a badlands in New Mexico. Note the film canister for scale. The view is from the drainage divide looking downslope

saturate the soil thickness and generate erosive overland flow, or the pore pressure necessary to initiate slope instability. Each of these relations allows models for channel initiation to be formulated in terms of thresholds related to the contributing drainage area. We propose that such a threshold-based view of channel initiation is most representative of field conditions in many landscapes.

Simple analytical models of channel initiation may be developed by assuming a steady state rainfall and associating channel initiation with the exceedence of erosional thresholds (see also Dietrich *et al.*, 1992). Drainage area may be considered to be a surrogate for surface or subsurface runoff in each of the cases discussed above. Models for channel initiation by Horton overland flow, saturation overland flow, and landsliding, as well as for channel head advance by sloughing, predict different relations between the contributing drainage area and the local slope at the channel head. The limitations of using the steady state assumption for modelling a transient process are obvious, but this assumption does allow useful conceptual relations to be drawn that are not possible for the transient case. These relations can be compared with published data for channel head locations to allow prediction of trends in landscape response to disturbance and environmental change and to examine linkages between the development of channel and valley networks. Perhaps equally important, none of the simple models that follow attempt to explain the size of a channel head. Hence these models are useful only in illustrating controls on where erosion due to runoff would tend to occur for various erosional processes.

### Horton Overland Flow

Horton overland flow controls the formation of gradual channel heads in semi-arid to arid landscapes and often results in the development of rills (channels that extend onto hillslopes) in arid, poorly vegetated, or disturbed lands (Figure 11.2). Rill development has been argued to be associated with a critical shear stress, a critical shear velocity, or a critical stream power (see review in Slattery & Bryan, 1992). However, sediment transport by overland flow does not always incise channels and may not always act to maintain valleys. For example, sediment transport by sheetflow may occur upslope of the point where diffusional processes dominate incisional processes, and thus an overland flow transport threshold may be exceeded on convex hillslopes. Furthermore, Dunne (1991) showed that if rainfall duration is short relative to the time required to transport sediment off of a hillslope, then even though rills may be incised, the entrained material will be redeposited lower on the slope. In this case, rills function as channels that act to maintain convex slope profiles.

Channel initiation by overland flow may occur under either laminar or turbulent flow regimes. While unchanneled overland flow generally is laminar (see discussions in Dunne & Dietrich, 1980; Reid, 1989), rill flow and rill initiation may be associated with turbulent or transitional conditions and the development of supercritical flow (Slattery & Bryan, 1992). Although flow over bare surfaces may be turbulent (Emmett, 1970), even high-velocity flow over grassy, or well-vegetated surfaces generally is laminar (see Moody diagrams in Dunne & Dietrich (1980) and Reid (1989)). Consequently, we develop models for overland flow erosion exceeding surface resistance by both turbulent and laminar flow.

Assuming a steady state rainfall intensity $R$ over a surface with uniform infiltration capacity, $I$, the discharge per unit contour length $q$ can be expressed as

$$q = (R - I)a \tag{3}$$

where $a$ is the drainage area per unit contour length. For turbulent open-channel flow, the hydraulic radius is approximately the flow depth. Using the Manning equation to calculate mean flow velocity, we can set the unit discharge equal to the product of the flow velocity and depth, and thus

$$q = (1/n)h^{5/3}S^{1/2} \tag{4}$$

where $n$ is the Manning resistance coefficient, $h$ is the flow depth, and $S$ is the water surface slope. The critical shear stress for incipient motion is determined by soil properties. A depth–slope product sufficient to generate a basal shear stress greater than the critical shear stress is required for sediment transport:

$$\tau_{cr} = \rho_w g (hS)_{cr} \tag{5}$$

where $\rho_w$ is the density of water and $g$ is gravitational acceleration. Thus the critical discharge ($q_{cr}$) can be found by substituting (5), rearranged in terms of $h$, into (4):

$$q_{cr} = \frac{\tau_{cr}^{5/3}}{(\rho_w g)^{5/3} n S^{7/6}} \tag{6}$$

Equating (6) and (3) and solving for the critical drainage area per unit contour length $a_{cr}$ necessary for erosion by overland flow yields

$$a_{cr} = \frac{\tau_{cr}^{5/3}}{(R - I)(\rho_w g)^{5/3} n S^{7/6}} \tag{7}$$

which describes an inverse loglinear source area–slope relation in which the contributing area per unit contour length for channel initiation by turbulent Horton overland flow is approximately proportional to the inverse of the local slope and is a function of surface roughness, critical shear stress, and the difference between rainfall intensity and infiltration capacity. According to equation (7), an increase in the critical shear stress or infiltration capacity of the soil will result in larger source areas, whereas an increase in rainfall intensity, surface roughness, or bulk density will result in smaller source areas (Figure 11.6A).

The turbulent flow assumption, however, may not be generally correct for channel initiation by overland flow. Moody diagrams for overland flow over grass surfaces (Dunne & Dietrich, 1980; Wilson, 1988; Reid, 1989) suggest laminar-type flow to Reynolds numbers in excess of 30 000. Consequently, even though rapid overland flow along hollow axes may appear turbulent, a laminar flow model may be more appropriate, especially where a well-vegetated grass surface protects the ground from surface runoff.

A model for channel initiation by laminar overland flow also may be constructed based on a critical shear stress argument. Assuming that a channel is initiated when a critical boundary shear stress is exceeded, we may again express the channel initiation criterion in terms of equation (5). Flow velocity $u$ may be expressed as

$$u = (2\,ghS/f)^{0.5} \tag{8}$$

where $f$ is a dimensionless friction factor, which itself may be expressed as

$$f = k\nu/q \tag{9}$$

where $k$ is a dimensionless surface roughness coefficient, $\nu$ is the kinematic viscosity, and $q$ is the unit discharge. Note that the unit discharge is equal to the product of the average flow velocity and depth, or that

$$q = uh \tag{10}$$

Consequently, substituting (10) into (9) and then (5), rearranged in terms of $h$, and (9) into (8) allows an expression for the critical discharge:

$$q_{cr} = \frac{2\tau_{cr}{}^3}{k\nu\rho_w{}^3 g^2 S^2} \tag{11}$$

During steady state runoff, the discharge is given by equation (3). Thus combining (11) and (3) results in an expression for the critical contributing area as a function of slope where

$$a_{cr} = \frac{2\tau_{cr}{}^3}{(R-I)k\nu\rho_w{}^3 g^2 S^2} \tag{12}$$

Equation (12) predicts an inverse loglinear relation where the contributing area is inversely proportional to the square of slope (Figure 11.6B).

Differences in the models for channel initiation for turbulent and laminar overland flow indicate that a change in hillslope flow hydraulics could have a pronounced impact on the extent of the channel network. The critical drainage area for turbulent overland flow is approximately proportional to the square of the critical shear stress, whereas the critical drainage area for laminar overland flow is a function of the cube of the critical shear stress. Moreover, channel initiation by turbulent overland flow is inversely proportional to the slope, whereas channel initiation by laminar overland flow is inversely dependent on the square of slope. The difference in the critical area between these two scenarios is expressed by the ratio of equations (12) and (7):

$$\frac{a_{crt}}{a_{crl}} \propto \frac{S^{5/6}}{\tau_{cr}^{4/3}} \tag{13}$$

where $a_{crt}$ is the critical area for turbulent flow and $a_{crl}$ is the critical area for laminar flow. This ratio indicates that the difference between the source area size under turbulent and laminar flow is approximately proportional to slope and is inversely proportional to the critical shear stress. Thus a change from laminar to turbulent flow, such as might accompany extreme changes in vegetation cover, would be expected to lead to significant expansion of the channel network.

## Seepage Erosion and Soil Saturation

Dunne (1990) developed an expression for the critical head gradient for seepage erosion and concluded that for cohesive material seepage gradients would have to be unrealistically high to cause liquefaction and thus initiate a channel. Consequently, seepage erosion is only a viable channel initiation mechanism in cohesionless materials or in materials in which weathering causes pronounced cohesion loss. In vegetated landscapes the root strength of plants provides an effective cohesion to the soil, suggesting that channel initiation by seepage forces is restricted to poorly vegetated landscapes. Abrupt channel heads in cohesive soils on low-gradient slopes, however, may be maintained by spalling of material from the seepage face at the channel head (Figure 11.3) (Bradford & Piest, 1985; Montgomery, 1991). Channel head advance at an abrupt channel head in the Tennessee Valley area of Marin County, California, for example, was correlated with peak piezometric levels at the base of the channel head, but stability calculations suggest that channel head spalling reflects loss of restraining pressure for the pedons exposed on the channel head (Montgomery, 1991). Overland flow over the channel head sufficient to remove the failed material is required to prevent buttressing and allow continued channel head advance, but does not directly contribute to erosion at the channel head. The burrowing activity of fossorial mammals appears to be the dominant sediment transport process upslope of the channel heads in this area (Black & Montgomery, 1991). Overland flow upslope of the channel head was not observed to transport sediment over the channel head and grain-by-grain seepage entrainment of material from the face of the channel head was not observed. Hence controls on the location of such channel heads involve a change in process. Maintenance of these abrupt channel heads may be modeled as dependent upon sufficient drainage area to saturate the soil profile in order to both fail the channel head and provide overland flow

**Figure 11.3**  Abrupt headcut in the Tennessee Valley area of Marin County, California. Note alcove formed by spalling of the seepage face at left side of headcut. Piezometers above headcut extend 0.5 m above ground surface

runoff to remove the failed material. This mechanism, however, is not a channel initiation mechanism. Disturbance or locally accelerated incision is necessary to develop abrupt channel head morphology (Leopold & Miller, 1956; Reid, 1989; Montgomery, 1991). Once initiated, however, such an abrupt channel head probably will advance until the drainage area becomes insufficient to generate either soil saturation to cause spalling or overland flow to remove the failed material (Montgomery, 1991). This indicates that stable locations for such abrupt channel heads are governed by a contributing drainage area sufficient to saturate the soil profile and generate overland flow. Consequently, for the present discussion we assume that the critical area for 'seepage' erosion occurs when the soil is saturated.

For the case where the infiltration capacity of the soil exceeds the rainfall intensity, but flow does not recharge to a bedrock groundwater table, the subsurface discharge due to flow through the soil is given by

$$q = Ra \qquad (14)$$

Further assuming that the hydrologic gradient is equal to the ground slope, Darcy's law can be expressed as

$$q = KzhS = ThS \qquad (15)$$

where $K$ is the saturated hydraulic conductivity, $z$ is the soil thickness, $h$ is the proportion of the soil column that is saturated, and $T$ is the transmissivity of the soil profile. Combining (14) and (15) the relative saturation of the soil profile may be expressed as

$$h = Ra/TS \tag{16}$$

For saturated areas $h = 1$ and thus the contributing area per unit contour length necessary to saturate the soil column $a_{cr}$ is given by

$$a_{cr} = TS/R \tag{17}$$

This approach is equivalent to those used by Beven & Kirkby (1979) and O'Loughlin (1986) to estimate the distribution of relative soil moisture in a catchment. Equation (17) shows that with greater rainfall a smaller contributing area per unit contour length is necessary to produce saturation at a given location. Equation (17) predicts a positive linear relation between the drainage area and slope (Figure 11.6C). This saturation threshold essentially defines a limit to potential channel head advance by seepage erosion and gullying.

### Saturation Overland Flow

Saturation overland flow controls the formation of gradual channel heads on many low-gradient slopes in humid, soil-mantled landscapes. Gradual channel heads in grasslands of Marin County, California, provide a well-studied example. A three-year monitoring program in this area did not reveal obvious sediment entrainment by overland flow upslope of channel heads even during high discharge events in which significant sediment transport is observed downstream of the channel head (Montgomery, 1991). Instead, sediment transport upslope of these channel heads is dominated by burrowing activity (Black and Montgomery, 1991). This change in sediment transport processes suggests a threshold condition that may be modeled using a critical shear stress. Sediment transport by overland flow is possible only after development of sufficient tractive force to overcome the resistance of the vegetation covering the soil surface, especially where interwoven grass shields the ground surface (Figure 11.4). In such areas, runoff occurs over this grass surface and sediment transport occurs only where runoff breaks through this protective mat. During extreme discharge events the channel head may extend upslope of its typical position and temporarily transport material until the channel is progressively infilled in the interval between extreme discharge events. Fire and grazing also can reduce the resistance of this material and allow upslope extensions of the channel head. Given sufficient time, a threshold-controlled channel head thus may intermittently extend significantly upslope of the typical channel head location.

Kirkby (1987) developed a model for channel initiation by saturation overland flow that predicts a positive source area–slope relation. This model considers erosion to be directly proportional to overland flow. Montgomery & Dietrich (1988), however, reported field data that indicated an inverse source area–slope relation from areas in which saturation overland flow controls channel heads on low-gradient slopes. Consequently, they (Montgomery & Dietrich, 1988) commented that theories for channel initiation by saturation overland flow required revision. Formulation of such a model in terms of an erosional threshold leads to a model more consistent with the available field data.

**Figure 11.4**   A mat of vegetation protects the ground surface from erosion by overland flow in the unchanneled valleys upslope of channel heads in many grassland catchments, such as the Tennessee Valley area of Marin County, California

For the case of steady state rainfall, the overland flow discharge at a point in a catchment is that portion of the rainfall ($Ra$) that cannot be accommodated by subsurface flow ($TS$). This is given by

$$q = Ra - TS \tag{18}$$

Assuming the surface flow to be laminar, equation (18) can be set equal to (11) to solve for the critical drainage area per unit contour length:

$$a_{cr} = \frac{2\tau_{cr}{}^3}{Rk\nu\rho_w{}^3g^2S^2} + \frac{TS}{R} \tag{19}$$

This is equivalent to equation (3) of Dietrich *et al.* (1992). For low-gradient slopes the left-hand term will dominate. With increasing slope the right-hand term will increase in importance and will dominate the relation for steep slopes. Thus equation (19) predicts a non-linear drainage area–slope relation for channel initiation by saturation overland flow that for low-gradient areas is similar to the Horton overland flow threshold, but that with increasing slope becomes asymptotic to the threshold for soil saturation (Figure 11.6D). On steep slopes, however, landsliding processes may dominate channel initiation.

## Landsliding

In many steep landscapes channel heads coincide with small-scale debris flow scars in topographic hollows (Figure 11.5) (Montgomery & Dietrich, 1988). In the Oregon Coast

**Figure 11.5**   Channel head associated with a debris flow at the base of a steep topographic hollow on Mettman Ridge near Coos Bay in the Oregon Coast Range

Range, for example, overland flow does not occur on steep slopes due to the high hydraulic conductivity of the low-density colluvial soils (e.g., Yee and Harr, 1977; Montgomery, 1991) and thus landsliding is a primary channel initiation mechanism. Although recurrent debris flows are a primary means of transporting sediment from hillslopes into downslope channels (Dietrich & Dunne, 1978), they typically excavate only a portion of the colluvium upslope of a channel head and only rarely, if ever, extend to drainage divides. Under a constant climatic regime, the frequency of debris flow excavation, and thus the long-term rate of sediment transport, probably tends to decrease from a relatively short recurrence interval low on the slope to essentially never at the drainage divide (Dunne, 1991). Other types of landsliding also may be associated with channel heads. For example, in some areas channels begin within and on the margins of earthflows. We will not consider such cases further, but will instead discuss debris-flow-controlled channel heads in well-defined topographic hollows, in part for simplicity of modeling and in part because of the long-term influence of debris flows on hollow form, which contrasts with the more ephemeral geomorphic expression of channels associated with other landslide types. The model outlined below (Dietrich *et al.*, 1986; Montgomery & Dietrich, 1989) is based on coupling a subsurface flow model (Iida, 1984) with a simple slope stability model.

As in the model for soil saturation, the steady state subsurface discharge per unit contour length may be expressed by equation (14). Assuming Darcy flow parallel to the ground surface and uniform saturated conductivity, this subsurface discharge also is given by

$$q = hzK\sin\theta\cos\theta \qquad (20)$$

where $K$ is the saturated hydraulic conductivity, $\theta$ is the ground surface slope and the proportion of the soil depth that is saturated, $h$, is measured vertically, rather than normal to the ground surface. Equations (14) and (20) allow an expression for the proportion of the soil thickness that is saturated as a function of the upslope contributing area:

$$h = RA/zK\sin\theta\cos\theta \qquad (21)$$

The simplest, and most appropriate stability model is the infinite slope model, which can be expressed as

$$h = \frac{C'}{\rho_w gz\cos^2\theta\tan\phi} + \frac{\rho_s}{\rho_w}\left[1 - \frac{\tan\theta}{\tan\phi}\right] \qquad (22)$$

where $C'$ is the effective cohesion of the soil, $\rho_s$ is the bulk density of the saturated soil, and $\phi$ is the angle of internal friction of the soil. Combining (21) and (22) yields an expression for the critical drainage area per unit contour length where

$$a_{cr} = \frac{zK\sin\theta\cos\theta}{R}\left[\frac{C'}{\rho_w gz\cos^2\theta\tan\phi} + \frac{\rho_s}{\rho_w}(1 - \frac{\tan\theta}{\tan\phi})\right] \qquad (23)$$

This equation predicts a highly non-linear relation between source area and slope (Figure 11.6E) and is only valid for

$$\tan\theta \geq [(\rho_s - \rho_w)/\rho_s]\tan\phi \qquad (24)$$

which for many soils may be approximated by $\tan\theta \geq \tan\phi/2$. For cohesionless soils, the steepest slope stable under this relation is given by the friction angle of the soil (i.e., $\tan\theta = \tan\phi$). For cohesive soils steeper slopes may be stable.

## DRAINAGE AREA–SLOPE PROCESS REGIMES

In a relatively uniform landscape, a single channel initiation process may be the dominant control on channel initiation. In landscapes with a variety of slopes and soil properties, however, all of these processes may influence channel initiation. In this case, we would

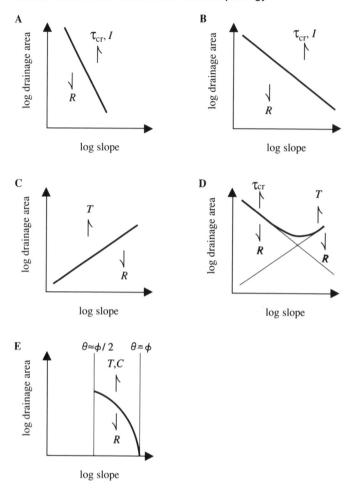

**Figure 11.6** Schematic illustration of the relation between drainage area and slope predicted from simple threshold theories of channel initiation by (A) turbulent Horton overland flow, (B) laminar Horton overland flow, (C) soil saturation, (D) saturation overland flow, and (E) shallow, small-scale landsliding. Arrows indicate effect of increases in rainfall intensity ($R$), critical shear stress ($\tau_{cr}$), infiltration capacity ($I$), transmissivity ($T$), cohesion ($C$), and angle of internal friction ($\phi$)

expect different portions of the landscape to be dominated by different processes. Channel initiation in steep areas would be dominated by landsliding, whereas in undisturbed, low-gradient areas overland flow would dominate. Seepage erosion and gullying would be expected to occur in response to local disturbance in low-gradient areas. In many real landscapes, therefore, a complex interplay between channel initiation processes will control channel head locations. Models of channel initiation processes can be used to interpret this interplay and allow prediction of those portions of a landscape in which different channel initiation processes provide the primary control on channel head locations. Consequently, we may divide the landscape into process regimes by plotting the erosional thresholds for saturation overland flow, soil saturation,

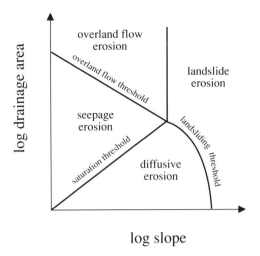

log slope

**Figure 11.7** Combining the thresholds depicted in Figure 11.6 allows division of the landscape into process regimes where different sediment transport and channel initiation mechanisms operate. Below the thresholds for transport by soil saturation, overland flow, and landsliding only diffusive sediment transport is effective and hence this region should correspond to hillslopes stable to channelization tendencies

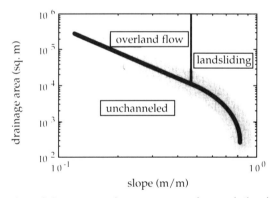

slope (m/m)

**Figure 11.8** Illustration of the expected source area–slope relation in a landscape with a range of slopes in which both overland flow and landsliding control channel initiation. The central tendency in the source area–slope relation (solid line) is governed by long-term changes in climate and the other factors illustrated in Figure 11.6. Specific channel head locations also define a distribution about the central tendency (shaded area) that reflects both spatial and short-term temporal variability in the processes influencing channel initiation

and landsliding on a graph of drainage area and slope (Figure 11.7) (Dietrich *et al.*, 1992). Portions of the landscape that plot above the overland flow threshold may be expected to support channels. Non-erosive overland flow occurs between the thresholds for soil saturation and overland flow erosion and this field essentially defines low- to moderate-gradient unchanneled valleys. Areas within this zone may be subject to channelization by seepage erosion in response to disturbance. Portions of the landscape plotting above the landslide threshold also will be subject to channelization. Together these thresholds surround a portion of the landscape within which none of these processes

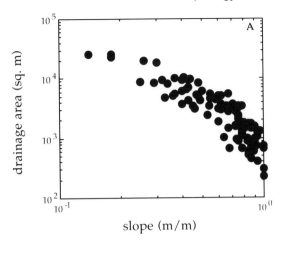

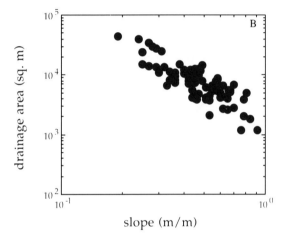

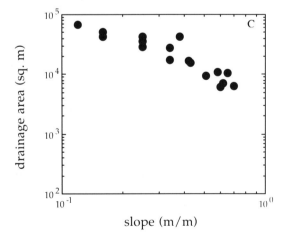

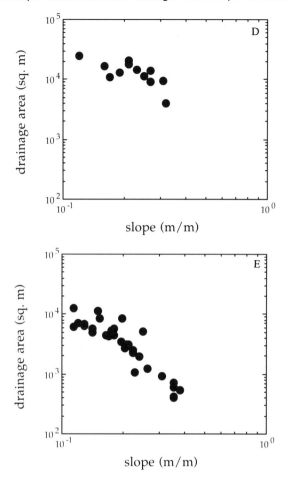

**Figure 11.9**  Plot of source area versus local slope at the channel head for studies areas in: (A) Coos Bay, Oregon, (B) Tennessee Valley, California, (C) Southern Sierra Nevada, California, (D) Stanford Hills, California, (E) northern Humboldt Range, Nevada. Channel initiation in the Coos Bay, Tennessee Valley, and Southern Sierra Nevada catchments is controlled primarily by overland flow on low-gradient slopes and landsliding on steeper slopes. The source area–slope relations for each of these areas reflect this transition, but it is best expressed in the Coos Bay data where landsliding is the dominant channel initiation process. Channel initiation in the Stanford Hills and Humboldt Range study areas is controlled by overland flow. The source area–slope relation in these catchments is consistent with the form predicted by an overland flow model. See text for sources of data

are effective at transporting sediment and in which diffusive processes dominate sediment transport [see also Dietrich *et al.* (1992)].

## OBSERVED SOURCE AREA–SLOPE RELATIONSHIPS

Few field data exist against which to test even the general form of predicted source area–slope relations. Abrahams (1980) measured source areas from Schumm's (1956) mapping of badlands and found that source area is inversely correlated with the average slope of first-order basins. He (Abrahams, 1980) also found that source-area size and average first-order basin slope were unrelated for another area where field inspection suggested that channel initiation was dominated by subsurface flow. Unfortunately, the use of the average first-order basin slope severely compromises comparisons of these data with predicted source area–local valley slope relations. Dietrich *et al.* (1986) reported an inverse relation between source area and average source-area slope for a limited number of field sites from several different regions. Dietrich *et al.* (1987) found that source-area size is slope-independent in catchments underlain by basalt where relatively impermeable dikes control groundwater flow paths. Montgomery & Dietrich (1988, 1989) subsequently found inverse source area–local valley slope relations in three study areas where channel initiation occurs by landsliding, overland flow, and to a lesser extent seepage erosion and channel head sloughing. Repka (unpublished data) also found an inverse source area–local slope relation for an area in which channel initiation occurs by Horton overland flow. Although we cannot at present constrain or fully characterize all of the parameters in the models presented above, we can test the general form of the predicted drainage area–slope relations for some of these models by comparison with available field data from areas in which these processes are the dominant channel initiation mechanisms.

Most of the available source area–slope data are from areas in which overland flow and/or landsliding are the dominant channel initiation mechanisms. In an area with both gentle and steep slopes, and in which both overland flow and landsliding occur, the threshold requiring the smallest drainage area should be reflected in the source area–slope relation. The models presented above predict that $a_{cr} \propto S^{-2}$ for low-gradient slopes dominated by laminar overland flow, and a more rapid decrease in source-area size with increasing slope for landslide-controlled channel intiation. Landsliding is only a relevant process when equation (24) is satisfied, a condition that commonly will be approximated by $\tan\theta > 0.5$. Conversely, overland flow is rare on steep hillsides in soil-mantled environments. Therefore, it is reasonable to expect different source area–slope relations for low and high-gradient slopes (Figure 11.8).

Montgomery & Dietrich (1988) reported inverse source area–slope relations across a wide range of slopes ($0.1 < \tan\theta < 1.0$) for three catchments in the western United States. They documented that landsliding dominated channel initiation on steeper slopes, whereas overland flow dominated channel initiation on low-gradient slopes and showed that a model for channel initiation by landsliding provided a reasonable fit to the observed data for $\tan\theta > 0.5$ for one of these areas (Montgomery & Dietrich, 1988, 1989). Willgoose *et al.* (1991) subsequently argued that the general inverse trend across the full range of these data indicates a single underlying channel initiation mechanism [an assertion directly contradicted by field observations (Montgomery & Dietrich, 1988, 1989)]. Examination of the available data confirms the expected generality of different source area–slope relations for low- and steep-gradient slopes. Furthermore, the observed data are consistent with the predicted form of the source area–slope relation for overland flow on low-gradient slopes and for landsliding on steeper slopes.

Nine source area–slope data sets are available at present. Five data sets report source area and local slope at the channel head (Figure 11.9), while four report average source-area gradient (Figure 11.10). On steep slopes gradient does not vary greatly and hence average source-area slopes are probably similar to local slopes at channel heads.

Together these data represent a broad range in geology, vegetation, and climate. Furthermore, observations pertaining to channel initiation mechanisms are available for each of these areas. The most intensively studied areas are a series of catchments in Coos Bay, Oregon, and the Tennessee Valley area of Marin County, California, in which all channel heads were mapped and local slopes at the channel head were measured in the field. Montgomery & Dietrich (1988, 1989, 1992) discuss the study areas, data collection methods, and field observations from these areas.

The Coos Bay area is underlain by nearly flat-lying Eocene sandstone, is covered by a managed coniferous forest, and receives an annual rainfall of approximately 1500 mm. Data from the catchments mapped in this area span a wide range of slopes (Figure 11.9A). On low-gradient slopes ($\tan\theta < 0.5$) the data followed an inverse trend consistent with $a_{cr} \propto S^{-2}$, while data for steeper slopes exhibit a more rapid decline in source area with increasing slope. Field observations from these catchments indicate that landsliding is a major channel initiation process on steep slopes (Montgomery & Dietrich, 1988; Montgomery, 1991), whereas overland flow occurs only on gentler slopes in this area due to the high conductivities of the colluvial soils.

A similar pattern of channel initiation by overland flow on low-gradient slopes and by landsliding on steeper slopes was observed in the Tennessee Valley catchments (Montgomery & Dietrich, 1988, 1989; Montgomery, 1991). A number of the channel heads in these catchments appear to be controlled by seepage erosion associated with the development of abrupt channel heads in response to historic overgrazing. These channel heads are discussed in greater detail elsewhere (Montgomery & Dietrich, 1989; Montgomery, 1991). The area is underlain by chert, greenstone, and greywacke, is covered by grassland and coastal prairie vegetation, and receives approximately 760 mm annual rainfall. Source area–slope data for this area are generally consistent with the relation predicted by overland flow for low-gradient slopes and exhibit an apparent steepening for slopes greater than 0.5 (Figure 11.9B). Montgomery & Dietrich (1988, 1989) reported that the data for the steeper slopes are reasonably modeled by a threshold-based equation for channel initiation by landsliding. Dietrich et al. (1992) show that the general inverse relationship is consistent with the saturation overland flow model [equation (19)].

Data from three other gentle-gradient study areas also reflect channel initiation by overland flow. The Southern Sierra data (Figure 11.9C) (Montgomery & Dietrich, 1988, 1992) are from an area underlain by deeply weathered, unglaciated granitic rocks covered by open oak forest and grasslands, which receives approximately 260 mm of rainfall annually. The Stanford Hills data (Figure 11.9D) are from an arena on the Stanford University campus south of San Francisco, California, underlain by basalts and sandstones and covered by open oak forest and grasslands. The data from Nevada (Figure 11.9E) are from a dissected Pleistocene alluvial fan in an arid region (Repka, unpublished data). Field observations suggest that channel head locations in each of these areas are dominated by laminar overland flow and the data from each of these areas exhibit an inverse relation consistent with the form of the predicted relation; $a_{cr} \propto S^{-2}$.

Three additional data sets using average source–area slope are available for steep environments. Data reported by Dietrich et al. (1987) from San Pedro Ridge, California (Figure 11.10A) suggest a steepening of the source area–slope relation at about $\tan\theta = 0.5$. The area is underlain by sandstone and is covered by a hardwood forest. Small-scale

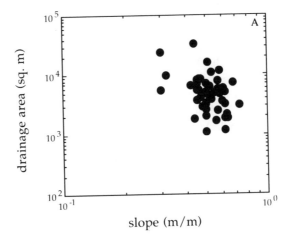

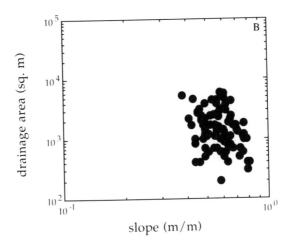

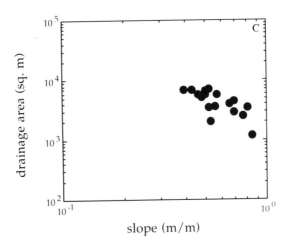

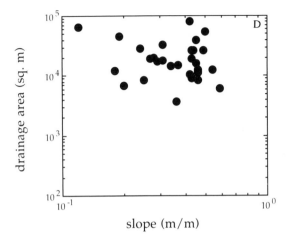

**Figure 11.10**  Plot of source area versus average source-area slope for study areas in (A) San Pedro Ridge, California, (B) San Dimas, California, (C) Japan, and (D) Rock Creek, Oregon. Channel initiation in the San Pedro Ridge, San Dimas, and Japan study areas is dominated by landsliding. The steep source area–slope relation in each of these areas is consistent with the form predicted by the landslide model. Channel head locations in the Rock Creek study area, however, are controlled by the location of impermeable dikes. In this case, the source areas are essentially slope independent on both gentle and steep slopes. See text for sources of data

landsliding is common at steep channel heads in this area. Data for the San Dimas area (Figure 11.10B) were derived from Maxwell's (1960) map of field surveyed channel networks in the Fern basin of the San Dimas National Forest, California. Landsliding is a common process on the steep slopes in this area and the data suggest a very steep source area–slope relation. Dietrich *et al.* (1987) reported data from Japan (Figure 11.10C) derived from figures in Tsukamoto *et al.* (1978). These data are from areas in which landsliding is an important sediment transport process. The form of the data from each of these steep landscapes conforms to the steeper relation expected from channel initiation by landsliding.

Dietrich *et al.* (1987) also reported data from Rock Creek, Oregon (Figure 11.10D), which do not exhibit a systematic source area–slope relation over low-gradient slopes ranging from $0.1 < \tan\theta < 0.6$. This area is underlain by gently dipping basalt flows and breccias cross-cut by fine-grained dikes. They (Dietrich *et al.*, 1987) found that fine-grained basalt was commonly exposed at the channel head and illustrated how such low-permeability zones in the underlying bedrock could strongly control channel head locations. This local bedrock hydrologic control on channel initiation may dominate source area–slope relations in areas with high bedrock conductivities and local low-permeability zones.

A plot of the available data indicate a steepening in the source area–slope relation at gradients greater than $\tan\theta = 0.5$ associated with a change in process dominance from overland flow on gentler slopes to landsliding on steeper slopes (Figure 11.11). Data from the northern Humboldt Range are excluded because they are from an arid environment and plot significantly below the range of the data from more humid environments. Data from Rock Creek, Oregon, also are excluded because of the strong

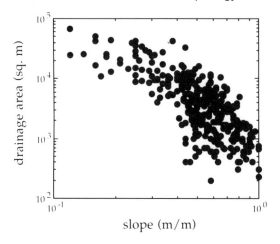

**Figure 11.11**   Plot of source area versus slope for all study areas except Rock Creek, Oregon, and northern Humboldt Range, Nevada

influence of bedrock dikes on channel head locations. The rapid decrease in source-area size on steeper slopes apparent in Figure 11.11 is consistent with the hypothesis that different channel initiation processes dominate different portions of a landscape (Figure 11.7), in this case reflecting the dominance of overland flow on low-gradient slopes and landsliding on steeper slopes. The approximately order-of-magnitude scatter about the central tendency of the composite data set reflects the variance of channel head locations both within and between study areas. The general agreement between the form of the observed source area–slope data and that predicted by models of relevant processes supports the general formulation of these models.

Source area–slope relations provide a conceptual framework within which to investigate landscape response to changes in the conditions affecting channel initiation. Slope may be thought of as the independent variable in the source area–slope relations due to the long response time and bedrock erosion required to significantly alter valley-long profiles. This leads to the question of what sets the valley slope and suggests that source-area size responds to changes in climate and process regimes through channel head migration along the valley axis. For all of the processes discussed above source area should decrease with either increasing rainfall or a decrease in infiltration capacity, hydraulic conductivity, or critical shear stress. The rainfall dependency suggests that channel network expansion should then accompany change to a wetter climate, whereas increased aridity would result in downslope retreat of channel heads and infilling of the upper reaches of first-order streams (Montgomery & Dietrich, 1988). However, the infiltration capacity, hydraulic conductivity, and critical shear stress may co-vary with changes in rainfall and the combined impact on channel head locations may be non-linear. For example, vegetation reduction in response to dramatically increased aridity may decrease the infiltration capacity, the critical shear stress, or possibly the flow resistance. Although flow resistance will tend to reduce flow depth and, consequently, the flow boundary shear stress, a relatively small reduction in critical boundary shear stress will greatly decrease the source area (Dietrich *et al.*, 1992). Similarly, land-use modifications that increase the effective rainfall intensity at the ground surface, or which decrease either the critical shear stress or the infiltration capacity of the soil would result in channel head advance, accelerated erosion, and channel network expansion. Furthermore, environmental perturbations resulting in a change in the controls on channel head locations would potentially destabilize the network of unchanneled valleys upslope of channel heads. As a whole then, these relations provide a quantitative framework within which to consider controls on channel initiation and thus channel network extent and landscape dissection.

## CHANNEL INITIATION THRESHOLDS AND LANDSCAPE DISSECTION

Channel heads are not static and this dynamism is apparent in the variance in the source area–slope data, which integrate both spatial variations in the physical properties of the soil (e.g., $\tau_{cr}$, $K$, $z$) and temporal variations in the position of individual channel heads (Figure 11.8). Over landscape-forming time scales, it is this range of channel head locations that defines the zone over which sediment transport occurs by channel processes. The variance about the general trend in the source area–slope data suggests that the threshold of channel initiation is best described as a zone of transition from channeled to unchanneled regions of the landscape. Such a transitional zone does not necessarily contradict models for valley stability based on the dominance of incisive and diffusive process. Rather this suggests the effect of both spatial and temporal dominance in maintaining the balance. Over long time scales, the extent of valley dissection represents an interplay between the frequency of channel head extention and the rate of diffusional infilling of the valley (e.g., Calver, 1978; Montgomery & Dietrich, 1992). In landscapes where the erosional threshold is significantly increased and the channel head is displaced far enough downslope, the range of channel head locations also will be displaced and the unchanneled valley network will expand downslope. In contrast, a decrease in the channel initiation threshold will result in expansion of the channel network and potentially the development of new valleys through incision into previously undissected hillslopes. This suggests a dynamic view of landscape development in which the relation between channel initiation and valley maintenance reflects both the spatial and temporal variance of channelization thresholds.

At the landscape scale, geomorphic response to environmental change may involve a shifting of the range over which channel heads oscillate. For example, a change from conditions favoring small source areas to conditions favoring larger source areas will result in a retraction of the zone defined by the short-term range in channel head locations. This will result in a decrease in the frequency that a channel will occupy any point upslope of the mean channel head location which, in turn, will allow diffusive transport to overwhelm channelization tendencies farther downslope from the drainage divide. Thus, a portion of the originally channeled valley will infill with colluvium, possibly to the point at which the original valley becomes obscured as a topographic feature. In contrast, a change from drier to wetter conditions (assuming that other factors, especially vegetation, do not co-vary with precipitation) will cause channel heads to advance upslope, excavate previously accumulated valley fills, and incise undissected hillslopes. The maximum distance upslope that channel extention may carve, and maintain, a valley depends upon the relation between the sediment transport function for extreme events and the rate of diffusive sediment transport. Therefore, the average channel head location will be significantly downslope of its long-term influence on valley maintenance. Consequently, the extent of landscape dissection may reflect the long-term interplay between a temporally variable channelization threshold and relatively continuous diffusive infilling.

## CONCLUSIONS

The extent of channel and valley networks provide identifiable and morphologically distinct descriptions of landscape dissection. Channel initiation is controlled by a variety of processes which may be modeled as threshold phenomena, whereas valley maintenance may be modeled as reflecting either a spatial transition in process dominance, or a temporal variance in the exceedence of a channel initiation threshold. Simple analytical models for channel initiation processes predict source area–slope relations that agree with the available field data. These relations define areas of the landscape in which incisive (slope and area-dependent) and diffusive (slope-dependent) processes operate. In essence, we suggest that in vegetated, soil-mantled landscapes the extent of the valley network reflects the spatial signature of a dynamic channel network controlled by channelization thresholds. Consequently, the Gilbert hypothesis of landscape dissection, in which valley development reflects a stable spatial transition in process dominance, is best viewed as relevant to geologic time scales (i.e., $10^4$–$10^6$ yr), whereas the Horton hypothesis of channel initiation and landscape dissection in response to exceedence of erosional thresholds is best viewed as relevant to geomorphic time scales (i.e., $10^2$–$10^3$ yr).

## ACKNOWLEDGEMENTS

This research was supported by NSF grant 8917467 and grant TFW FY92-010 from the CMER committee of the Washington State Timber/Fish/Wildlife agreement. We thank Kelin Whipple for comments on drafts of this manuscript.

## REFERENCES

Abrahams, A. D. (1980). Channel link density and ground slope. *Annals of the American Association of Geographers*, **70**, 80–93.

Beven, K. J. and Kirkby, M. J. (1979). A physically based variable contributing area model of basin hydrology. *Hydrological Sciences Bulletin*, **24**, 43–69.

Black, T. A. and Montgomery, D. R. (1991). Sediment transport by burrowing mammals, Marin County, California. *Earth Surface Processes and Landforms*, **16**, 163–172.

Bradford, J. M. and Piest, R. F. (1985). Erosional development of valley-bottom gullies in the upper midwestern United States. In Coates, D. R. and Vitek, D. (eds), *Thresholds in Geomorphology*, Allen & Unwin, London, pp. 75–101.

Calver, A. (1978). Modelling drainage headwater development. *Earth Surface Processes*, **3**, 233–241.

Dietrich, W. E. and Dunne, T. (1978). Sediment budget for a small catchment in mountainous terrain. *Zeitschrift für Geomorphologie*, Supplement, **29**, 191–206.

Dietrich, W. E. and Dunne, T. (1993). The channel head. In Beven, K. and Kirkby, M. J. (eds), *Channel Network Hydrology*, Wiley, Chichester, pp. 175–219.

Dietrich, W. E., Wilson, C. J. and Reneau, S. L. (1986). Hollows, colluvium and landslides in soil-mantled landscapes. In Abrahams, A. D. (ed.), *Hillslope Processes*, Allen & Unwin, Boston, pp. 362–388.

Dietrich, W. E., Reneau, S. L. and Wilson, C. J. (1987). Overview: 'Zero order basins' and problems of drainage density, sediment transport and hillslope morphology. In Beschta, R. L., Blinn, T., Grant, G. E., Ice, G. G. and Swanson, F. J. (eds), *Proceedings of the International Symposium on Erosion and Sedimentation in the Pacific Rim*, International Association of Hydrological Sciences Publication 165, pp. 27–37.

Dietrich, W. E., Wilson, C. J., Montgomery, D. R., McKean, J. and Bauer, R. (1992). Erosion thresholds and land surface morphology. *Geology*, **20**, 675–679.

Dunne, T. (1980). Formation and controls on channel networks. *Progress in Physical Geography*, **4**, 211–239.

Dunne, T. (1990). Hydrology, mechanics, and geomorphic implications of erosion by subsurface flow. In Higgins, C. G. and Coates, D. R. (eds), *Groundwater Geomorphology; the Role of Subsurface Water in Earth-Surface Processes and Landforms*, Geological Society of America Special Paper, **252**, pp. 1–28.

Dunne, T. (1991). Stochastic aspects of the relations between climate, hydrology and landform evolution. *Transactions, Japanese Geomorphological Union*, **12**, 1–24.

Dunne, T. and Black, R. D. (1970). Partial area contributions to storm runoff in a small New England watershed. *Water Resources Research*, **6**, 1296–1311.

Dunne, T. and Dietrich, W. E. (1980). Experimental investigation of Horton overland flow on tropical hillslopes. 2. Hydraulic characteristics and hillslope hydrographs. *Zeitschrift für Geomorphologie, N.F., Supplement*, **35**, 60–80.

Emmett, W. W. (1970). The hydraulics of overland flow on hillslopes. *US Geological Survey Professional Paper*, 662-A, 68 pp.

Gilbert, G. K. (1877). Geology of the Henry Mountains. *US Geographical and Geological Survey*, 160 pp.

Gilbert, G. K. (1909). The convexity of hill tops. *Journal of Geology*, **17**, 344–350.

Gregory, K. J. and Walling, D. E. (1968). The variation of drainage density within a catchment. *Bulletin of the International Association of Scientific Hydrologists*, **13**, 61–68.

Horton, R. E. (1932). Drainage basin characteristics. *Transactions of the American Geophysical Union*, **13**, 350–361.

Horton, R. E. (1945). Erosional development of streams and their drainage basins; Hydrophysical approach to quantitative morphology. *Geological Society of America Bulletin*, **56**, 275–370.

Iida, T. (1984). A hydrological method of estimation of the topographic effect on the saturated throughflow. *Transactions of the Japanese Geomorphological Union*, **5**, 1–12.

Kirkby, M. J. (1971). Hillslope process-response models based on the continuity equation. *Institute of British Geographers Special Publication* No. 3, pp. 15–30.

Kirkby, M. J. (1980). The streamhead as a significant geomorphic threshold. In Coates, D. R. and Vitek, J. D. (eds), *Thresholds in Geomorphology*, Allen & Unwin, London, pp. 53–73.

Kirkby, M. J. (1987). Modelling some influences of soil erosion, landslides and valley gradient on drainage density and hollow development. In Ahnert, F. (ed.), *Geomorphological Models*, *Cantena Supplement*, **10**, pp. 1–11.

Kirkby, M. J. and Chorley, R. J. (1967). Throughflow, overland flow and erosion. *Bulletin of the International Association of Scientific Hydrology*, **12**, 5–21.

Leopold, L. B. and Miller, J. P. (1956). Ephemeral streams – hydraulic factors and their relation to the drainage net. *US Geological Survey Professional Paper*, 282-A, 37 pp.

Loewenherz, D. S. (1991). Stability and the initiation of channelized surface drainage: A reassessment of the short wavelength limit. *Journal of Geophysical Research*, **96**, 8453–8464.

Maxwell, J. C. (1960). Quantitative geomorphology of the San Dimas Experimental Forest, California: Technical Report 19, Office of Naval Research project NR 389-042, Department of Geology, Columbia University, New York.

Montgomery, D. R. (1991). Channel initiation and landscape evolution. Unpublished Ph.D. Dissertation, University of California, Berkeley, 421 pp.

Montgomery, D. R. and Dietrich, W. E. (1988). Where do channels begin? *Nature*, **336**, 232–234.

Montgomery, D. R. and Dietrich, W. E. (1989). Source areas, drainage density, and channel initiation. *Water Resources Research*, **25**, 1907–1918.

Montgomery, D. R. and Dietrich, W. E. (1992). Channel initiation and the problem of landscape scale. *Science*, **255**, 826–830.

O'Loughlin, E. M. (1986). Prediction of surface saturation zones in natural catchments by topographic analysis. *Water Resources Research*, **22**, 794–804.

Reid, L. M. (1989). Channel incision by surface runoff in grassland catchments. Unpublished Ph.D. Dissertation, University of Washington, Seattle, 135 pp.

Reneau, S. L., Dietrich, W. E., Dorn, R. I., Berger, C. R. and Rubin, M. (1986). Geomorphic and paleoclimatic implications of latest Pleistocene radiocarbon dates from colluvium-mantled hollows, California. *Geology*, **14**, 655–658.

Schaefer, M. G. (1979). The zero order watershed. Unpublished Ph.D. Dissertation, University of Missouri-Rolla, Rolla, Missouri, 69 pp.

Schumm, S. A. (1956). Evolution of drainage systems and slopes in badlands at Perth Amboy, New Jersey. *Geological Society of America Bulletin*, **67**, 597–646.

Slattery, M. C. and Bryan, R. B. (1992). Hydraulic conditions for rill incision under simulated rainfall: A laboratory experiment. *Earth Surface Processes and Landforms*, **17**, 127–146.

Smith, T. R. and Bretherton, F. P. (1972). Stability and the conservation of mass in drainage basin evolution. Water Resources Research, **8**, 1506–1529.

Tarboton, D., Bras, R. L. and Rodriguez-Iturbe, I. (1992). A physical basis for drainage density. *Geomorphology*, **5**, 59–76.

Tsukamoto, Y., Matsuoka, M. and Kurihara, K. (1978). Study on the growth of stream channels. (VI) Landslides as the process of erosional development of basin morphology. *Journal of the Japanese Erosion Control Engineering Society*, **17**, 25–32.

Willgoose, G. R., Bras, R. L. and Rodriguez-Iturbe, I. (1991). A coupled channel network growth and hillslope evolution model: 1. Theory. *Water Resources Research*, **27**, 1671–1684.

Wilson, C. J. (1988). Runoff and pore pressure development in hollows. Unpublished Ph.D. Dissertation, University of California, Berkeley, 284 pp.

Yee, C. S. and Harr, S. D. (1977). Influence of soil aggregation on slope stability in the Oregon Coast Ranges. *Environmental Geology*, **1**, 367–377.

# 12 Influence of Slope/Stream Coupling on Process Interactions on Eroding Gully Slopes: Howgill Fells, Northwest England

**A. M. HARVEY**
*Department of Geography, University of Liverpool, UK*

## ABSTRACT

Process interactions on eroding gullies include on-slope process interactions and those involving slope/stream coupling. Processes on experimental plots on eroding gullies in the Howgill Fells, northwest England, have been monitored by sequential photography. The plots have been selected to represent varying styles of coupling between eroding slopes and the stream system. Seasonal process interactions are evident on all plots, with overland flow processes dominant during summer and mass movement during winter. The style of interaction varies with the strength of coupling, with fluvial processes strongly influenced by local base-level controls and more effective where the coupling is strong. As erosion continues over the longer term, there is a progressive decoupling of the system and a switch away from fluvial dominance. Deposition begins to exceed erosion and vegetation colonisation eventually leads to stabilisation of the eroded hillslopes.

## INTRODUCTION

Although the importance of coupling between the component parts of geomorphic systems has long been acknowledged (e.g. Brunsden and Thornes, 1979), there is relatively little work that has attempted to characterise its role, either in the context of the sediment flux through the system, or in the context of landform adjustment (see Newson, 1992). The strength of coupling is central to the functioning of the system. In strongly coupled systems, sediment throughput may be rapid and spatially continuous, allowing the downstream transmission of catchment or hillslope-induced perturbations in water and sediment supply. The whole system may exhibit sensitive and more or less synchronous adjustment to environmental change (e.g. Macklin *et al.*, 1992). In weakly coupled systems, sediment transport may be temporally and spatially intermittent between major storage zones, with little downstream transmission of hillslope or catchment-induced changes through the system. Sedimentation zones may act as buffers within the system (e.g. Harvey, 1989). The system as a whole may be much less sensitive in the short term, and landform adjustment may be local rather than basin-wide.

*Process Models and Theoretical Geomorphology*. Edited by M. J. Kirkby
©1994 John Wiley & Sons Ltd

Coupling within fluvial systems has both spatial and temporal characteristics. Spatially, coupling may be important on hillslopes, between hillslopes and channels, and downstream within the channel system. Temporally, differing styles of coupling may produce a range of sediment flux/landform adjustment relationships, from continuity to cyclicity to episodicity.

There have been very few studies of coupling *per se* within fluvial systems. This paper addresses the temporal and spatial characteristics of the coupling between hillslope gully systems and basal stream systems in the Howgill Fells, Cumbria, northwest England.

The rates and styles of development of hillslope gullies are often controlled, not by one single process, but by interactions between multiple processes. These interactions may be fundamentally related to the coupling characteristics of the system as a whole. Furthermore, as morphology progressively changes during gully development, the style of process interactions may also change as the coupling between the component parts of the system changes. This will result in dynamically variable, rather than constant, process/form relationships.

Spatially, process interactions may occur between on-slope processes, e.g. weathering, mass movement, surface and subsurface erosion, and may exhibit temporal variations ranging from event-related short-term to progressive long-term timescales (Calvo-Cases *et al.*, 1991). Process interactions commonly exhibit seasonality (Schumm, 1956a,b; Harvey, 1987a) or cyclicity (Calvo-Cases *et al.*, 1991; Calvo & Harvey, in press). There may also be an interaction between on-slope and basal stream processes, involving varying degrees of slope/stream coupling, and producing a periodicity in sediment accumulation/flushing relationships (Harvey, 1977; Wells & Gutierrez, 1982; Faulkner, 1988).

This paper builds on previous work on gully systems in the Howgill Fells, which has demonstrated both on-slope process interactions between seasonally variant surface erosion and mass movement processes (Harvey, 1974, 1987a), and slope/stream coupling interactions, producing a cyclicity of basal sediment accumulation and removal, with a periodicity of several years (Harvey, 1977, 1987b, 1988, 1992; Harvey *et al.*, 1979). Over the longer term, progressive gully development itself appears to modify the nature of the slope/stream coupling, as gully heads retreat and the influence of basal removal becomes more remote. This ultimately results in the stabilisation of the gully slopes (Harvey, 1988, 1992) in a manner not unlike that described by Beaty (1959) or that by Wells *et al.* (1991).

This paper deals with process interactions on three gullies at Carlingill in the Howgill Fells, adjacent to the long-term monitored site at Grains Gill (Figure 12.1). They were specifically selected to examine the influence of slope/stream coupling on process interactions during the stages of gully development, from gully initiation to ultimate stabilisation. The first results of the monitoring of on-slope process interactions alone have been described elsewhere (Harvey & Calvo, 1991). This paper builds on those preliminary results and extends consideration from the eroding surfaces to the whole gully system.

## STUDY SITES

The three study sites are located on the north side of Carlingill valley in the western Howgills (Figure 12.1). They are cut into Pleistocene glacial and periglacial sediments.

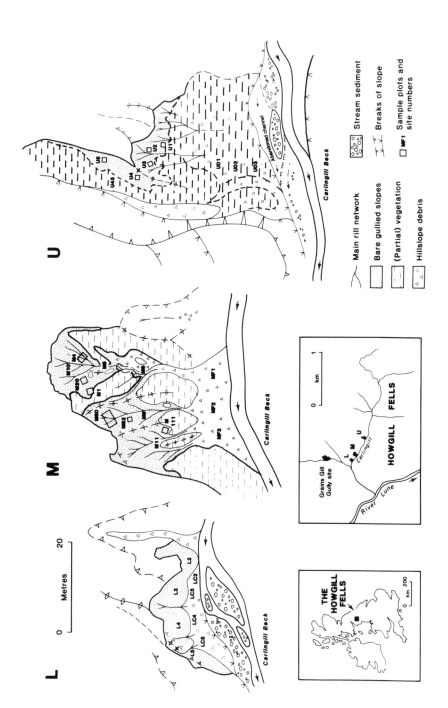

**Figure 12.1** Study site locations. Numbers refer to plot locations (see text and Table 12.1)

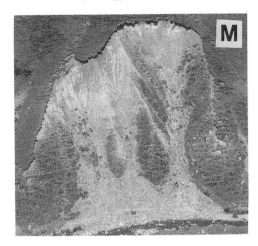
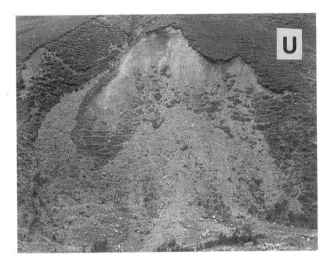

**Figure 12.2**  Carlingill gully sites L, M and U

The three gullies have been selected to represent the range of slope/stream coupling characteristic of the progressive evolution of gully systems in the Howgills (Harvey, 1990, 1992). Gully L (Figure 12.2) is a relatively young streamside scar, not visible in 1948 air photographs, with strong and direct coupling to the stream channel. Steep, rapidly eroding slopes feed sediment to the stream margins. The sediment builds up in debris cones or debris aprons which are periodically removed by stream floods. Gully M (Figure 12.2) is a larger, more mature system, where the active gully heads have cut back into the hillslope, and are connected to the basal stream by linear gully channels. Sediment is fed from the eroding slopes down the gully channels to debris cones at the stream margin. Periodically these are scoured by stream floods but the strength of coupling is less than on Gully L. The linear gully channels and the debris cones partially buffer the active slopes from basally induced stream erosion. At Gully U (Figure 12.2) the eroding slopes are completely disconnected from basal stream activity. The gully was apparently initiated by basal stream erosion by a now abandoned channel of Carlingill Beck. Lichenometry from fluvial deposits adjacent to this channel suggests abandonment approximately 50–60 years ago (Harvey *et al.*, 1984; Harvey, 1992). Since then, sediment derived by gully erosion has accumulated on the debris slope, and partial stabilisation of the gully has occurred (Harvey, 1990, 1992).

The three gully sites therefore represent three degrees of slope/stream coupling, related to three stages in the progressive development of the Howgill gully systems from initiation as streamside scars to eventual stabilisation of the gully slopes.

## RESEARCH DESIGN

In order to identify the relative importance of the different processes involved in the process interactions, direct measurement of the processes at the scale desired would have been virtually impossible. However, previous work in a variety of environments had demonstrated the potential of sequential photography for recording the effects of different processes, and for providing a satisfactory basis for the derivation of quasi-quantitative measures to express the relative importance of different processes (see Suwa & Okuda, 1988; Harvey & Calvo, 1991; Calvo & Harvey, in press). The design was therefore to establish plots for repeated photography, located to represent the range of slope/stream coupling styles within the three gully systems (Figure 12.3), including both erosional and transportational/depositional zones.

Four categories of eroding slope were distinguished (Table 12.1); basal slopes, feeding sediment directly to ephemeral streamside debris cones or to the main stream itself;

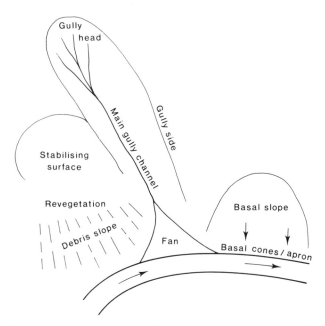

**Figure 12.3** Characteristic gully morphology showing the various styles of slope/stream coupling used as a basis for a siting of sample plots

**Table 12.1**   Monitored sites (for locations see Figure 12.1)

| Type of site | Site Nos. (parentheses indicate multiple sites) | Notes |
|---|---|---|
| *Erosional Zones* | | |
| Basal sites | L2 L3 L4 L5 | Slopes 50–65°, feeding basal cones or directly into Carlingill Beck. |
| Gully-head sites | M4 M10 M20 M60 | Slopes 40–60°, convergent rill networks feeding main gully channels. |
| Gully-side sites | M62 U2 | Slopes *c.* 50°, planar slopes feeding laterally into gully channels. |
| Stabilising sites | M1 M111 U1 U3 U4 U5 | Slopes 45–50°, all based by revegetating slopes. |
| | | |
| *Transport and Depositional Zones* | | |
| Basal cones | LC2 LC3 LC4 LC5 | Ephemeral cones, fed by eroding sites L2–L5. |
| Gully channels and fans | M5 M9 (3)     MF1 (3) | Three separate channel–fan systems |
| | M8 (2)     MF2 (3) | |
| | M11 (3)     MF3 (3) | |
| Debris slopes | UO1 UO2 UO3 U45 | Depositional sites |

gully-head slopes, with convergent rill networks feeding sediment into a main gully channel; gully-side slopes, planar slopes feeding sediment laterally into main gully channels; and stabilising slopes, based by a revegetating debris slope and therefore with little or no slope/stream coupling. Sixteen photo plots were established on eroding slopes.

Three categories of transport and depositional zone were identified (Table 12.1): ephemeral basal cones (Gully L), main gully channels feeding streamside fans or cones (Gully M), and debris slopes with no obvious connection to the stream channel (Gully U). On these, 13 photo plots were established, some of which yielded multiple point observations giving a total of 26, and a grand total of 42 observational sites altogether.

For each plot, sequential photographs were taken, on each occasion from a fixed point using the same camera lens setting. A5 prints were later analysed for surface changes and evidence of erosional/depositional regime and dominant process (see below).

The first plots were established in October 1987 and recorded again in April 1988 when most of the remaining plots were established. Since then observations have been maintained at approximately eight-week intervals. Preliminary results from a sample of the eroding sites over the period to 1989 have been presented elsewhere (Harvey & Calvo, 1991). This paper extends cover to the transportational/depositional zones over the four-year observational period to 1991. This period covers a wide range of weather types and includes two important stream floods and the snowy winter of 1990/91 (see Table 12.2).

## METHODOLOGY

The sequential photographs were analysed in a manner similar to that used by Harvey & Calvo (1991), Calvo & Harvey (in press) and not unlike that used by Suwa & Okuda

**Table 12.2** Rainfall sequence during the study period. (Data derived from Grayrigg, 5 km west of Carlingill)

| Period (dates to) | Weeks | Rainfall total (mm) | Days > 10 mm | Total (mm) at > 10 mm/ day | Max. day (mm) | Notes |
|---|---|---|---|---|---|---|
| 1 Sept. 87 | | | | | | |
| to 24 Oct. 87 | 7 | 353 | 13 | 245 | 33 | very wet |
| 12 April 88 | 23 | 762 | 35 | 437 | 41 | fairly wet, some frost, two minor snows |
| 12 July 88 | 12 | 238 | 8 | 134 | 27 | fairly dry |
| 15 Oct. 88 | 12 | 516 | 19 | 342 | 45 | wet, very little frost |
| 21 Dec. 88 | 9 | 174 | 7 | 95 | 18 | fairly dry, very little frost, one minor snow |
| 19 March 89 | 12 | 474 | 16 | 289 | 40 | fairly wet, little frost, some snow |
| 22 April 89 | 4 | 156 | 5 | 94 | 32 | fairly wet |
| 17 June 89 | 8 | 67 | 1 | 10 | 10 | very dry |
| 17 Aug. 89 | 8 | 175 | 7 | 123 | 33 | some heavy rains |
| 28 Sept. 89 | 5 | 73 | 1 | 20 | 20 | very dry |
| 20 Oct. 89 | 3 | 82 | 3 | 60 | 27 | — |
| 11 Jan. 90 | 11 | 379 | 16 | 258 | 46 | very wet, frost, some snow |
| 10 March 90 | 8 | 511 | 20 | 372 | 50 | very wet, snow |
| 19 April 90 | 5 | 76 | 1 | 13 | 13 | very dry |
| 8 June 90 | 7 | 129 | 3 | 51 | 20 | fairly dry |
| 31 Aug. 90 | 11 | 209 | 6 | 99 | 32 | fairly dry, some storms |
| 19 Oct. 90 | 7 | 187 | 6 | 106 | 22 | — |
| 29 Dec. 90 | 10 | 317 | 15 | 233 | 21 | wet, cold, heavy snow |
| 17 March 91 | 11 | 289 | 7 | 154 | 29 | very heavy snow, very cold |
| 9 May 91 | 7 | 154 | 4 | 101 | 42 | wet |

(1988). For the plots on eroding gully slopes two features were examined; changes in the rill network, and evidence for different processes in the inter-rill areas. For the rill network a composite index of rill development was derived by scoring each photograph, on the basis of pre-determined criteria, to a scale ranging 0–8 points for each of the following characteristics: rill clarity/continuity, depth of incision, rill extent/density, network complexity/bifurcation. The index for each photo, i.e. for each period at each site, was the mean of these point scores. Care was taken to cross-check within and between sites.

For the inter-rill areas two successive photographs of each plot were examined together. Each was subdivided into (normally) six units, and within each unit evidence for the dominant processes during the period covered by the two photographs was recorded in one of the following classes: erosion by surface wash, mass movement, erosion of stones, no change, deposition of stones, deposition of fine sediment. This yielded data on the relative importance of each surface process at each site for each period.

For the transport/depositional zones a similar technique was used, but classifying the changes/dominant processes individually as follows: on the main gully channel sites and basal fans on Gully M and the debris cones on Gully L; basal stream erosion (where appropriate), major/minor channel incision, transport of individual stones but no morphological change, no change (the latter two classes were later amalgamated), major/minor deposition; and on the debris slopes below Gully U, major/minor sedimentation by debris flow deposition, fine sediment deposition, stone movement, no change, major/minor changes by incision. Again these observations provided the basis for assessment of the importance of erosional/depositional changes at each site for each period.

## RESULTS

### Erosional Zones

The results for the rills index show three important trends (Figure 12.4): strong seasonality, some year to year variability, and contrasts related to slope type, reflecting coupling/network connectivity characteristics. Seasonality is very clear, reflecting a classic 'Perth Amboy' (Schumm, 1956b) type of rill development in summer and destruction by winter frosts, and similar to that previously identified at the neighbouring Grains Gill gully site (Harvey, 1974, 1987a). Rill index minima are reached normally in late winter, then development takes place until the following autumn, before degradation of the rill network by frost and snow occurs. Contrasts from year to year reflect weather variations, such as the generally slow growth of the network and lower maximum values of the index during and following the dry early summer of 1989 and the generally less than complete degradation of the rills during the mild almost frost-free winter of 1989/90. Clear contrasts are apparent between the various slope types. Rills are generally much better developed on the basal and gully-head sites and least well developed on the stabilising plots, where in some cases they disappear completely during winter. The seasonality is well marked in most cases but is clearest, with maximum synchroneity, on the gully-head plots.

For the inter-rill areas the individual site data have been amalgamated to give the relative importance of the various surface processes within each slope type (Figure 12.5). Again a strong seasonality, limited year to year variability and clear contrasts between slope types are evident. In winter, erosion is dominated by mass movement and stone fall, and is more active than in summer, when it is dominated by surface wash. Of the limited year to year variability the most obvious trend appears to be a slight reduction in erosional area and increase in deposition, evident on all slope types except the basal plots. Although the results relate to spatial extent of erosion rather than directly to erosion rates, the contrasts between the slope types clearly indicate higher rates of erosion on the basal plots, followed by gully-head, gully-side and finally by stabilising plots.

### Transport and Depositional Zones

The results from the plots in the erosional zones confirm what was expected; both spatially, with rapid erosion indicated on basal and gully-head plots; and temporally, with higher winter than summer erosion rates indicated. Sediment derived from this zone is fed into the transport and depositional zones. On Gully L it is fed either directly into the main stream or into ephemeral debris cones at the base of the slopes. On Gully M it is fed into the linear gully channels and hence to the streamside fans, and on Gully U it is fed onto the debris slopes. In all cases but Gully L, where the processes are more directly related to stream activity, subsequent sediment movement depends on the power of the transport mechanism downgully, to move the sediment supplied from the hillslopes. This is equivalent to the relationship between critical and actual power as defined by Bull (1979) in the context of larger stream systems, and here depends primarily on the relative supplies of water and sediment from the gully slopes.

Periods of high runoff production would be expressed by marked increases in the rill index, and perhaps by a dominance of wash over other inter-rill processes. Periods of high relative sediment production would be expressed by degradation of the rills and important stone movement or mass movement from the slopes. Table 12.3 represents an attempt to classify the observation periods on the basis of importance for runoff or sediment supply to the transport and depositional zones. As would be expected there is a general pattern of seasonality with wet summer periods (see Table 12.2) indicating major runoff production, and wet, especially snowy winter periods indicating major sediment production. There is general accordance with the rainfall totals (Table 12.2). During dry periods there is little indication of either water or sediment production. How the transport zones respond depends not only on water and sediment production on the gullies but also on basal stream activity, though by definition, the style of this influence varies between the three gullies. During the study period there were several periods with stream floods, notably March 1990 which created major erosional changes at the bases of Gullies M and L (see below and Figure 12.6). This event had similar effects to those described earlier on Grains Gill gullies (Harvey, 1987a), rejuvenating the base of the gullied slopes and adding massive amounts of coarse sediment to the stream.

Erosion and deposition within the transport/depositional zones, over the study period is summarised on Figure 12.7. On Gully L the behaviour is simple. Major periods of slope erosion feed sediment directly to the debris apron/basal cone zone (sites LC 2–5, Figure 12.7). Widespread removal took place once during the study period, as the result of the major flood in March 1990, with minor trimming and removal by smaller floods in October and December 1990. A similar cyclicity of cone aggradation and removal as that observed at Grains Gill (Harvey et al., 1979) appears to characterise this site. On Grains Gill such events have a return period of c. 2–6 years (Harvey et al., 1979), but show a range of from c. 2 years for the flushing of linear gully channels, to c. 3 years for the scouring of debris aprons, to c. 6 years for the removal of large debris cones (Harvey, 1992). Gully L is similar to the debris aprons along Grains Gill, and major scour of the debris cone/apron zone once every c. 3 years accords with prior field knowledge of this site. Major cone-removing floods have occurred on a number of occasions since 1970.

On Gully U (Figure 12.7) the pattern is even simpler. Without the complication created by basal removal, sedimentation is a simple response to sediment production on the gullied slopes. Deposition is quite frequent on the upper part of the debris slopes (site UO1) but much rarer downslope (site UO3). The style of deposition varies, with debris flow deposition important in top and mid-slope locations with occasional flushing out

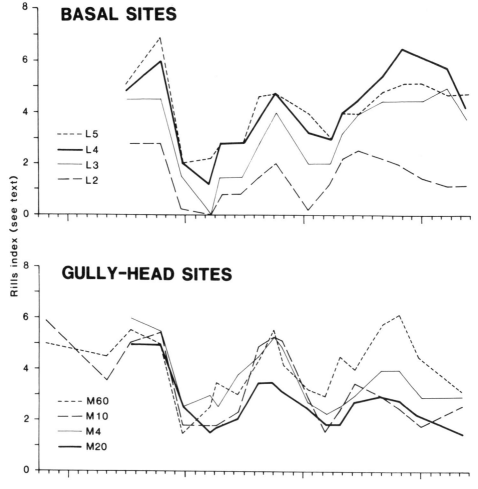

**Figure 12.4** Rills index for 1987–1991, for each site, grouped by slope type. For explanation of the rills index see text. Data for plot *M*1, a stabilising site, is omitted because of the virtual absence of any rill development

of fines to basal locations. Only once during the study period, in response to a major snowmelt in March 1991, did deposition by debris flow take place over the whole slope.

The pattern on Gully M is more complex (Figure 12.7). Three separate channels feed fans MF1–3. The results of monitoring the feeder channels and fans show some degree of consistency but also within-system and between-system variability. Processes within these gully systems are influenced not only by water and sediment supply and by basal stream activity, but also by local factors. In general, during the major periods of sediment production on the hillslopes (e.g. December 1988, March 1989, March 1991), sediment is fed to the gully channels and causes aggradation there. Only occasionally is it fed down the channel system to the fan to cause synchronous sedimentation throughout the system. Quite often, the behaviour of the two zones is out of phase. The response to runoff production can be variable. In some cases (e.g. July 1988, August 1989, MF1

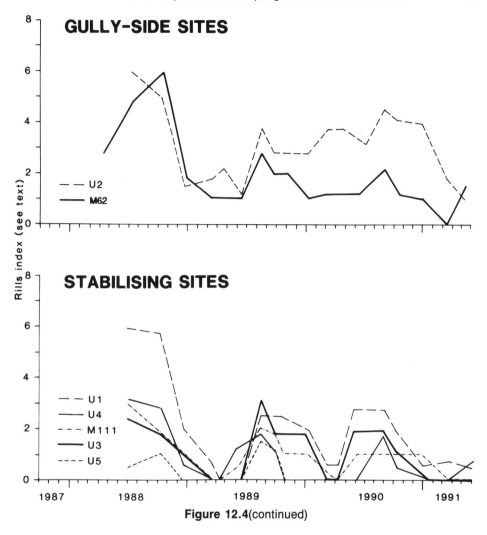

**Figure 12.4**(continued)

system) the response in mid-gully is channel scour, but this does not persist downchannel to the fan. On other occasions the general response to periods of runoff production tends to be minor within-channel sedimentation. On some occasions, associated with little apparent sediment movement from the gully slopes (e.g. April and June 1990), minor but fairly widespread sedimentation occurred within the gully channels. Such events may represent delayed responses to earlier sediment-producing events.

On several occasions stream floods caused erosion of the fan toe zones, resulting in trenching of the channels through the fans. This occurred on all three fans during the flood of March 1990, but trenching was limited to the fans themselves, while sedimentation occurred further up the gully channels. Even on MF1 system, where trenching in the major gully channel occurred, two trenching zones were separated by a zone of sedimentation. On two other occasions (October 1988, December 1991) basal incision of fan MF3 occurred, with only limited trenching elsewhere. There is clearly a pattern of discontinuity, related to storage within the gully channels.

Percentage observations

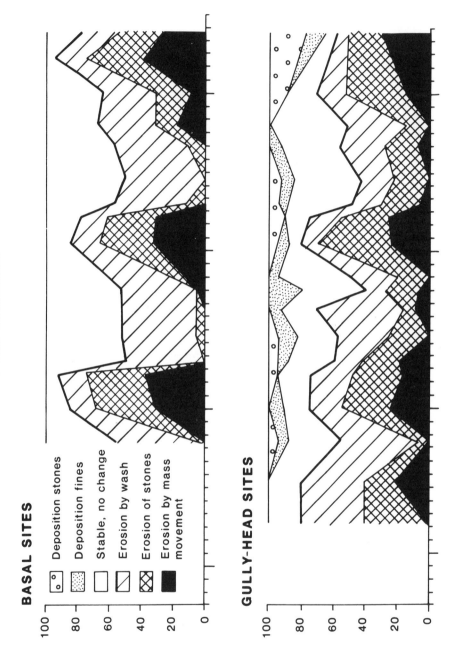

**BASAL SITES**

- Deposition stones
- Deposition fines
- Stable, no change
- Erosion by wash
- Erosion of stones
- Erosion by mass movement

**GULLY-HEAD SITES**

259

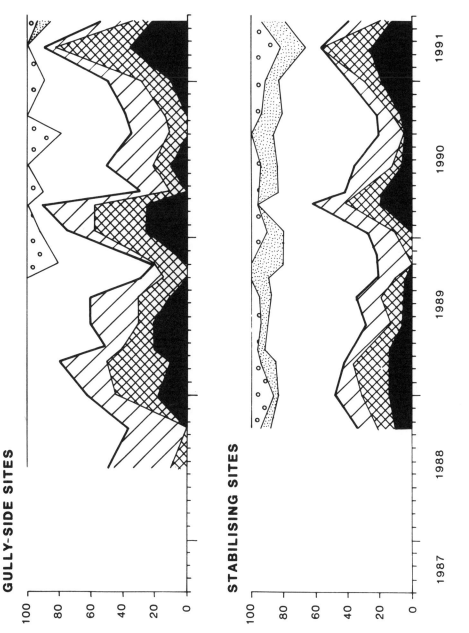

**Figure 12.5** Relative importance of the various surface processes in inter-rill areas 1988–1991, by slope type

**Table 12.3** Classification of observation periods on the basis of major/minor runoff and sediment supply on the hillslopes. Parentheses indicate minor importance

| Period dates (to) | Data availability (no. of sites) | Rills. No. of sites showing major (minor) | | Inter-rill areas. No. of sites showing major (minor) | | | Periods important for | |
|---|---|---|---|---|---|---|---|---|
| | | Rill development | Rill degradation | Wash | Stone movement | Mass movement | Runoff | Sediment |
| 24 Oct. 87 | 2 | 2 – | – | 2 – | – | – | (+?) | – |
| 12 April 88 | 3* | – | – (1) | – | – | – | + | ? |
| 12 July 88 | 15** | 10 (4) | – (1) | – (1) | – (1) | – | + | – |
| 15 Oct. 88 | 15*** | 5 (10) | – | 3 (5) | – | – | + | – |
| 21 Dec. 88 | 15 | – | 6 (5) | – (2) | 2 (7) | 2 (6) | – | + |
| 19 March 89 | 15 | – | 7 (7) | – (4) | 3 (4) | 2 (5) | – | + |
| 22 April 89 | 15 | – | 1 (12) | – (5) | – (2) | – (2) | – | (+) |
| 17 June 89 | 15 | – (3) | – (9) | – (7) | – (2) | – (1) | – | – |
| 17 Aug. 89 | 15 | 11 (4) | – | 2 (6) | – (1) | – | + | – |
| 28 Sept. 89 | 15 | 6 (9) | – | – (4) | – (1) | – | + | – |
| 20 Oct. 89 | 7 | – (4) | – | – (2) | – | – | – | – |
| 11 Jan. 90 | 15 | – (1) | 2 (6) | – (2) | 4 (3) | 3 (3) | – | + |
| 10 March 90 | 15 | 1 (1) | 6 (6) | – (6) | 2 (6) | 1 (5) | (+) | + |
| 19 April 90 | 15 | – (6) | 1 (8) | – (5) | – (1) | – | – | – |
| 8 June 90 | 15 | 3 (6) | – (4) | – (5) | – | – | (+) | – |
| 31 Aug. 90 | 15 | 7 (6) | – (1) | – (6) | – (1) | – | + | – |
| 19 Oct. 90 | 15 | 2 (9) | – (3) | – (6) | – (2) | – (2) | (+) | – |
| 29 Dec. 90 | 13 | – (5) | 2 (4) | – (6) | 3 (2) | – (3) | – | (+) |
| 17 March 91 | 11 | 1 (2) | 3 (5) | – (2) | 6 (4) | 5 (4) | (+) | + |
| 9 May 91 | 15 | – (3) | 2 (9) | – (5) | – (7) | 1 (7) | – | (+) |

*No data for inter-rill sites.
**Inter-rill data available for only two plots.
***Inter-rill data available for only eight plots.

**Figure 12.6**  Basal scour on Gully L, March 1990

The aggradation and dissectional behaviour of the channel and fan systems on Gully M represents an interaction between water and sediment supplied by the gully systems and basal erosion induced by stream floods. This general picture is complicated by several factors. In many cases feeder channels and fans behave differently. Secondly, there are clear differences between the three gully systems, which may result from different water:sediment mix fed from the gully slopes, different basal behaviour, or random within-channel differences related to individual slugs of sediment moving only short distances but locally producing a juxtaposition of scouring and filling sites. Finally, there are complications related to timelags between sediment supply and movement through the system. The data on Figure 12.7 accord in a general sense with the magnitude/frequency relations observed on Grains Gill (Harvey, 1992). Sediment flushing within the linear gully channels occurs fairly frequently, albeit with discontinuity, basal scour of the cones occurs less frequently. Not once since 1970 has complete cone removal taken place.

## DISCUSSION

Two styles of process interaction have been identified and monitored on the Carlingill gullies. These are (a) on-slope process interactions and (b) interactions involving coupling between the gully slopes and the basal stream. These two types are themselves related and appear to progressively change during gully evolution.

The on-slope process interactions involve mass movement and runoff-related processes and show a clear seasonality. Both the style of interaction and the seasonality vary with

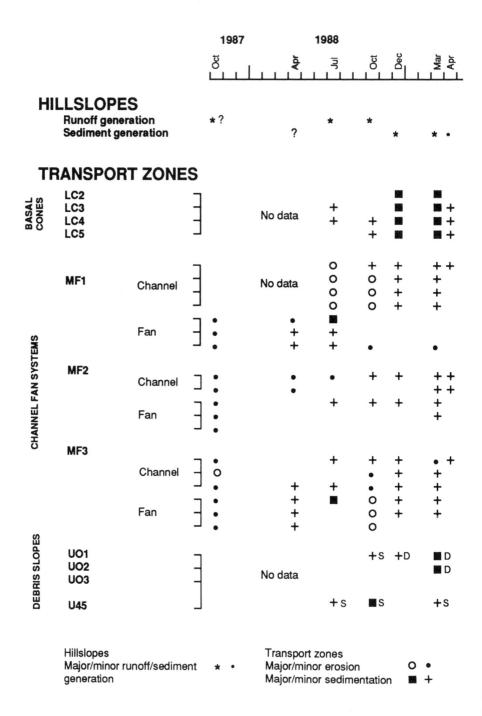

**Figure 12.7** Erosion and deposition in transport and depositional zones, 1987–1991

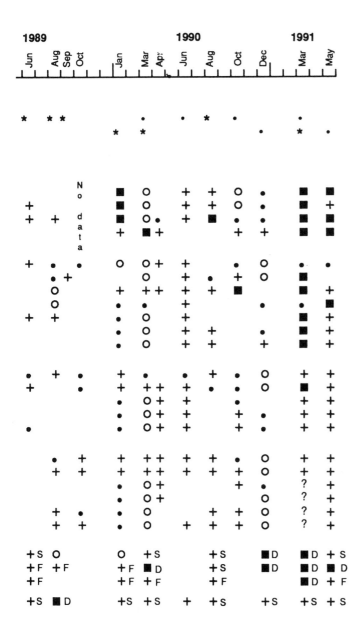

For Gully U only
Debris flows             D
Stone deposition         S
Fine sediment deposition F

**Figure 12.7**(continued)

**Table 12.4**  Summary results of rainfall simulation experiments, Carlingill gullies (Alexander and Harvey, unpublished data). Plot size 0.24 m², rainfall applied 40–60 mm/hr for 20 minutes, plot slopes 39–52°

| Plot type | Bare | Partial veg. | Well veg. |
|---|---|---|---|
| Vegetation cover | < 5% | 15–30% | > 50% |
| Number of runs | 4 | 3 | 3 |
| Mean runoff rate (ml/min) | 185 | 170 | 135 |
| Mean sediment concentration (g/l) | 41 | 14 | 1.1 |
| Mean erosion rate (g/min) | 6.9 | 2.0 | 0.13 |

the type of slope. Similarly, erosion rates could be expected to differ considerably between the different slope types. In addition to the evidence presented here, sediment trap data from Grains Gill gullies, show greater winter dominance in sediment production on unrilled as opposed to rilled slopes (Harvey, 1987b), and rainfall simulation experiments carried out on Carlingill gullies suggest much lower erosion rates from partially vegetated stabilising slopes, than from bare and rilled slopes (Alexander & Harvey, unpublished data and Table 12.4). Strong and dynamic process interactions and higher erosion rates appear to characterise basal and gully-head sites, rather than gully-side or stabilising sites. Local base level appears to be important, in that active incision prevents the basal accumulation of sediment and subsequent slope stabilisation.

Coupling between slope and stream varies between the three gullies. Coupling on Gully L is immediate, with no within-gully sediment storage. Sediment is fed directly to streamside cones which are periodically removed by stream floods. As long as the frequency of stream erosion is similar to that observed on Grains Gill, i.e. once every c. 2–6 years, stabilisation of the gully slopes is unlikely to occur. On Gully M, the coupling is buffered by the gully channel/fan system. Although the fans are periodically scoured by the stream, causing local trenching, rarely does this incision proceed far up the gully channels. Not once since 1970 have the cones been completely eroded. Continued erosion on the gully-head slopes is related to on-slope processes rather than to gully-channel incision; however, it is permitted only by the transmission of the sediment down the main gully channel. On Gully U all coupling with the main stream has been severed since the channel migrated away, and although headward retreat of the gully has continued, this appears to be part of a long-term stabilisation trend.

The style of coupling changes with gully evolution, as does the distribution of the various types of gully slope. A photographic survey carried out in 1970 allows a comparison of slope type over a 20 year period to be made on the three gullies (Figure 12.8). An older gully just to the east of Gully L has become revegetated and stabilised since 1970. Otherwise on Gully L, slope recession, coupled with repeated removal of debris cones, has allowed the continued development of basal slopes supplying sediment directly to streamside locations. Two small re-entrant gully heads have also developed. On Gully M the gully-head slopes have cut back into the hillslope, but are reduced in size. There has been considerable vegetation colonisation of former gully-side and stabilising slopes (Figure 12.9). This process is even more dramatic on Gully U, where large areas of what were active, but stabilising slopes in 1970, are now vegetated and stable. There has also been considerable vegetation growth on the debris slope. What appears to be taking place is natural stabilisation by vegetation colonisation, when erosion rates fall below a critical amount. Although so far we know little of the rates of vegetation colonisation, some idea of its rapidity can be gained by simple counts of the number of identifiable plant clusters within a selection of the photo plots (Table 12.5). In each case an increase in plant cover is recorded between 1988 and 1990, decreasing slightly after the heavy winter of 1990/91.

Process dominance, process interactions, and the style of coupling, progressively change during gully evolution (Figure 12.10). The gullies are basally initiated, by stream erosion triggering a small-scale slope failure (Harvey, 1992). If early stabilisation does not take place, a streamside scar of Gully L type develops. At this stage, erosion rates on the slopes are high, involving interactions between surface erosion and mass-movement processes. Basal processes are also active and coupling is strong, allowing little or no sediment accumulation. As scars develop into linear gullies of type M, overall rates of erosion appear to decline, though there is a strongly seasonal surface erosion/mass movement interaction. Basal removal becomes less frequent and the coupling becomes weaker, allowing some buildup of sediment within the system. Erosion rates continue to decline, and the style of interaction changes. Basal coupling becomes much weaker, more sedimentation occurs within the system, and vegetation colonisation may begin to stabilise the slopes. If the coupling with the basal stream becomes severed, either by stream migration or by basal stabilisation (gully type U), overall stabilisation of the system takes place. During this sequence, slope erosion rates rapidly decline, and process interactions are further modified. Within-system sedimentation initially accelerates, but later declines with the ultimate decline in erosion rates, and vegetation-induced stabilisation accelerates. This process has been an important mechanism in the evolution of the steep valley-side slopes, cut into the Pleistocene glacial and periglacial sediments, adjacent to Carlingill. These are characterised by at least two generations of now-stabilised gully systems Harvey, 1992).

There are two important implications of this simple schematic model; one relating to experimental work in geomorphology and the estimation of erosion rates; the other relating to timescales of gully evolution and slope development. In the first case, field experiments designed to measure erosion rates must take into account variability induced by process interactions, both on-slope and involving slope/stream coupling. Monitoring needs to be carried out over sufficiently long periods to account not only for seasonality, but also for any cyclicity associated with basal storage/flushing relationships, and for longer term progressive trends in landform evolution.

In the second case, and specifically related to gully development in environments similar to that of the Howgill Fells, an evolution of gully form can be identified. This involves a progressively weaker gully/stream coupling (i.e. from Gully L to Gully M), which through decoupling (Gully U), may lead to ultimate stabilisation. The implications are that in any particular context, gully development will have finite limits, and take place over finite timescales. A model has been developed for the Howgill gullies, based on the Grains Gill site (Harvey, 1992). This suggests that, under present environmental conditions, gullies grow to a maximum length of something over 40 m, over a total period of 150–200 years, before stabilisation takes place. The Carlingill gullies, dealt with in this paper, confirm that general model. Gully L is less than c. 40 years old, and is still developing rapidly. Gullies M and U are considerably older. The largest channel on Gully M appears to have reached something like its maximum size, and stabilisation appears to have begun. Gully U decoupled c. 60 years ago,

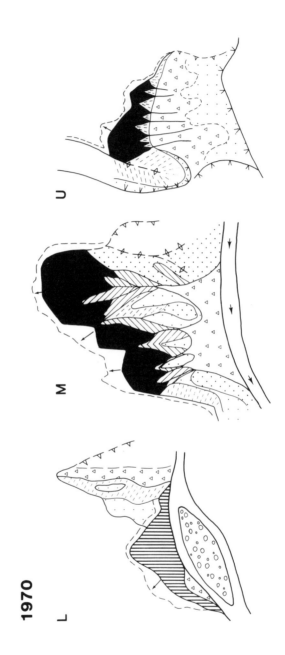

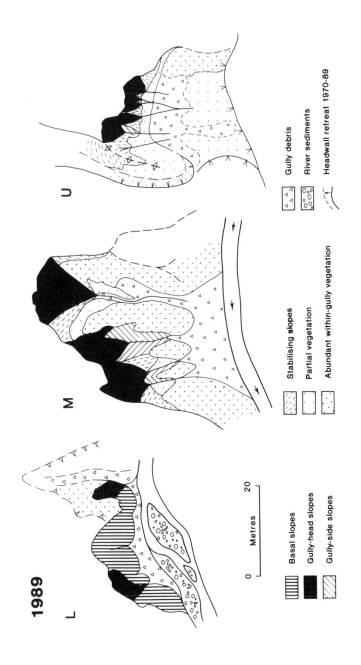

**Figure 12.8** Longer term progressive change on Carlingill gully slopes, 1970–1989: status of slope and surface types (Break of slope symbols as on Fig. 12.1)

**Figure 12.9**  Gully M, 1970 and 1988. Note: 1, headwall recession, and 2, vegetation colonisation and stabilisation of previously active slopes

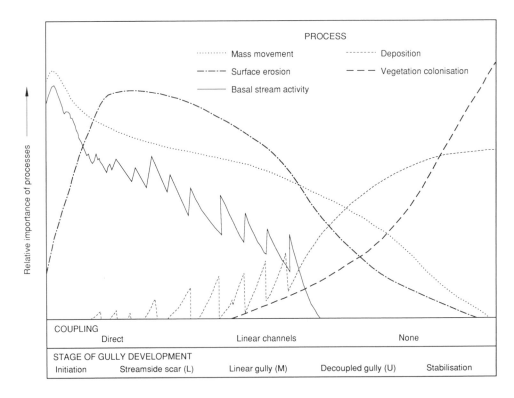

**Figure 12.10**  Schematic model to illustrate the changing processes, process interactions, and style of coupling, during gully evolution

**Table 12.5** Vegetation growth in selected plots 1988–1991: number of individual plant clusters (mostly grasses)

| Plot | U1 | U2 | U3 | U4 | U5 | M4 | M20 |
|------|----|----|----|----|----|----|-----|
| July 1988 | 13 | 6 | 8 | 7 | 3 | 4 | 8 |
| June 1989 | 36 | 12 | 30 | 14 | 11 | 31 | 48 |
| June 1990 | 60 | 26 | 65 | 21 | 13 | 42 | 55 |
| May 1991 | 45 | 12 | 55 | 18 | 26 | 35 | 68 |

and, given the rates of change since 1970, now appears to be rapidly reaching total stabilisation.

The coupling between on-slope, gully-channel and basal stream processes, appears to be the major control of process interactions on eroding gullies in this environment. Understanding the coupling relationships is fundamental to understanding process/form relationships during longer term landform evolution.

## ACNOWLEDGEMENTS

I am grateful to the drawing office and photographic sections of the Department of Geography, University of Liverpool, for producing the diagrams, and to Dr Adolfo Calvo of the University of Valencia for assistance during the initial establishment of the project.

## REFERENCES

Beaty, C. R. (1959). Slope retreat by gullying. *Bulletin Geological Society of America*, **70**, 1479–1482.

Brunsden, D. and Thornes, J. B. (1979). Landscape sensitivity and change. *Transactions Institute of British Geographers*, New Ser. **4**, 463–484.

Bull, W. B. (1979). Threshold of critical power in streams. *Bulletin Geological Society of America*, **90**, 453–464.

Calvo, A. and Harvey, A. M. (1993). Morphology and development of selected badlands in southeast Spain. *Earth Surface Processes and Landforms* (in press).

Calvo-Cases, A., Harvey, A. M. and Paya Serrano, J. (1991). Process interactions and badland development in S.E. Spain. In Sala, M., Rubio, J. L. and Garcia-Ruiz, J. M. (eds), *Soil Erosion Studies in Spain*, Geoforma Ediciones, Logrono, Spain, pp. 75–90.

Faulkner, H. (1988). Gully evolution in response to both snowmelt and flash flood erosion, Western Colorado. In Gardiner, V. (ed.), *International Geomorphology 1986 Vol. I*, Wiley, Chichester, pp. 947–969.

Harvey, A. M. (1974). Gully erosion and sediment yield in the Howgill Fells, Westmorland. In Gregory, K. J. and Walling, D. E. (eds), *Fluvial Processes in Instrumented Watersheds*, Institute of British Geographers, Special Publication No. 6, pp. 45–58.

Harvey, A. M. (1977). Event frequency in sediment production and channel change. In Gregory, K. J. (ed.), *River Channel Changes*, Wiley, Chichester, pp. 301–315.

Harvey, A. M. (1987a). Seasonality of processes on eroding gullies, a twelve year record of erosion rates. In Goddard, A. and Rapp, A. (eds), *Processes et Mesure de l'Erosion*, CNRS, Paris, pp. 439–454.

Harvey, A. M. (1987b). Sediment supply to upland streams, influence on channel adjustment. In Thorne, C. R., Bathurst, J. C. and Hey, R. W. (eds), *Sediment Transport in Gravel-Bed Rivers*, Wiley, Chichester, pp. 121–150.

Harvey, A. M. (1988). Seasonal and longer term development of rill systems on eroding slopes (Abstract). In International Geographical Union COMTAG Symposium; Amsterdam/Leuven 1988, Abstracts of Papers.

Harvey, A. M. (1989). The occurrence and role of arid-region alluvial fans. In Thomas, D. (ed.), *Arid-Region Geomorphology*, Belhaven, London, pp. 136–158.

Harvey, A. M. (1990). Gully development in the Howgill Fells: the role of process interactions (Abstract). *IBG Annual Conference, January 1990, Glasgow*, Abstracts of Papers, 13.

Harvey, A. M. (1992). Process interactions, temporal scales and the development of hillslope gully systems: Howgill Fells, northwest England. *Geomorphology*, 5, 323–344.

Harvey, A. M. and Calvo, A. (1991). Process interactions and rill development on badland and gully slopes. *Zeitschrift für Geomorphologie*, Suppl. 83, 175–194.

Harvey, A. M., Hitchcock, D. H. and Hughes, D. J. (1979). Event frequency and morphological adjustment of fluvial systems in upland Britain. In Rhodes, D. D. and Williams, G. P. (eds), *Adjustments of the Fluvial System*, Kendall/Hunt Publ. Co., Dubuque, Iowa, pp. 139–167.

Harvey, A. M., Alexander, R. W. and James, P. A. (1984). Lichens, soil development and the age of Holocene valley floor landforms, Howgill Fells, Cumbria. *Geografisca Annaler*, 66A, 353–366.

Macklin, M. G., Rumsby, B. T. and Newson, M. D. (1992). Historical floods and vertical accretion of fine-grained alluvium in the lower Tyne valley, northeast England. In Billi, P., Hey, R. D., Thorne, C. R. and Tacconi, P. (eds), *Dynamics of Gravel-Bed Rivers*, Wiley, Chichester, pp. 573–589.

Newson, M. D. (1992). Geomorphic thresholds in gravel-bed rivers – Refinement for an era of environmental change. In Billi, P., Hey, R. D., Thorne, C. R. and Tacconi, P. (eds), *Dynamics of Gravel-Bed Rivers*, Wiley, Chichester, pp. 3–20.

Schumm, S. A. (1956a). The role of creep and rainwash on the retreat of badland slopes. *American Journal of Science*, 254, 693–700.

Schumm, S. A. (1956b). Evolution of drainage systems and slopes in badlands at Perth Amboy, New Jersey. *Bulletin Geological Society of America*, 67, 597–646.

Schumm, S. A. (1964). Seasonal variations in erosion rates on hillslopes in Western Colorado. *Zeitschrift für Geomorphologie*, Suppl. 5, 215–238.

Suwa, H. and Okuda, S. (1988). Seasonal variations of erosional processes in the Kamikamihori Valley of Mt. Yakedale, Northern Japan. In Harvey, A. M. and Sala, M. (eds), *Geomorphic Processes in Environments with Strong Seasonal Contrasts Vol. II Geomorphic Systems*, Catena Suppl. 13, 61–77.

Wells, N. A., Andriamihaja, B. and Solo Rakotovololona, H. F. (1991). Patterns of development of Lavaka, Madagascar's unusual gullies. *Earth Surface Processes and Landforms*, 16, 189–206.

Wells, S. G. and Gutierrez, A. A. (1982). Quaternary evolution of badlands in the southeastern Colorado Plateau, USA. In Bryan, R. B. and Yair, A. (eds), *Badland Geomorphology and Piping*, Geobooks, Norwich, pp. 239–258.

# 13 Hydrogeomorphology Modelling with a Physically Based River Basin Evolution Model

**GARRY WILLGOOSE**
*Department of Civil Engineering and Surveying, University of Newcastle, Australia*

**RAFAEL L. BRAS**
*Department of Civil and Environmental Engineering, Massachusetts Institute of Technology, USA*

and

**IGNACIO RODRIGUEZ-ITURBE**
*Instituto International de Estudios Avanzados, Caracas, Venezuela*

## ABSTRACT

A model of catchment evolution that explicitly describes the evolution of the catchment will be discussed. Catchment elevations are modelled from sediment transport continuity within the basin. The evolution of the channel network is modelled by a threshold mechanism which is based on the runoff mechanisms of the hillslopes; e.g. when velocity exceeds a threshold a channel is formed. These elevations interact with the channels through the assumption that different erosion processes occur in the hillslopes and channels. The channels interact with the elevations through the threshold dependence on hillslope properties such as the contributing area and slope. The runoff processes modelled include both the Hortonian and subsurface saturation mechanisms. The structure of the model will be discussed and some of its assumptions highlighted. Results from simulations using the computer model based on this conceptualisation, SIBERIA, will be shown and their implications discussed. In particular, we will discuss (1) channel network thresholds, particularly addressing the significance of the channel head and (2) a physical explanation, that follows from the continuity of sediment transport in catchments, for an observed correlation between area and slope.

## INTRODUCTION

The catchment form determines the flood and erosion response of the catchment and the flood and erosion response, over geological time, determines the catchment form. Such is a fundamental truism of fluvial geomorphology and hydrology. Yet, because of the difficulty of the problem, until recent years, research into catchment form and flood response has proceeded independently. For many years hydrologists have fitted

*Process Models and Theoretical Geomorphology.* Edited by M. J. Kirkby
©1994 John Wiley & Sons Ltd

hydrograph parameters to geomorphological statistics such as slopes and Strahler ratios (reported in Strahler, 1964; Rodriguez-Iturbe & Valdes, 1979). Geomorphologists have likewise fitted geomorphology statistics without regard for the dominant hydrological processes that have formed the landscape (Strahler, 1964; Shreve, 1966; Mock, 1971) though this has not always been the case (Gilbert, 1909; Horton, 1945; Schumm, 1956). Recently there has been a resurgence of interest in the interactions between catchment processes, form and temporal changes (Kirkby, 1971, 1987; Ahnert, 1976; Dunne, 1980; Huggett, 1988). The main stumbling block to understanding these interactions are the long timescales associated with the evolution of catchments; the difficulty of observing changes; and the inability to categorically attribute differences between catchments to differences in age or process because of spatial and temporal heterogeneity. Physically based computer models of catchment development are important tools in the understanding of the interactions between hydrological process and response, primarily because of their ability to explore temporal trends and sensitivity of physical inputs, such as erodibility, tectonic uplift and runoff, to change, without the difficulties of identification and generalisation associated with the heterogeneity encountered in field studies.

This paper describes the development and application of a process response model of the erosional development of catchments and their channel networks. A crucial component of this model is that it explicitly incorporates the interaction between the hillslopes and the growing channel network based on physically observable mechanisms. The elevations within the catchment – both hillslope and channel – are simulated by a mass transport continuity equation applied over geologic time. Mass transport processes considered include fluvial sediment transport, such as modelled by the Einstein–Brown equation, and mass movement mechanisms such as creep, rainsplash or landslide. An explicit differentiation between the processes that act on the hillslopes and in the channels is made. The growth of the channel network is governed by a physically based threshold mechanism, where if a function (called the channel initiation function) is greater than some predetermined threshold then channel head advance occurs. The channel initiation function is primarily dependent on the discharge and slope at that point, and the channel initiation threshold is dependent on the resistance of the catchment of channelisation. Channel growth is thus governed by the hillslope form and processes that occur upstream of the channel head, but in a way that can be independent of channel growth stability arguments (Smith and Bretherton, 1972). The elevations on the hillslopes and the growing channel interact through the different transport processes in each regime and the preferred drainage to the channels that results. The interaction of these processes produces the long-term form of the catchment. The preferential erosion in the channels results in the familiar pattern of hills and valleys with hillslope flow being towards the channel network in the bottoms of the valleys.

A computer implementation of this conceptual model of catchment evolution (SIBERIA) has been used to generate synthetic catchments and to examine their sensitivity to changes in exogenous forces, such as temporal variation in tectonic uplift. The numerical model allows the examination of temporal changes in both the channel network form and the elevation properties of the basin and allows the development of a conceptual understanding of the complicated large-scale interactions between the channels and the hillslopes that occur in the field. One important issue that can be examined is the effect of variation in the tectonic uplift and the implications for the attainment of dynamic equilibrium. There have been numerous arguments through the years about dynamic equilibrium and the form of catchments when they reach maturity (Davis (1899) and Penck (1924) in Chorley *et al.*, 1984; Gilbert, 1909; Strahler, 1964; Scheidegger, 1970). The determination of the tectonic history of field catchments and the spatial distribution of mass transport are extremely difficult and prone to error so that computer modelling can be a useful tool for examining the effect of variations in these physical processes on catchment characteristics (e.g. Ahnert, 1976).

In particular, the model here allows us to explicitly define the conditions for 'dynamic equilibrium', a term interpreted in different ways by different authors. We will show by examination of the governing equations that dynamic equilibrium is not possible in the absence of tectonic uplift. If tectonic uplift varies in time, dynamic equilibrium can only occur in an approximate sense and then only if the timescales of changes in the tectonic uplift are longer than the timescales of elevation adjustment.

## GEOMORPHOLOGY MODEL

The geomorphology model to be used here has two main elements. The first element is a model of elevation variation and the second is where the channels are formed in the catchment. The channels develop in response to changes in the elevation and, in turn, the elevations change in response to the channels.

The first component of the model, the elevation within the catchment, is simulated by a mass transport continuity equation applied over geologic time. If more material enters a region than leaves it then the elevations rise, and vice versa. The mass transport processes considered include both fluvial sediment transport and a conceptualisation of diffusive mass movement mechanisms such as creep, rainsplash and landslide. The model averages these processes in time so that the elevations (and the channel network) are indicative of the average with time of the full range of erosion events; the elevations simulated are average elevations with time. The model explicitly differentiates between the transport processes that act on the hillslope and in the channel.

The model's second component, the channel network, is simulated by an equation that initiates the advance of the channel heads into the surrounding hillslopes. Catchments start with an initial pattern of channelisation, or no channelisation at all, and channel head advance occurs when a channel initiation function, non-linearly dependent on the local slope and discharge, exceeds a threshold, characteristic of the landscape. Conceptually this threshold can represent overland flow velocity or shear stress, subsurface flow criteria or criteria based on local landsliding at the channel head.

The first component of the model, the governing equation of the elevations in the catchment model, is expressed as

$$\frac{\partial z}{\partial t} = c_0 + \frac{\boldsymbol{\nabla} \cdot \boldsymbol{q}_s}{\rho_s (1-n)} + D \left( \frac{\partial^2 z}{\partial x^2} + \frac{\partial^2 z}{\partial y^2} \right) \tag{1}$$

where $x$ and $y$ = horizontal directions, $z$ = elevation, $t$ = time, $c_0$ = tectonic uplift, $\boldsymbol{q}_s$ = vector sediment transport per unit width (mass/time), $\rho_s$ = density of the sediment, $n$ = porosity of the sediment, and $D$ = diffusivity.

Variables whose variation in space and time is dependent on the form of the catchment, and thus change as the simulated catchment evolves in time, are highlighted here in bold.

All other parameters, though they may vary in space and time, are considered independent of the evolving form of the catchment. The behaviour of the equation will not be discussed in detail here as it has been dealt with adequately elsewhere (Willgoose et al., 1991a, b, c, d). However, we will discuss those features that impinge on later sections.

The differential equation for elevation, equation (1), is a continuity equation in space for sediment transport. It is an average equation that models the average sediment transport over many erosion events to give the average elevations with time. The first term in the elevation equation is the rate of tectonic uplift (positive upwards). This term may be time varying. The third term in the elevation equation represents diffusive mechanisms occurring in certain mass transport processes, such as creep, rainsplash and landsliding (Culling, 1963; Dunne, 1980; Andrews & Bucknam, 1987). The rate of these processes is governed by the diffusivity $D$. Both the diffusivity and tectonic uplift may vary over the catchment but are not dependent on the form of the catchment. Other workers (e.g. Ahnert, 1976) have used more sophisticated diffusive processes in models of hillslope evolution. Their processes (e.g. viscous and plastic flow) were spatially variable and dependent on soil depth. In this model we do not model the chemical and physical processes associated with weathering and soil formation. We currently do not have the capability to model the spatial and temporal variation of soil and regolith depth. Accordingly we cannot model these processes that depend on soil depth, though we believe the spatial variability of these processes to be of second order importance.

The sediment transport process, $q_s$, modelled by the second term in equation (1), can be parameterised in any way that is believed to reflect the processes occurring in the catchment. Willgoose et al. (1989) and others (e.g. Kirkby, 1971) suggest that a realistic formulation is

$$q_s = \mathrm{f}(Y)Tq^{m_1}S^{n_1} \tag{2}$$

where $q_s$ = sediment transport per unit width (mass/time), $q$ = discharge per unit width, $S$ = slope in the steepest downstream direction, $\mathrm{f}(Y)$ = a sediment transport coefficient dependent on the pattern of channelisation (discussed below), $T$ = a function dependent of the runoff process modelled (discussed below), and $m_1, n_1$ = sediment transport coefficients.

This fluvial sediment transport term is one that has been commonly used by geomorphologists and hydrologists (Kirkby, 1971; Smith & Bretherton, 1972; Julien & Simons, 1985; Moore & Burch, 1986) to represent a transport-limited process. It can be directly related to generally accepted instantaneous sediment transport physics, such as Einstein–Brown, by averaging over the range of flood events. Briefly, when modelling the instantaneous sediment transport rate the appropriate discharge to use is the instantaneous discharge at that time. However, here the equation is used to model the mean sediment transport, so the appropriate discharge to use is the mean peak discharge derived from a frequency analysis of runoff events (Willgoose et al., 1989). Note that we describe as fluvial erosion any sediment transport process that results from surface runoff, whether it be on the hillslope, as sheet or rill flow, or in the channels.

The function $T$ indicates what *proportion of storms* saturate that point and thus generate surface runoff and is a conceptualisation of the subsurface saturation runoff generation mechanism. It is only during those storms when surface runoff is generated that fluvial erosion occurs. The calculation of this parameter is discussed in further detail below. Briefly, for Hortonian runoff it may be assumed that $T = 1$. For subsurface saturation generated runoff, smaller storms saturate a smaller proportion of the catchment than do larger storms. Thus for subsurface saturation runoff $T$ is less than 1 and largest in those parts of the catchment saturated most frequently (Figure 13.1b).

The $T$ factor is an adjustment on the sediment transport rate to reflect what proportion of storms result in saturation at that point and thus cause overland flow and fluvial sediment transport. If the rainfall rate during the storm triggers saturation at that point then fluvial erosion is considered to occur at that point. If saturation does not occur then overland fluvial erosion cannot occur at that point. Averaging of the overland fluvial sediment transport then occurs over the population of storms that trigger saturation.

The deterministic saturation criteria used to derive $T$ (Beven and Kirkby, 1979) is

$$\beta_3 > \beta_6 \ln \left[ \frac{MS}{A} \right] \tag{3}$$

$$T = \int_{\beta_6 \cdot \ln \left( \frac{MS}{A} \right)}^{\infty} f_{\beta_3}(\beta_3) d\beta_3 \tag{4}$$

where $f_{\beta_3}(\beta_3)$ = frequency distribution of the peak runoff rate during storms, $\beta_6$ = saturation capacity of the soil, $M$ = a saturation criterion, and $T$ = frequency factor between 0 and 1.

Again bold indicates those variables whose variation in space and time is dependent on the form of the catchment. A value of $T = 0$ indicates that that point never becomes saturated and surface runoff never occurs; a value of $T = 1$ indicates that that point is always saturated and surface runoff, and thus fluvial erosion, always occurs. The definition of where subsurface saturation occurs within the catchment outlined above is consistent with previous definitions (Beven & Kirkby, 1979; O'Loughlin, 1981) if $m_3 = 1$ and $n_3 = 0$ so that discharge is proportional to the incident rainfall $\beta_3$ and the area.

The slope in the fluvial sediment transport equation is determined directly from the catchment elevations and the direction of steepest downhill drainage. The discharge relationship, dependent on area and slope, can be formulated to reflect the processes that occur in the field. However, it is important to note that if the sediment transport equation is to model the *long-term average* sediment transport equation then the discharge per unit width, $q$, should be interpreted as the mean annual peak discharge, analogous to the idea of a dominant discharge (Willgoose et al., 1989), so that

$$Q = \beta_3 A^{m_3} S^{n_3} \tag{5}$$

where $Q$ = discharge in the channel, $\beta_3$ = runoff rate constant, $S$ = slope in the steepest downstream direction, $A$ = area per unit width, and $m_3, n_3$ = coefficients.

This empirical relationship accounts for runoff routing effects within the catchment and the spatial correlation of rainfall (Leopold et al., 1964; Pilgrim, 1987; Huang & Willgoose, 1992). The discharge, $Q$, is converted to discharge per unit width, $q$, using empirical channel morphology relationships (Henderson, 1966).

$$q = \frac{Q}{w} \tag{6}$$

$$w = \beta_4 Q^{m_4} \tag{7}$$

where $Q$ = discharge in the channel, $w$ = width of the channel, $q$ = discharge per unit width, and $\beta_4, m_4$ = channel width coefficients.

A crucial feature of this model is its ability to explicitly model the extension of the channel network and to allow for different sediment transport processes in the channels and on the hillslopes. A variable is defined in space, $Y$, that identifies where channels exist ($Y \approx 1$) versus where the catchment is hillslope ($Y \approx 0$). Initially a catchment can either have no channels or it can have a predefined channel network and drainage pattern. The extension of the network occurs when a function, non-linearly dependent on contributing area slope, called the channel initiation function $a$, exceeds a threshold value called the channel initiation threshold $a_t$. The exact means by which the transformation from hillslope to channel occurs appears to be unimportant, though Willgoose and coworkers have extensively used one that results in channel being permanently formed at a point once the threshold has been exceeded at that point. More important is the functional dependence of the channel initiation function on discharge and slope which Willgoose and coworkers have formulated as

$$a = \beta_5 q^{m_5} S^{n_5} \tag{9}$$

where $a$ = channel initiation function, and $\beta_5, m_5, n_5$ = coefficients.

Again, within the conceptual framework of the model, the form of the channel initiation function can be formulated as seen fit in light of physical processes observed in the field. The formulation above results from consideration of surface-flow-driven channel formulation processes where it has been postulated that channel formation occurs when a critical velocity or tractive force is exceeded by overland flow or where the head grandient in the groundwater exceeds a specified piping threshold (Willgoose *et al.*, 1989). It is consistent with field data collected by other workers (Patton & Schumm, 1975; Montgomery & Dietrich, 1988).

The channel network calculated by the model is used to determine the rate at which fluvial sediment transport occurs as

$$f(Y) = \begin{cases} \beta_1 & Y \approx 0 \text{ (hillslope)} \\ \beta_1 O_t & Y \approx 1 \text{ (channel)} \end{cases} \tag{9}$$

where $Y$ = channel indicator variable, $\beta_1$ = erosion rate constant, and $O_t$ = ratio of hillslope to channel erosion rate.

The transport rate $\beta_1$ can be spatially variable in any predefined way; structural controls due to the differential erodibility of strata can be easily modelled. However, the sediment transport rate $\beta_1$ is not varied as a result of the evolving catchment's hills and valleys so that differential sediment transport rates in valleys and interfluves cannot be modelled other than with a crude area or slope dependence. Very few data are available to calibrate such a dependence. The parameter $O_t$ is generally assumed to be somewhat less than 1 and this reflects the increased velocities, and thus transport rates, in channels over those occurring on the hillslopes. Diffusive transport is assumed to occur at the same rate on both hillslopes and channels.

Note that the sole use of the channel network *within the model* is in determining the differential rates of erosion that occur in the channel and on the hillslopes. No field interpretation is made regarding whether a channel head is an abrupt or gradual transition, only that the fluvial transport rate changes abruptly. Willgoose and coworkers have normally assumed that the actual channel network observed in the field and the channel network postulated in the model are synonymous.

In summary, the important feature of the presented model – the one that distinguishes the networks it generates from other, stochastic, network generation models (Leopold & Langbein, 1962; Howard, 1971) – is that the network extension process is governed by physical conditions; the drainage pattern on the hillslopes and the local slopes in the hillslopes around the channel head. That channels can be assumed to erode faster than the hillslopes facilitates the natural tendency towards convergence of flow on the hillslopes around the channel heads. The pattern of pre-existing channels governs the valley erosion, which in turn governs the drainage pattern of the hillslopes and their slopes, and, thus, the spatial pattern of the channel initiation function. This complicated interaction of flow and sediment transport in both the channels and hillslopes over long timescales is central to the channel network extension process and it gives catchments their form.

The hydrogeomorphology perspective is *how does the geomorphology respond to and determine the hydrological response of the catchment*? This work has been triggered by the observation that a catchment's hydrology has, over geological time, shaped the catchment. The hope is, if we understand how this evolution proceeds, and what the physical dependencies are, we can then use this geomorphologic information as the basis of models of hydrological response. Does, for instance, the catchment evolve to some form that is optimal in passing water and sediment? Our inquiries are necessarily biased by this interest, though the overlap with traditional research in geomorphology is obvious. For instance, it has been noted that catchment drainage density has a significant influence on the shape of the hydrograph generated by a catchment (Wyss *et al.*, 1990), so that questions of drainage density determination are important. Furthermore, it has often been asserted (e.g. Pilgrim, 1977) that the flood wave velocity in catchment is constant because the balance of erosional forces should lead to this result. Preliminary work by the first author using simulated catchments suggest that the flood wave velocity may vary systematically within the catchment; variations in the flood wave velocity within the catchment will change the hydrograph shape while variations from storm to storm will change the peak discharge. It is to these types of questions that hydrogeomorphologic research addresses itself. It is to some of these questions that this paper addresses itself. First, however, we digress slightly to discuss important issues of notation and the interpetation of models of catchment evolution.

## Timescales in Catchment Evolution

Since catchments evolve with time we need to consider the timescales of the catchment. We distinguish between those changes that occur over short timescales and those that occur over the longer timescales of catchment evolution. Short-term changes are those that result from random events in time, or which from the perspective of the longer timescales appear to be random. Here we model only the long-term changes – the short-term variations are explicitly averaged out. For instance, a landslide is a short timescale process since it occurs very quickly compared with the time that it takes for a catchment to evolve. Similarly any single flood and erosion event is only a minor fluctuation in the life of a catchment which is exposed to the full range from small to large events.

The underlying philosophy of this approach is that the full range of flood and erosion events shapes the catchment. An alternative philosophy, apparently not compatible with our approach, is *catastrophism* (or neocatastrophism) (Thorn, 1988). Here certain very large events cause catastrophic and irrevocable changes in the form of the catchment, and every event larger than the last largest historic event wipes out all changes before it, so that the catchment always reflects the last large event rather than the complete history of events. We will not discuss catastrophism here but refer the reader elsewhere (e.g. Leopold *et al.*, 1964).

A seductive, but we believe erroneous, idea is that a single effective or characteristic discharge can effectively replace the range of flood and erosion events. This idea appears to be supported by the form of the governing equations, with their dependence on a single discharge at a point, $q$, in the sediment transport and channel initiation functions, equations (2) and (8). This interpretation appears to be correct for the Hortonian runoff mechanism (when $T = 1$) but is incorrect for the subsurface saturation mechanism. The factor $T$ reflects the region of saturation for the range of storms; regions that are only saturated for infrequent events have a low value of $T$; they are saturated for short periods of time. Regions with large $T$ are saturated frequently (Figure 13.1). Considering variation in $T$ alone, points on the hillslope with large $T$ have greater average fluvial erosion than those with small values, because they are saturated more often. The value of $T$ at any point in the catchment reflects the statistical distribution of runoff. Moreover the complete distribution of rainfall, runoff, and thus $T$, is important since different parts of the catchment are saturated for different proportions of time; the distribution of the level of $T$ over the catchment reflects the complete probability distribution on runoff. This means that not only the mean peak discharge, $q$, is important but also its variance, and higher order moments, as well (equation 4). Stochastic effects in the runoff influence the distribution of the erosion within the catchment so that the catchment elevations

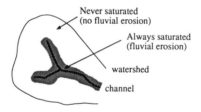

Never saturated
(no fluvial erosion)

Always saturated
(fluvial erosion)

watershed

channel

(a) Saturated area based on effective discharge

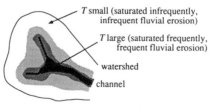

$T$ small (saturated infrequently,
infrequent fluvial erosion)

$T$ large (saturated frequently,
frequent fluvial erosion)

watershed

channel

(b) Saturated area factors based on range
of discharges

**Figure 13.1**    Schematic diagram showing the variation of the subsurface saturation function $T$

are dependent on the frequency distribution of the runoff events rather than simply a single *effective* discharge.

An important concept in geomorphology is that of dynamic equilibrium. While a 'mature' catchment may exhibit short-term, episodic, fluctuations in form, catchments in dynamic equilibrium exhibit a balance between uplift and erosion over long timescales. When mathematically modelling landscape evolution a formal definition is required. The definition we adopt follows from equation (1) and is that the (mean) elevations and channel networks are invariant with respect to time (Hack, 1960) so that the governing equations are

$$\frac{\partial z}{\partial t} = 0 \tag{10}$$

$$\frac{\partial Y}{\partial t} = 0 \tag{11}$$

This definition, that requires the spatial form of the catchment to remain the same, we call *deterministic dynamic equilibrium*. Deterministic because every aspect of the catchment must remain the same. Contrast this with other definitions below. Note that this definition does not specifically prohibit variations with time of the tectonic uplift, only that the catchment must adjust to the new regime faster than the tectonic uplift changes so that the catchment is always in equilibrium; this involves difficult questions related to timescales that have not yet been fully resolved (Ahnert, 1987).

A less restrictive definition, implicitly assumed by many workers, is that the statistics describing the geomorphology are invariant with time. This we call *statistical dynamic equilibrium*; any catchment in deterministic dynamic equilibrium is clearly in statistical dynamic equilibrium. This definition is implicitly assumed, for instance, when it is asserted that a catchment is in dynamic equilibrium because the mean depth of sediment loss from the catchment equals the tectonic uplift (Ahnert, 1970). Sediment yield is just one statistic describing the catchment; the catchment can change while the sediment yield remains the same. A major operational problem with this definition is that while statistical dynamic equilibrium means that the statistics are invariant with time the inverse argument cannot be made. How then do we observe and assert that a catchment is in statistical dynamic equilibrium? We don't have an answer to this question.

Commonly the decline of hillslopes has been discussed in the terms of a characteristic form of the hillslope (Kirkby, 1971; Smith & Bretherton, 1972; Ahnert, 1976). Consider a catchment which is declining with time but whose form (e.g. hillslope profile) remains the same when judged relative to the vertical scale of the catchment. If this catchment form, or non-dimensional characteristic profile, remains the same then we say that the catchment is in *non-dimensional dynamic equilibrium*, either statistical or deterministic as the case may be. The formal definition is as in equations (10) and (11) except that non-dimensional properties such as non-dimensional elevation and horizontal distances are used (see Willgoose *et al.*, 1991c, for a detailed discussion of non-dimensionalisation). If the catchment is in non-dimensional statistical dynamic equilibrium then two examples are that the Strahler slope ratio (the ratio of two slopes; Strahler, 1964) and hypsometric curve are constant with time. Willgoose (1993) has discussed in detail the case of non-dimensional dynamic equilibrium for a catchment declining to a peneplain (i.e. Davisian landscape).

Not all catchments are in dynamic equilibrium. In fact many catchments never reach dynamic equilibrium either because there is no tectonic uplift leading to dynamic equilibrium or because tectonic uplift changes so quickly that catchments are always converging to a new equilibrium slope (Ahnert, 1970), though a case can be made that most catchments are close to dynamic equilibrium (Ahnert, 1987). However, even in these circumstances it is possible to make quantitative statements about catchment geomorphology (Willgoose, 1992, 1993). It is important, however, to understand what dynamic equilibrium is so that it is clear when catchments are not in dynamic equilibrium. This distinction is important when predicting the equilibrium states of the landscape, and will be further discussed in a later section of this paper.

## THE CHANNEL INITIATION THRESHOLD

One of the most important issues in hydrogeomorphology, and one of the most obvious characteristics of a catchment, is the channel or gully network and its drainage density. Variations in drainage density are important for determining the hydrologic response of the catchment; catchments with higher drainage density respond more quickly and thus have higher peak discharges because of the reduced length of hillslope.

When a catchment is in dynamic equilibrium with respect to its channels the channel heads are neither receding nor advancing with time. At the channel head, the channel initiation function on the hillslope is equal to the threshold so that we may write

$$TA_h = \beta' \frac{A_s{}^{m_3 m_5 (1 - m_4)} S_s{}^{n_5 + n_3 m_5 (1 - m_4)}}{a_t} = 1 \qquad (12)$$

where $TA_h$ is a non-dimensional number that is the magnitude of the channel initiation function divided by the channel initiation threshold. $TA_h$ and other non-dimensional numbers have been discussed by Willgoose *et al.* (1991c). These other non-dimensional numbers will not be discussed here. This relationship shows that the area draining to the channel head, called the source area, $A_s$, is related to the hillslope slope at the channel head, $S_s$, and is related to the channel initiation threshold. The parameter $\beta'$ is a function of the parameters $\beta_1$, $\beta_2$, $\beta_3$, $m_1$, $n_1$, $m_3$, $n_3$, and $m_4$.

Everything else being equal, if the threshold is decreased then the source area decreases and the drainage density increases. This behaviour can be clearly seen by comparing Figure 13.2 with Figure 13.6c. The sole difference between these two simulated catchments is that the channel initiation threshold used in Figure 13.2 is 50% higher than that used in Figure 13.6.

The exact form of this equation relating area, slope and the channel initiation function follows from the proposed surface-runoff-driven mechanism that triggers the channel formation process. The more general principle of a physically based threshold below which channel formation cannot occur is a powerful concept, and other workers have published examples of different mechanisms using this concept (Dunne, 1980; Dietrich *et al.*, 1986; Montgomery & Dietrich, 1988, 1989). The exact form of the relationship between area and slope is different in these latter cases but the threshold behaviour leading to equation (1) remains the same.

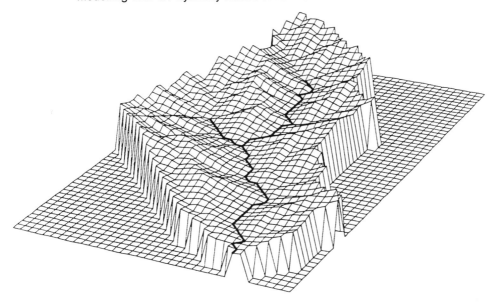

**Figure 13.2** Sample catchment with channel initiation threshold 50% higher than in Figure 13.6; simulation time = 60 000

Willgoose *et al.* (1990) compare the area–slope relationship at channel heads predicted by equation (12) with that obtained in a field study by Montgomery and Dietrich (1988). Figure 13.3 shows their data for a site in Marin County, California, together with their threshold criteria based on a one-dimensional slope stability model, as well as a fitted curve of the form of equation (12). At least for low slopes equation (12) seems to provide a satisfactory fit. Montgomery (pers. comm.) noted that the runoff processes triggering gully growth appear to vary among the channel heads ranging from stream sapping to surface runoff. It is thus unlikely that in their data a single area–slope line will perfectly

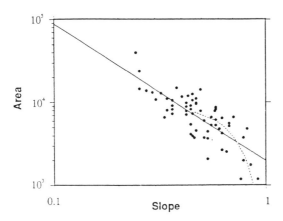

**Figure 13.3** The area contributing to a channel head versus the slope at the channel head (data from Montgomery & Dietrich, 1989), the threshold mechanism of Montgomery & Dietrich (dotted line) and the Willgoose *et al.* (1989) surface flow mechanism (solid line)

fit the data. Indeed it is likely that their slope stability criteria will dominate for high slopes but for low slopes the surface flow threshold will dominate; the data seems to support this interpretation.

Willgoose *et al.* (1990) also noted that their data for this site and two other sites in Oregon and California were consistent with a runoff dependence in the threshold, because the sites with a higher rainfall had smaller source areas for the same slope. Such behaviour is consistent with equation (12) since $\beta'$ is positively correlated with the runoff rate, $\beta_3$.

Such behaviour does not necessarily follow from other channel formation criteria (Smith & Bretherton, 1972; Kirkby, 1987), which have been used to postulate that a channel is formed when a small perturbation can grow unstably. Apart from possibly giving a different trend between area and slope for channel heads, these latter criteria are not directly related to the runoff rate; rather they are dependent on the second derivatives of the landscape surface. Source areas are only reduced when increased runoff rates are able to modify the landscape surface and so are not related to runoff over short timescales. Furthermore, Loewenherz (1990) suggests that Smith & Bretherton's criteria will lead to a system of rills at an infinitesimal distance apart unless a basic length scale is built into the equations. Introducing this realistic scale of separation is conceptually consistent with the threshold behaviour postulated here.

The authors suggest that the stability criteria of Smith and Bretherton may, in fact, define the uppermost portion of the hillslope that can be channellised, and no matter how high the value of the channel initiation function becomes, by increasing, for instance, runoff and thus $\beta'$, channel network extension cannot proceed upstream of this point of the catchment. This idea is consistent with anecdotal evidence of active gully erosion after catchment disturbance by forest clearing, climate change, etc.

Moreover, the dependence of the channel initiation equation on the channel initiation threshold reflects the resistance of the catchment to channel formation. More resistant catchments will have larger source areas, and thus lower drainage density, which is a generally observed property of catchments (Ruhe, 1952; Hack, 1957).

## EQUILIBRIUM STATES OF THE LANDSCAPE

The velocity of a flood wave is a function of the discharge and the slope. If the discharge is related to area and slope, as, for instance, postulated in equation (5) above, we may relate the flood wave velocity to the area draining through that point and the slope at that point by use of Mannings equation and kinematic wave theory. These properties may be obtained directly from the catchment geomorphology.

Willgoose *et al.* (1989, 1991d) and Willgoose (1993) discuss the usefulness of the model for describing the long-term form of catchments. For instance, they were able to explain a widely observed log–log correlation between slope and area in field catchments (Flint, 1974; Tarboton *et al.*, 1989b) by arguing that a catchment in dynamic equilibrium tends to an equilibrium form where the erosional loss of soil equals the tectonic uplift at that point. If the sediment transport and runoff equations are of the form outlined above then this leads to a relationship between area, slope and catchment geologic and climatic conditions of the form (Willgoose *et al.*, 1991d)

$$\overline{c_0} = -D' \cdot A_o{}^{\gamma_1} S_o{}^{\gamma_2} - \beta_1' \cdot A_o{}^{\gamma_3} \cdot S_o{}^{\gamma_4} \tag{13}$$

where $\overline{c_0}$ is the areal average tectonic uplift for the catchment of area $A_0$, with slope at the outlet of the catchment $S_0$. The parameters $D'$ and $\beta_1'$ are primarily functions of the diffusivity and rate of fluvial sediment transport but also are dependent on other parameters in the model as are the parameters $\gamma_1$, $\gamma_2$, $\gamma_3$, and $\gamma_4$. Willgoose *et al.* (1991d) noted that if $c_0$ is spatially constant and if only one transport mechanism (i.e. fluvial transport or diffusion) is dominant then for the Hortonian case ($T = 1$) equation (13) reduces to a log–log linear relationship between contributing area and slope of the form

$$A^\gamma \cdot S = \text{constant} \tag{14}$$

where $\gamma$ is a function of the process dominating the catchment, which can be easily derived from equation (13) by setting $D' = 0$ (for fluvial dominated catchments) or $\beta_1' = 0$ (for diffusion dominated catchments). If both diffusion and fluvial transport are important in a single catchment then the area–slope relationship is asymptomatically log–log linear for large (fluvial transport dominated) and small (diffusion dominated) areas with a log–log non-linear transition region at intermediate areas (Figure 13.4). A sample catchment converging to dynamic equilibrium, described by equation (13), is shown in Figure 13.5.

Willgoose (1993) extended this result to describe the long-term form of catchments where the long-term tectonic uplift is not equal to the erosion. These are catchments that may have been initially uplifted but because of the lack of continuous uplift they are declining to a peneplain. For these declining catchments the tectonic uplift, $c_0$, is zero. The final equilibrium form of the catchment will then be the Davisian peneplain with the elevations of the catchment asymptomatically approaching the datum elevation. Willgoose (1993) showed that the declining catchment converged to a constant non-dimensional form (i.e. non-dimensional dynamic equilibrium). For the case of either a catchment dominated by fluvial sediment transport alone (where diffusion is insignificant) or a catchment dominated by diffusive transport alone the catchment is described by

$$\frac{A_0^{\gamma_5/\gamma_6} S_0}{\overline{z_0}^{1/\gamma_6}} = \text{constant} \tag{15}$$

where $\gamma_5$ and $\gamma_6$ are parameters dependent on the runoff and erosion equations and $\overline{z_0}$ is the mean elevation of the catchment. Note that the values of $\gamma_5$ and $\gamma_6$ are different

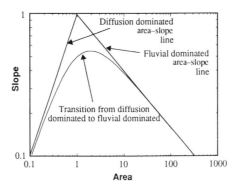

**Figure 13.4**  Schematic diagram showing the area–slope correlation with catchments for (a) diffusion dominated transport, (b) fluvial dominated transport, and (c) a composite of both transport processes

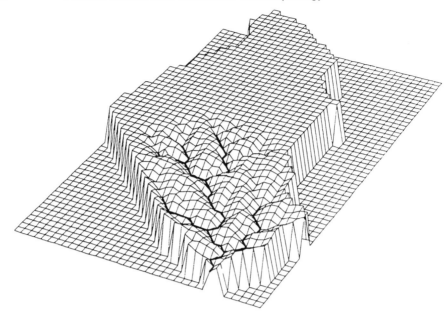

(a) simulation time = 30 000

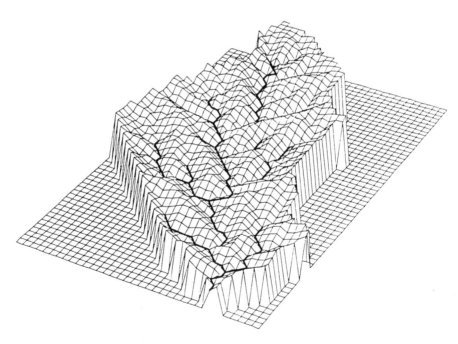

(b) simulation time = 46 000

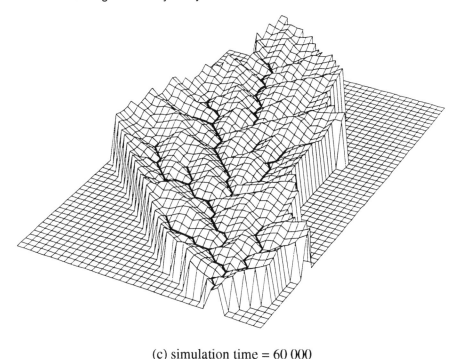

(c) simulation time = 60 000

**Figure 13.5** Sample catchment simulation with time. Continuous tectonic uplift with time. The catchment is elevated above the background to highlight the catchment boundaries

for the fluvial and diffusion dominated cases with $\gamma_5$ being negative for diffusive transport processes and positive for fluvial transport processes. Note that if this equation is multiplied by $\overline{z}_0^{1/\gamma_6}$ it has a similar form to the equation for dynamic equilibrium. The difference is that the right-hand side constant declines to zero as the catchment erodes to the Davisian peneplain. That the catchment is in non-dimensional dynamic equilibrium can be clearly seen in the series of pictures showing a simulated catchment declining to a peneplain (Figures 13.6 and 13.7). Figure 13.6 shows the actual simulated catchment while Figure 13.7 shows the non-dimensionalised catchment (scaled by the catchment relief) converging on non-dimensional dynamic equilibrium. Other statistics, such as the hypsometric curve, confirm this convergence to non-dimensional dynamic equilibrium with time (Willgoose, 1993).

The area–slope and area–slope–elevation relationships also indicate the relative importance of the diffusive and fluvial processes in determining catchment form. Where slope increases with increasing area diffusion is dominant and where slope decreases with increasing area fluvial transport is dominant. Figure 13.8 shows a field catchment from southeast Australia studied by Willgoose (1991, 1993)). Figure 13.9 shows the area–slope–elevation curve he fitted to this catchment. The right-hand side of the figure (i.e. large areas) is dominated by fluvial erosion while the left-hand side (i.e. areas less than about 50 m$^2$ per metre width) is dominated by a diffusive mechanism. It has proved very difficult in practice to calibrate the magnitude of $D$ in the diffusive mechanism from the slope–area–elevation relationships because of the poor resolution of most digital

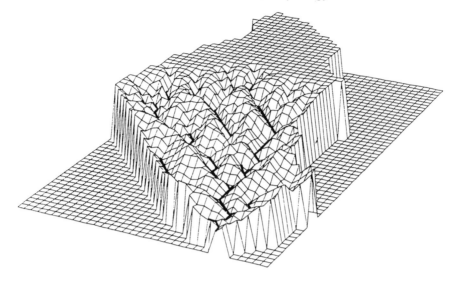

(a) simulation time = 10 000

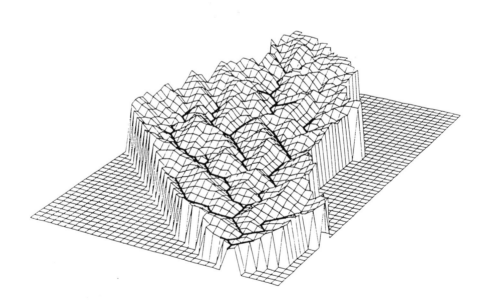

(b) simulation time = 20 000

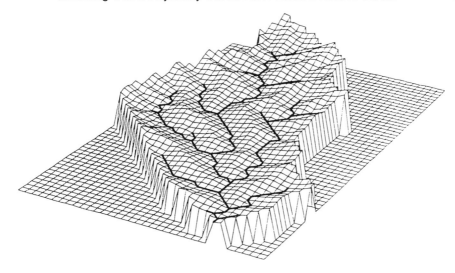

(c) simulation time = 60 000

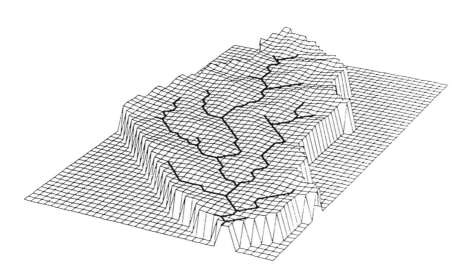

(d) simulation time = 170 000

**Figure 13.6**  Sample catchment simulation with time. Initial uplift followed by decline to peneplain with continuing fluvial erosion

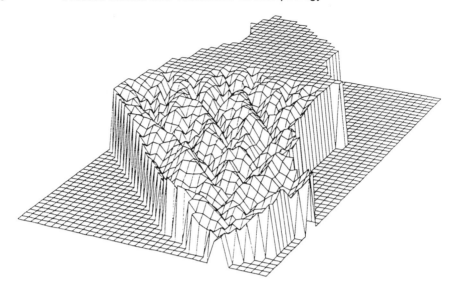

(a) simulation time = 10 000

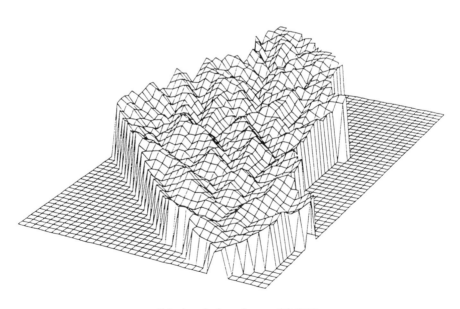

(b) simulation time = 20 000

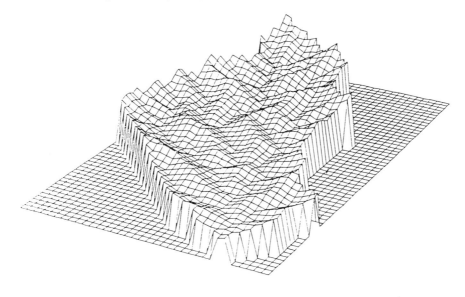

(c) simulation time = 60 000

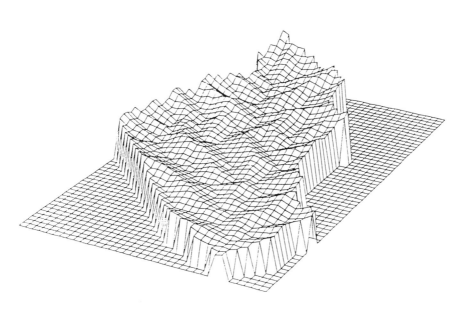

(d) simulation time = 170 000

**Figure 13.7** Sample catchment simulation of Figure 13.6 with time. Elevations non-dimensionalised by the maximum relief of the catchment

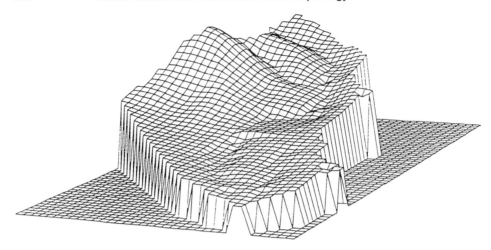

**Figure 13.8**   Pokolbin #1 field site (grid spacing 20 m)

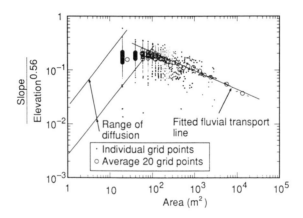

**Figure 13.9**   Pokolbin #1 field site area–slope–elevation diagram. Dots are the values for individual nodes in the catchment, circles are averages of 20 nodes

terrain data; the fluvial component is much easier. the difficulty is in simultaneously obtaining the large-scale data necessary for the fluvial processes, and the small-scale hillslope data for the diffusive processes from readily available digital terrain data. Work is proceeding at the field site to alleviate this problem.

If more complicated diffusive mechanisms, such as the plastic or viscious flow mechanisms studies by Ahnert (1976) (referred to above), are used in the model then the calibration problems are significantly increased because of the increased number of parameters. Tarboton *et al.* (1989a) indicate that, as for the linear diffusion and fluvial processes, these mechanisms will plot on a single line in the slope–area curve though the trend of slope with area varies depending on the process. Despite this it appears that the spatial variability in the diffusive transport (as indicated by the scatter in Figure 13.9, and other work by Tarboton *et al.*, 1989a) is such that more accurate representation of the processes is not warranted at this stage.

# CONCLUSIONS

This paper has highlighted aspects of a process response model of catchment geomorphology that has been used to study problems in hydrogeomorphology. The model is based on a conceptualisation of the important transport processes occurring in the field over the timescales of the catchment evolution. Plots of the area draining through a point and the slope at that point suggest that, in particular, a more sophisticated representation of the hillslope processes is unwarranted at this stage even though it is possible to construct a more complicated model.

A key feature of the presented model is that it explicitly distinguishes between the transport processes on the hillslope and in the channel and that it models the development of the channel network in response to the changing catchment geomorphology. Different transport processes are postulated for the hillslope and channel regimes. The channel network extension is a threshold phenomenon where a channel forms if a physically based threshold (e.g. overland flow velocity) is exceeded.

A key requirement for this model is the need to mathematically describe the boundary conditions and exogenous inputs to be used for modelling exercises. Firstly it was noted that, in general, it is not possible to think in terms of an effective discharge in terms of modelling catchment development; rather the complete range of runoff events is needed. A variety of equilibrium conditions were discussed. A mathematical definition of dynamic equilibrium was proposed. A non-dimensional version of this dynamic equilibrium definition was proposed that is applicable to catchments declining after a tectonic uplift event.

Simulated catchments produced by a computer implementation of this model (SIBERIA) were presented. These catchments were for a variety of tectonic conditions and physical parameters and illustrated some of the most important characteristics of the model, such as the ability to model changes in drainage density due to material properties.

It has been observed by other workers using digital-terrain-map-based rainfall–runoff models that changes in the drainage density assumed for the channel network change the hydrologic response. Simulating accurately the extent of the channel network and its slope–area characteristics thus appears to be of key importance for hydrogeomorphologists. Some of the key issues in determining the extent of the channel network have been discussed in this paper, together with previously published field data that appear to support the threshold behaviour postulated in the model presented here. The mechanism used in this paper results in the position of the channel head being a function of the area draining to the channel head, the slope at the channel head, and the runoff upstream of the channel head; this appears to be consistent with field data.

Closed form solutions for catchment form have been presented. These relationships between area, slope and elevation are primarily dependent on the long-term sediment transport processes occurring in the catchment. The form is also dependent on the tectonic regime active in the region. Equations were presented for (1) the dynamic equilibrium case where sediment loss is in balance with the tectonic uplift, and (2) the declining catchment case where an initially uplifted catchment is gradually declining to a peneplain under the action of erosion. In both cases there is an equilibrium form that can be simply related to the governing climate, geology and transport processes. A field catchment was shown to demonstrate the principles of this type of analysis.

The authors believe that this paper has demonstrated that a model of catchment evolution explicitly modelling the interactions between the hillslopes and the dynamically varying channel network can be a useful tool to describe important features of field catchments. We firmly believe that this interaction between the channel network and the hillslopes is important in determining the shape of catchments, and modelling of this interaction is key to understanding the interaction between hydrology and geomorphology. The authors cannot claim to have fulfilled the aim of hydro-geomorphology – to quantitatively link hydrology and geomorphology – at this stage. However, we firmly believe that the SIBERIA model is a significant step towards satisfying those aims.

## ACKNOWLEDGEMENTS

The work reviewed here has benefited at various times from discussions with a number of people, most notably William Dietrich, David Montgomery and David Tarboton. This work results from multiple efforts funded, in part, by the National Science Foundation of the USA, a cooperative agreement with the University of Florence (through the National Research Council of Italy) and the National Weather Service (Cooperative Research Agreement NA86AA-H-HY123) and the Australian Water Research Advisory Council through support of the first author by an AWRAC Research Fellowship.

## REFERENCES

Ahnert, F. (1970). Functional relationships between denudation, relief, and uplift in large mid-latitude drainage basins. *American Journal of Science*, **268**, 243–263.
Ahnert, F. (1976). Brief description of a comprehensive three-dimensional process-response model for landform development. *Zeitschrift für Geomorphologie N. F. Supplement*, **25**, 29–49.
Ahnert, F. (1987). Approaches to dynamic equilibrium in theoretical simulations of slope development. *Earth Surface Processes and Landforms*, **12**, 3–15.
Andrews, D. J. and Bucknam, R. C. (1987). Fitting degradation of shoreline scarps by a nonlinear diffusion model. *Journal of Geophysical Research*, **92** (B12), 12857–12867.
Beven, K. J. and Kirkby, M. J. (1979). A physically based, variable contributing area model of basin hydrology. *Hydrological Sciences Bulletin*, **24** (1), 43–69.
Chorley, R. J. Schumm, S. A. and Sugden, D. E. (1984). *Geomorphology*, Methuen, London.
Culling, W. E. H. (1963). Soil creep and the development of hillside slopes. *Journal of Geology*, **71** (2), 127–161.
Dietrich, W. E., Wilson, C. J. and Reneau, S. L. (1986). Hollows, colluvium, and landslides in soil-mantled landscapes. In Abrahams, A. D. (ed.), *Hillslope Processes*, Binghampton Series on Geomorphology, pp. 361–368.
Dunne, T. (1980). Formation and controls of channel networks. *Progress in Physical Geography*, **4**, 211–239.
Flint, J. J. (1974). Stream gradient as a function of order, magnitude and discharge. *Water Resources Research*, **10** (5), 969–973.
Gilbert, G. K. (1909). The convexity of hillslopes. *Journal of Geology*, **17**, 344–350.
Hack, J. T. (1957). Studies of longitudinal stream profiles in Virginia and Maryland, USGS, Professional Paper 294-B, pp. 45–97.
Hack, J. T. (1960). Interpretation of erosional topography in humid temperate regions. *American Journal of Science*, **258-A**, 80–97.

Henderson, F. M. (1966). *Open Channel Flow*, Macmillan, New York.

Horton, R. E. (1945). Erosional development of steams and their drainage basins; hydrophysical approach to quantitative morphology. *Bulletin of the Geological Society of America*, **56**, 275–370.

Howard, A. D. (1971). Simulation of stream networks by headward growth and branching. *Geographical Analysis*, **3**, 29–50.

Huang, H. Q. and Willgoose, G. R. (1992). *Numerical Analyses of Relations between Basin Hydrology, Geomorphology and Scale*, Research Report 075.04.1992, Department of Civil Engineering and Surveying, The University of Newcastle, Australia.

Huggett, R. J. (1988). Dissipative systems: implications for geomorphology. *Earth Surface Processes and Landforms*, **13**, 45–49.

Julien, P. Y. and Simons, D. B. (1985). Sediment transport capacity of overland flow. *Transactions of the American Society of Agricultural Engineers*, **28** (3), 755–762.

Kirkby, M. J. (1971). Hillslope process-response models based on the continuity equation. In *Slopes: Form and Process*, Institute of British Geographers Special Publications, 3, London, pp. 15–30.

Kirkby, M. J. (1987). Modelling some influences of soil erosion, landslides and valley gradient on drainage density and hollow development. In Ahnert, F. (ed.), *Geomorphological Models: Theoretical and Empirical Aspects*, Aachen, Germany, 1986.

Leopold, L. B. and Langbein, W. B. (1962). *The Concept of Entropy in Landscape Evolution*. US Geological Survey Professional Papers 500-A, USGS.

Leopold, L. B., Wolman, M. G. and Miller, J. P. (1964). *Fluvial Processes in Geomorphology*, Freeman, London.

Loewenherz, D. (1990). Lengthscale-dependent material transport and the initiation of surface drainage: a modified linear stability theory, American Geophysical Union Meeting, Baltimore, May, 1990, *EOS*, **71** (17), 515–516.

Mackin, J. H. (1948). Concept of the graded river. *Bulletin of the Geological Society of America*, **59**, 463–512.

Mock, S. J. (1971). A classification of channel links in stream networks. *Water Resources Research*, **7** (6), 1558–1566.

Montgomery, D. R. and Dietrich, W. E. (1988). Where do channels begin? *Nature*, **336**, 232–234.

Montgomery, D. R. and Dietrich, W. E. (1989). Source areas, drainage density and channel initiation. *Water Resources Research*, **25** (8), 1907–1918.

Moore, I. D. and Burch, G. J. (1986). Sediment transport capacity of sheet and rill flow: Application of unit stream power theory. *Water Resources Research*, **22** (8), 1350–1360.

O'Loughlin, E. M. (1981). Saturation regions in catchments and their relations to soil and topographic properties. *Journal of Hydrology*, 53, 229–246.

Patton, P. C. and Schumm, S. A. (1975). Gully erosion, northern Colorado, a threshold phenomena. *Geology*, 3, 88–90.

Pilgrim, D. H. (1977). Isochrones of travel time and distribution of flood storage from a tracer study on a small watershed. *Water Resources Research*, **13** (3), 587–595.

Pilgrim, D. H. (1987). Estimation of peak flows for small to medium sized rural catchments. In Pilgrim, D. H. (ed.), *Australian Rainfall and Runoff*, The Institution of Engineers, Canberra, pp. 93–116.

Ruhe, R. V. (1952). Topographic discontinuities of the Des Moines Lobe. *American Journal of Science*, **250**, 46–56.

Scheidegger, A. E. (1970). *Theoretical Geomorphology*, Springer-Verlag, Berlin.

Schumm, S. A. (1956). Evolution of drainage systems and slopes in badlands at Perth Amboy, New Jersey. *Bulletin of the Geological Society*, **67**, 597–646.

Smith, T. R. and Bretherton, F. P. (1972). Stability and the conservation of mass in drainage basin evolution. *Water Resources Research*, **8** (6), 1506–1529.

Strahler, A. N. (1964). Quantitative geomorphology of drainage basins and channel networks. In Chow, V. T. (ed.), *Handbook of Applied Hydrology*, McGraw-Hill, New York, pp. 4-39–4-76.

Tarboton, D. G., Bras, R. L. and Rodriguez-Iturbe, I. (1989a). The Analysis of River Basins and Channel Networks using Digital Terrain Data, TR 326, Ralph M. Parsons Laboratory, Department of Civil Engineering, MIT, Boston.

Tarboton, D. G., Bras, R. L. and Rodriguez-Iturbe, I. (1989b). Scaling and elevation in river networks. *Water Resources Research*, **25** (9), 2037–2052.

Thorn, C. E. (1988). *Introduction to Theoretical Geomorphology*, Unwin Hyman, London.

Valdes, J. B., Fiallo, Y. and Rodriguez-Iturbe, I. (1979). A rainfall–runoff analysis of the geomorphologic IUH. *Water Resources Research*, **15** (6), 1421–1434.

Willgoose, G. R. (1991). Field application of river basin evolution model. In *Challenges for Sustainable Development*, International Hydrology and Water Resources Symposium, Perth, Australia, 1991.

Willgoose, G. R. (1992). Geomorphologic modelling and interactions between active tectonics and topography. In *Chapman Conference on Tectonics and Topography*, August 31–September 4, Snowbird, Utah.

Willgoose, G. R. (1993). A physical explanation for an observed area–slope–elevation relationship for declining catchments. *Water Research* (in press).

Willgoose, G. R., Bras, R. L. and Rodriguez-Iturbe, I. (1989). *A Physically Based Channel Network and Catchment Evolution Model*, TR 322, Ralph M. Parsons Laboratory, Department of Civil Engineering, MIT, Boston.

Willgoose, G. R., Bras, R. L. and Rodriguez-Iturbe, I. (1990). A model of river basin evolution. *Transactions of the American Geophysical Union*, **71** (47), 1806–1807.

Willgoose, G. R., Bras, R. L. and Rodriguez-Iturbe, I. (1991a). Results from a new model of river basin evolution. *Earth Surface Processes and Landforms*, **16**, 237–254.

Willgoose, G. R., Bras, R. L. and Rodriguez-Iturbe, I. (1991b). A physically based coupled network growth and hillslope evolution model: 1. Theory. *Water Resources Research*, **27** (7), 1671–1684.

Willgoose, G. R., Bras, R. L. and Rodriguez-Iturbe, I. (1991c). A physically based coupled network growth and hillslope evolution model: 2. Applications. *Water Resources Research*, **27** (7), 1685–1696.

Willgoose, G. R., Bras, R. L. and Rodriguez-Iturbe, I. (1991d). A physical explanation of an observed link area–slope relationship. *Water Resources Research*, **27** (7), 1697–1702.

Willgoose, G. R., Bras, R. L. and Rodriguez-Iturbe, I. (1991e). The relationship between catchment and hillslope properties: implications of a catchment evolution model. *Geomorphology*, **5** (1/2), 21–38.

Wyss, J. Williams, E. R. and Bras, R. L. (1990). Hydrologic modelling of New England river basins using radar rainfall data. *Journal of Geophysical Research (Atmospheres)*, **95** (3), 2143.

# 14  Thresholds and Instability in Stream Head Hollows: A Model of Magnitude and Frequency for Wash Processes

**MIKE KIRKBY**
*School of Geography, University of Leeds, UK*

## ABSTRACT

The Smith and Bretherton instability criterion and the Horton/Willgoose tractive stress criterion are combined in a comprehensive theory of stream head location. The theory is given for a combination of wash and rainsplash processes only. Both processes are disaggregated on a storm basis to provide explicit long-term integrals over the frequency distribution, as well as a magnitude and frequency interpretation of stream head behaviour. The theory takes account of tractive stress thresholds, subsurface flow and storms which are brief relative to overland flow travel times.

This theory forecasts a constant stream head location where instability is dominant, and a strong inverse relationship between distance to stream head and gradient where threshold behaviour is dominant. There is a continuous transition between these two types. Over the distribution of storms, there is also a transition of types along the length of channels. The different behaviours are also associated with a morphological classification of channel types.

## INTRODUCTION

The scale of fluvially eroded landscapes is determined mainly by the spacing of significant valleys. Where valleys are very closely spaced, at 5 to 50 m apart, we see a characteristic badland topography. At the other extreme, temperate valleys may be more than 1 km wide, so that there are significant variations over at least two orders of magnitude. In any network, half of the channel links are external links (i.e. first order streams). Since links tend to have similar lengths, approximately half of the total channel length lies in these headwaters, and factors which control the position of stream heads therefore have a considerable effect on total channel density and therefore landscape scale. This paper is concerned with the location of stream heads within their unbranched headwater valleys, which are commonly described as stream head hollows. It is important to understand how stream heads are located, both for explaining static spatial patterns of variation, and also for potential sensitivity to land use and to global climate change.

The reasons for the wide range of variation in channel spacing appear to be related in some way to the balance between hillslope and channel processes for water and sediment transport. Closely spaced active channel networks are generally associated with

---

*Process Models and Theoretical Geomorphology.* Edited by M. J. Kirkby
©1994 John Wiley & Sons Ltd

sparse vegetation and high rates of hillslope runoff and sediment loss. Under natural conditions, badland channel densities are commonly found in semi-arid areas on impermeable shales or marls, where sediment yields are highest. Melton (1957) has shown a strong dependence of drainage density on infiltration capacity and proportion of bare (unvegetated) areas for small catchments in the American South West.

In humid temperate areas, channel densities are generally low in association with continuous soil and vegetation cover, but empirical studies generally fail to show strong dependence of density on simple external causal variables, such as infiltration capacity or rainfall intensity. A number of studies in California and Oregon (e.g. Dietrich *et al.*, 1986; Dietrich and Dunne, 1993) have, however, shown a strong inverse dependence of stream head area on valley gradient, which may partially explain the apparent lack of dependence on external factors.

Two theoretical views have been proposed to account for the location of channel heads. One is based on stability theory and the second on erosion thresholds. The first is based on a mathematical stability analysis of the relevant process laws (Carson & Kirkby, 1972; Smith & Bretherton, 1972). If a small random perturbation is made to indent a smooth landscape surface, stability analysis examines the conditions under which the perturbation will grow to macroscopic size, or shrink until it disappears in the original smooth surface. The analysis can, in principle, be made for any continuously varying process or combination of processes. An indentation in the surface concentrates both water and sediment flows by convergence of the flow lines. Unstable growth of the perturbation occurs if the convergence of flow allows more sediment to be transported out of the indentation than the convergence of sediment flow has brought into it. This happens if sediment transport increases more than linearly with water discharge, so that the increased flow is able to carry out more than the increased sediment inflow.

In the alternative, threshold-based view, channel heads are defined by an erosion threshold, where tractive stress exceeds material resistance. This is the basis of Horton's (1945) 'belt of no erosion' and $x_c$ distance. It is also the basis used by Willgoose *et al.* (1991; this volume, Chapter 13), in the form of an 'activator function'. This recent formulation generalises the condition for a stream head to cover a range of possible processes. This approach appears to refer to behaviour at a dominant flow condition, and has not yet incorporated magnitude and frequency concepts, so that there is no explicit spatial or temporal pattern associated with the area around a channel head. In Willgoose's analysis the activator function separates zones to which different sediment transport laws are applied: stream transport laws below the channel head, and slope process laws above it, although this distinction is not a necessary part of the threshold view.

Here we argue that these views need not conflict, but may be combined to give a more general theory. Over a spectrum of conditions which are related to flow regimes and critical erosion thresholds, there is seen to be a progressive change in response from one dominated by stability conditions to one dominated by threshold behaviour, and these correspond loosely to the semi-arid and humid-temperate cases described above. Here we present the detail for the case corresponding to stream heads created by rainsplash and overland flow. There may be corresponding generalisations to other processes, in particular to stream heads associated mainly with mass movements or to subsurface piping.

## FLOW GENERATION

To understand the way in which threshold and instability criteria apply, across the magnitude and frequency spectrum, we require an explicit model for generating overland flow and sediment transport by the flow and as rainsplash. To be reasonably workable, the model should be simple enough to be integrated over the frequency distribution for long periods, and realistic enough to be relevant. The model used here is based on daily rainfalls, which roughly represent individual single storms, and for which there is widespread observational data.

In the simple view taken here, the soil has a storage capacity $h$, and any storm rainfall greater than $h$ produces overland flow. A useful and workable approximation to the infiltration equations in the absence of detailed intensity data is to assume that a proportion $\beta$ of additional rainfall flows overland, so that the total contribution to overland flow from a storm rainfall $r$ is $q(r) = \beta(r - h)$. If $r$ is taken as the daily rainfall total, then the distribution of daily rainfalls can be fitted adequately to the distribution:

$$N(r > r_1) = N_0 \exp(-r_1/r_0) \tag{1}$$

where $r_0$ is the mean rain per rain day, and $N_0$ is the annual number of rain days ($= R/r_0$) where $R$ is the total annual rainfall.

If, as is commonly the case, the data are not a perfect fit to this inverse exponential distribution, then the parameters $N_0$, $r_0$ and $R = N_0 r_0$ should be fitted to it over the relevant range of storm rainfalls (de Ploey et al., 1991).

Summing over the distribution of rain days, the total overland flow produced at a point is:

$$q_* = \beta \int_h^\infty (r - h) n(r) dr = \beta r \exp(-h/r_0) \tag{2}$$

where $n(r)$ is the frequency of days with rainfall $r[ = 1/r_0 \exp(-r/r_0)]$, $R$ is the total annual rainfall, and $q_*$ is the local average annual rate of flow production.

During a period of constant overland flow production, overland flow depth initially increases linearly with time. Local discharge then builds up gradually towards a steady state in which discharge is equal to unit catchment area times intensity. We will first take the simplest case in which flows reach equilibrium everywhere. The total collecting area for overland flow at a point is delimited by neighbouring flow streamlines, ignoring kinetic effects. Defining the equivalent catchment length, $a$, as the area drained per unit contour length, we can write the topographic identities:

$$a = \frac{1}{w} \int_0^x w \, dx'$$

$$\frac{da}{dx} = 1 + \frac{a}{\rho} \tag{3}$$

$$\frac{1}{\rho} = \frac{1}{w} \frac{dw}{dx}$$

where $\rho$ is the contour radius of curvature ($+$ in hollows; $-$ on spurs), $x$ is horizontal distance measured along the flow direction and $w$ is the width of the elementary flow strip.

The total overland flow is then given as:

$$Q_* = q_* a = \beta R a \exp(-h/r_0) \tag{4}$$

where $Q_*$ is the average annual overland flow discharge per unit contour width.

Our second case is for overland flow which does not reach this equilibrium, because of the short duration of rainfall bursts. The time taken to reach this equilibrium stage is short near to the divides, and increases downslope. The overland flow hydrograph may be obtained as a kinematic cascade, using Manning's equation or another appropriate flow law (e.g. Wooding, 1965). The equilibration time at any given point is approximately the time taken for overland flow to travel to the point in question from the divide. Flow velocity depends on gradient and surface roughness, and commonly lies in the range 0.01 to 0.1 m s$^{-1}$. Equilibration at 5 m from the divide therefore occurs after 1 to 8 minutes, but at the base of a 100 m slope requires rainfall for 0.3 to 3.0 hours. Comparing with the duration of bursts of intense rainfall, it can be seen that the lower parts of many slopes will not reach equilibrium, particularly under arid and semi-arid conditions. After the end of a burst of intense rain, while intensities are lower or rain has stopped, some of the overland flow previously generated re-infiltrates into the soil. For a single burst of rain, there is a clear transition point between equilibrium conditions and steady state conditions. When overland flow is summed over a distribution of rainstorms varying in duration, the transition is blurred, although the linear increase with distance is preserved near the divide, and the steady state is preserved at long distances downslope. A convenient generalisation for our second case, of non-equilibrium overland flow, takes the form:

$$Q_* = q_* a \left[ 1 - \left( \frac{a}{x_0} \right)^b \right] \bigg/ \left[ 1 - \left( \frac{a}{x_0} \right)^{b+1} \right] \tag{5}$$

where $x_0$ is the distance travelled by overland flow during the dominant storm and $b$ is a positive exponent, usually in the range 1 to 4.

It can be seen that $Q_*$ behaves like $q_* a$ close to the divide, and tends towards an upper limit of $q_* x_0$ far from the divide, with a transition in the region of $a = x_0$. For actual slopes, lithological differences, surface roughness and topography may modify this general form, often considerably, through different rates of runoff production and flow velocities. The extent of such local variations, and the general trend illustrated by equation (1) is clearly seen in the work of Dunne and his co-workers in Kenya (e.g. Dunne & Aubry, 1986).

Our third case considers overland flow which reaches equilibrium, but in which previous rainfall sets up a downslope pattern of soil moisture, modifying the deficit $h$ in equation (2) in a systematic fashion. 'Saturation' overland flow is produced when the soil can be brought to saturation by the addition of storm rainfall, even of low intensity. Where rainfall exceeds evapotranspiration rates over a period of days or weeks, subsurface flow within the regolith or rock may persist between storms, so that, even at the beginning of a storm, downslope areas and hollows are wetter than divides and spurs. These wetter areas then become saturated during the course of a storm, producing overland flow per unit area at a rate which depends sensitively on topography and soil thickness. Other things being equal, the pattern of overland flow production is concentrated along the margins of channels, especially where there is some floodplain development; and around valley heads. Under extreme storm conditions, this area expands, particularly in hollows (areas of flow convergence) and on concave foot slopes.

The effect of antecedent conditions may be seen for a given hillside profile. For a net rate, $j$, of antecedent subsurface runoff, and an exponential soil moisture store, specified by deficit $D$, the steady state flow is given (Beven and Kirkby, 1979) as:

$$Q_s = ja = q_1 \Lambda \exp(-D/m) \tag{6}$$

where $j$ is the net rate of antecedent subsurface runoff, $q_1$ is the saturated soil flow on unit gradient, $\Lambda$ is slope gradient and $m$ is a soil moisture parameter with units of depth.

Down the length of the slope, there are consistent changes in $a$ and $\Lambda$, which generally reduce the saturation deficit $D$ downslope. Solving for $D$ over a particular topography, and replacing $h$ by $D$ in equation (2) gives the overland flow production at every point, and this may be integrated, in principle, down the length of the flow strip. For a single rainstorm of $r$, we obtain:

$$Q(r) = \frac{\beta}{w} \int w[r - D(a)] \, dx = ra - E(a) \tag{7}$$

where $Q(r)$ is the total overland flow discharge from a single rainfall of r and $E(a) = 1/w \int wD(a) dx$ is the accumulated deficit downslope.

Integrating over the rainfall distribution, the total annual overland flow discharge is then:

$$Q_* = \beta Ra \exp\left[ -\frac{E(a)}{r_0 a} \right] \tag{8}$$

This third case is commonly relevant in humid temperate conditions, where there is significant subsurface flow. Overland flow increases much more than linearly with distance, but only becomes significant in amount within the slope concavity.

## SEDIMENT TRANSPORT

The sediment transport processes considered here are rainsplash, which is driven by rainfall intensity and gradient, and wash processes, which are driven by overland flow discharge, usually to a square power. A threshold for erosion is also included. For both rainsplash and wash, we assume that sediment is carried at its full capacity, that is that sediment removal is transport or flux limited.

The sediment transport model used here for rainsplash is:

$$S_R = \alpha r^2 \Lambda \tag{9}$$

where $S_R$ is the rate of rainsplash sediment transport per unit width in a storm of rainfall $r$ and $\alpha$ is an empirical rate constant.

This expression assumes that net sediment transport by splash is proportional to the square of rainfall intensity and to gradient. Integrating over the frequency distribution of rainfalls, we obtain the average annual transport as:

$$S_{R*} = \alpha \frac{N_0}{r_0} \int_0^\infty r^2 \exp(-r/r_0) dr = 2\alpha R r_0 \Lambda = \varkappa \Lambda \tag{10}$$

where $\varkappa$ has been written for the overall diffusion rate coefficient $[= 2\alpha R r_0]$, measured on an annual basis.

For wash transport, we assume that sediment detachment is proportional to unit flow power in excess of a threshold, and to discharge:

$$S_F = \zeta Q(r) [Q(r)\Lambda - \Theta_c] \tag{11}$$

where $S_F$ is the sediment carried by the flow in a single storm of $r$, $\Theta_c$ is the power threshold for sediment detachment and $\zeta$ is an empirical rate constant.

This can be integrated in principle over the frequency distribution, to give the average annual wash transport:

$$S_{F*} = \zeta \frac{N_0}{r_0} \int_{r_c}^\infty Q(r) [Q(r)\Lambda - \Theta_c] \exp(-r/r_0) dr \tag{12}$$

where $r_c$ is the rainfall required to overcome the threshold.

In practice, the integration can readily be completed only for the first case, where the soil deficit $h$ is constant, and overland flow reaches equilibrium. Then we have $r_c = h + \Theta_c/(a\Lambda)$, and:

$$S_{F*} = 2\beta^2 \zeta a^2 R r_0 \Lambda \exp(-h/r_0)(1 + \phi) \exp(-2\phi) \tag{13}$$

where $\phi = \Theta_c/(2a r_0 \Lambda)$ is a threshold dependent function, which takes the value zero when $\Theta = 0$.

In the case of a zero threshold, the combination of splash and wash may be expressed in the simple form:

$$S_* = \varkappa \Lambda \left[ 1 + \left( \frac{a}{u} \right)^2 \right] \tag{14}$$

where $u$ is a parameter with units of length.

Comparison with equations (10) and (13) give the distance parameter:

$$u = \frac{\exp(h/2r_0)}{(2R r_0 \beta^2 \zeta / \varkappa)^{1/2}} \tag{15}$$

The parameter $u$ may be interpreted as the distance from the hillslope divide at which wash becomes greater than rainsplash. Empirically this is found to range from 1 m to 1000 m.

For the more general case, with a threshold, equation (13) may be re-written in the form:

$$S_* = \varkappa\Lambda[1+(a/u)^2(1+\phi)\exp(-2\phi)]$$ (16)

where $u$ has the same value as above.

In the arid case of brief storms, the distance $a$ should be replaced in equation (16) by: $a[1-(a/x_0)^b]/[1-a/x_0)^{b+1}]$, following equation (5) above. For the case of humid temperate slopes, equation (16) remains valid, provided that $h$ in equation (15) is replaced by $E(a)/a$ (from equation (7) above) to give the relevant value for $u$.

The effect of these slope processes may be seen for a ridge in equilibrium with a constant rate, $T$, of tectonic uplift. The total sediment transport is then necessarily $S_* = Ta$. In the simplest case, with equilibrium overland flow, no subsurface flow and a zero threshold, we may write down an explicit expression for the slope profile obtained. We obtain:

$$\Lambda = \frac{Tx}{\varkappa[1+(x/u)^2]}$$

$$z = \frac{u^2T}{2\varkappa}\ln[1+(x/u)^2]$$ (17)

where $x$ ($=a$ for a ridge) is distance from the divide, $\Lambda$ is the local gradient and $z$ is the fall from the divide.

This profile is convexo-concave in form, with its steepest gradient at $x=u$. The distance parameter, $u$, thus has a clear morphological expression. For non-zero thresholds, we can write down two explicit relationships for distance, $x$, and gradient, $\Lambda$, in terms of the parameter $\phi$ (which is inversely proportional to the distance gradient product):

$$\left(\frac{u}{x}\right)^2 = \frac{\phi u^2}{\varkappa'\Theta'} - (1+\phi)\exp(-2\phi)$$

and

$$\Lambda = \frac{\Theta'}{\Theta x}$$ (18)

where $\Theta' = \Theta_c/(2r_0)$ and $\varkappa' = \varkappa/T$ each have the dimensions of length.

The effect of increasing the erosion threshold is seen in Figure 14.1, which is a log–log plot of gradient against distance. The initial linear rise represents a convex divide. For zero threshold, the maximum gradient occurs at $u$ (10 m in this example). As the threshold is increased, the point of inflexion (or maximum gradient) is displaced downslope, due to the lower efficacy of wash. The transition from convex to concave slopes also becomes slightly more abrupt, while the maximum gradient is almost unchanged.

The effect of short storms is to also extend the convexity. For certain parameter values, it is theoretically possible to infer a convexo-concavo-convex profile, but many reasonable

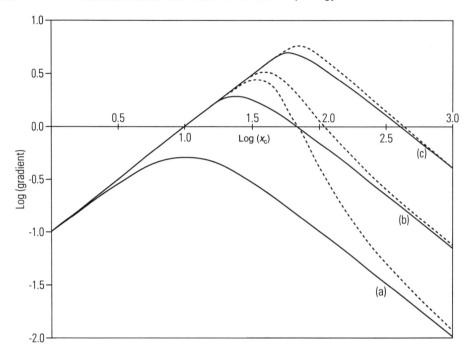

**Figure 14.1** Example slope profiles for conditions of constant downcutting, shown as log–log plots of gradient against distance from divide. The conditions are for strong subsurface flow ($u_0 = 10$ m; $m/r_0 = 10$; $a_0 = 15\,000$ m). The three pairs of curves are for traction thresholds of (a) 0, (b) 100 and (c) 1000 m. Solid lines are for conditions of Hortonian overland flow; broken lines for conditions with appreciable subsurface flow and saturation overland flow

parameter values lead to very long gentle convexities of the type described by Dunne (Dunne & Aubry, 1986) for Kenya. Appreciable subsurface flow also lengthens the convexity, but terminates it with very short and abrupt concavity in the slope base region where saturated conditions become dominant.

## CHANNEL HEADS

We now use the sediment transport model above to apply the stream head criteria of exceeding the threshold or reaching instability. This is presented both for the individual storm event and for the integrated average effect. The threshold criteria has already been implicitly presented in the form of the wash transport law (equation (11) above). Since only wash, and not rainsplash, is able to cut a channel, it is clear that the power threshold, $\Theta_c$, must be exceeded in an individual event to cut a new channel, and on average to cut a permanent channel. Below the threshold, no incision can occur.

The instability criterion for the establishment of a local channel is taken to be that for the growth of an initial small hollow (Carson & Kirkby, 1972, p. 394; Smith & Bretherton, 1972). The continuity equation for sediment transport along a flow strip, of variable width, $w$, and ignoring aerial input of dust, etc., is:

$$\frac{\partial z}{\partial t}+\frac{1}{w}\frac{\partial(wS)}{\partial x}=0 \qquad (19)$$

Rewriting equation (19), making use of the identities in equations (3):

$$-\frac{\partial z}{\partial t}=\frac{\partial S}{\partial x}+\frac{S}{w}\frac{\mathrm{d}w}{\mathrm{d}x}+\frac{\partial S}{\partial\Lambda}\frac{\mathrm{d}\Lambda}{\mathrm{d}x}$$

$$=\frac{\partial S}{\partial a}\frac{\mathrm{d}a}{\mathrm{d}x}+\frac{S}{w}\frac{\mathrm{d}w}{\mathrm{d}x}+\frac{\partial S}{\partial\Lambda}\frac{\mathrm{d}\Lambda}{\mathrm{d}x}$$

$$=\frac{\partial S}{\partial a}\left(1+\frac{a}{p}\right)-\frac{S}{p}+\frac{\partial S}{\partial\Lambda}\frac{\mathrm{d}\Lambda}{\mathrm{d}x}$$

$$=\frac{\partial S}{\partial a}+\frac{1}{p}\left(a\frac{\partial S}{\partial a}-S\right)+\frac{\partial S}{\partial\Lambda}\frac{\mathrm{d}\Lambda}{\mathrm{d}x} \qquad (20)$$

In the neighbourhood of a small irregularity, it is argued that contour curvature is initially changed appreciably, while gradient and unit area are changed to a much smaller extent. The direction of change of erosion rate is therefore given by the sign of the second term on the last line of the right hand side of equation (20). That is to say that small hollows ($1/p$ positive) will grow, by increasing their local erosion rate relative to adjacent points, if and only if:

$$\frac{\partial S}{\partial a}>\frac{S}{a} \qquad (21)$$

where the partial differentiation keeps gradient constant.

This criterion can be directly applied for simple types of transport-limited sediment transport, such as the splash plus wash expression of equation (14) above, which is applicable for semi-arid areas with storms of appreciable duration. Unstable hollow growth occurs, if and only if $a>u$. For the more complex patterns of humid or arid overland flow production, or if there are significant erosion thresholds ($\Theta_c$ in equations (11) and (12)), then the stability criterion is no longer independent of gradient. Nevertheless, the criterion, although applying to the two-dimensional form of the hillside contours, can readily be evaluated for a one-dimensional slope profile.

There is a strong connection between the form of slope profile (in vertical section) and the stability criterion, and slopes which are unstable with respect to hollow growth are generally concave in profile. To make this relationship more explicit, equation (20) may be rewritten in the form:

$$\frac{\partial S}{\partial\Lambda}\frac{\mathrm{d}\Lambda}{\mathrm{d}x}=\left(-\frac{\partial z}{\partial t}+\frac{S}{\rho}\right)-\frac{\partial S}{\partial a}\left(1+\frac{a}{\rho}\right) \qquad (22)$$

Writing the local rate of slope lowering $-\partial z/\partial t=T$, and its weighted average down the flow strip to the point of interest as $U=(1/a)\int Tw\,\mathrm{d}x=S/a$, gives:

$$\frac{\partial S}{\partial \Lambda} \frac{d\Lambda}{dx} = \left(T + \frac{Ua}{\rho}\right) - \frac{\partial S}{\partial a}\left(1 + \frac{a}{\rho}\right) \tag{23}$$

Since $S$, $\partial S/\partial \Lambda$ and $a$ are always positive, the slope profile is concave ($d\Lambda/dx < 0$) if and only if:

$$\frac{\partial S}{\partial a}\left(1 + \frac{a}{\rho}\right) > \frac{S}{a}\left(\frac{T}{U} + \frac{a}{\rho}\right) \tag{24}$$

which should be compared with the stability condition in equation (21) above.

Normally $(1 + a/\rho)$ is positive, although there are rare exceptions. For the normal case, equation (24) leads to a series of relationships which clarify the relationship between hollow loction and profile concavity. For the special case of a constant downcutting slope ($T = U$), the conditions are identical, so that, on a profile of constant downcutting, concave sections of the profile are unstable and convex sections stable. For more general circumstances, only more restricted statements can be made, according to the value of the ratio $T/U$. For mature slopes, with rates of lowering decreasing downslope, then the ratio is less than unity. All convexities must then be stable, and the unstable zone must begin strictly within the concavity. For youthful slopes at an early stage of incision into an uplifted block, $T/U$ is typically greater than unity. In this state all concavities must be unstable, and the unstable zone may extend somewhat into the convexity above. In pronounced hollows with strong flow convergence, where $a/\rho$ is high, the sensitivity of the condition for concavity to local rates of lowering is much less than on a ridge or a nose (spur), so that profile concavity and instability should usually roughly coincide. This analysis can refer meaningfully only to the average position of the instability, since slope profile form changes only slowly over time.

The effect of an erosion threshold, and the role of magnitude and frequency in stream head location may be seen from the single storm version of the sediment transport equations ((9) and (11) above), and their integral (equation (16)). Applying the stability criterion to equation (16), substituting $E(a)/a$ for $h$ to allow for subsurface flow:

$$\frac{\partial S}{\partial a} = \varkappa\Lambda\left[\frac{2a}{u^2}(1 + \Phi)\left(1 - \frac{D-h}{r_0}\right) - \frac{a^2}{u^2}(1 + 2\Phi)\frac{d\Phi}{da}\right]\exp(-2\Phi) \tag{25}$$

where $D$ is the local saturation deficit and $h$ is the averaged deficit $E(a)/a$ from equation (7).

The critical stable equivalent catchment length, $a_c$ is given by applying the stability criterion (21) and substituting $-\Phi/a$ for $d\Phi/da$, to give:

$$\frac{a_c}{u} = \left[(1 + \Phi)\left(1 - \frac{D-h}{r_0}\right) + \Phi(1 + 2\Phi)\right]^{-1/2}\exp(\Phi) \tag{26}$$

The significance of equation (26) will first be examined for the case of a fixed soil water deficit $h$, where subsurface flow is relatively unimportant. Bearing in mind that $\Phi$ has been defined as $\Theta_c/(2ar_0\Lambda)$ in equation (13) above, and contains terms in both distance $a$ and gradient $\Lambda$, equation (26) may thus be read as an expression relating equivalent length to gradient at the critical point, and this expression is drawn in Figure 14.2. It may

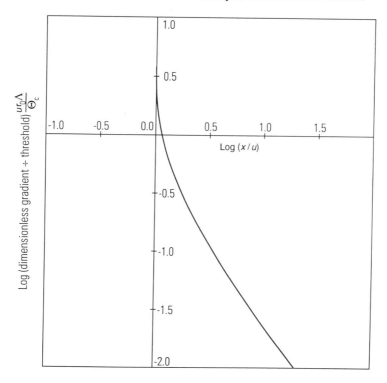

**Figure 14.2** The theoretical relationship between average stable (unchannelled) valley length, $x_c$, and local slope gradient, $\Lambda$ in the presence of a slope threshold. The curve is drawn from equation (26) for conditions of splash and wash erosion, without significant saturation overland flow

be seen that, where erosion thresholds are low or gradients high, then the previous model of a fixed stream head distance, $a_c$ remains appropriate, and stream head position is determined primarily by stability considerations. However, where thresholds are large and/or gradients low, there is a strong inverse relationship between $a_c$ and gradient. Beyond the constant $a_c$ region, this behaves approximately like a power law $\Lambda \propto x_c^{-p}$, with an exponent, p, which falls asymptotically towards 1.0. This relationship broadly matches Dietrich's empirical evidence for a constant product of catchment area and gradient for northern California. No exact comparison is possible in general since equivalent catchment length, $a$ and catchment area at the stream head are not identical measures.

The critical ratio of gradient to dimensionless threshold required for a gradient dependent stream head may be seen from Figure 14.2 to be around 1.0. The critical threshold required is therefore $\Theta_c > r_0 u \Lambda$, which is the flow power expended at the soil surface during an average rain day, at distance $u$ from the divide. It may be seen that, as the threshold strength falls, gradient dependence occurs only at lower and lower slope angles. Thus typical stream head gradients are in the constant $a_c$ region where surfaces are weak and thresholds low, and in the gradient dependent region where surfaces are strong. Arguably, low threshold conditions are related to unvegetated surfaces, at least partially covered with easily erodible fine material. High thresholds are then associated with coarse debris, and even more with strong turf cover. Optimum conditions for strong threshold control are therefore argued to be moderate slopes, a sub humid to humid climate year round to provide a strong turf cover, and some intense rainstorms to erode it, corresponding closely to the areas described by Dietrich in northern California. Semi-arid conditions, such as those of the Colorado Plateau described by Melton (1957), are similarly thought to lie in the constant $a_c$ domain, with low erosion thresholds. Using climatic parameters from such areas, it is suggested that the critical value of $\Theta_c$ should lie close to $0.1 \, m^2 \, day^{-1}$.

Where subsurface flow is important, an exact analysis of equation (26) depends on a complete knowledge of the slope profile. For a linear ridge (parallel contours), one relevant solution is to look at conditions of constant downcutting, for which the unstable position is equivalent to the point of inflexion in the slop profile. Under these conditions there are just two additional parameters which describe the behaviour, both related to the subsurface flow regime. The first is the ratio of the soil moisture parameter, $m$ (in equation (6)), to the mean rain per rain day, $r_0$. Ratios greater than 1.0 are associated with areas where subsurface flow is significant, and for humid-temperate areas like Britain this may reach values of up to 10. The second relevant parameter is the distance, $a_0$, at which subsurface flow seeps out at the surface on unit slope gradient. High values of $a_0$ ($> 1000 \, cm$) are associated with highly permeable deep soils and low (but positive) mean net rainfall. High values of $m/r_0$ and $a_0$ give conditions which might favour a positive relationship between hollow area and hollow gradient, since lower gradients enhance overland flow, but Figure 14.3 shows that in practice this effect is likely to be very weak, and only evident at realistic gradients where traction thresholds are low. The strongest effect of subsurface flow is on the form of the slope profile concavity, as may be seen in Figure 14.4, which shows the strong concavities associated with low thresholds.

We can obtain some further insight into the stream head relationships by disaggregating the sediment transport into daily rainfall components. Equations (7) and (11) provide an adequate basis for disaggregating the rill transport, and equation (9) for the splash term in the sediment transport equation. Using the same notation as above, the total sediment transport in a day with rainfall $r$ is:

$$S = \left( \frac{x}{2Rr_0} \right) \left\{ \frac{(r-h)x}{\beta^2 u^2} \exp(h/r_0)\,[r-h)x\Lambda - \Theta_c] + \Lambda r^2 \right\} \tag{27}$$

where the first term is for rillwash (with zero replacing negative values) and the second is for rainsplash.

Applying the stability criterion as before, instability should occur where two conditions are met: first the instability criterion $x\,\partial S/\partial x > S$, and second the threshold condition for non-zero rillwash $(r-h)x\Lambda > \Theta_c$. For the case of constant deficit $h$, the instability criterion simplifies to: $x/u > r/(r-h) \exp(-h/2r_0)$. These two expressions are sketched in Figure 14.5. In general they cross at the storm rainfall $r = (\Theta_c/u\Lambda) \exp(h/2r_0)$, and this intersection is in the relevant range (i.e. $r > h$) if $\Theta_c > u\Lambda h \exp(-h/2r_0)$. Plainly both conditions must be met for a channel to form, so that the upper curve is the control at any given storm rainfall. In other words there are two regimes where thresholds are appreciable, and only one, controlled by the instability criterion, for low thresholds. In both cases the critical distance falls with storm rainfall, supporting the intuitive view that channels incise headwards in severe storms.

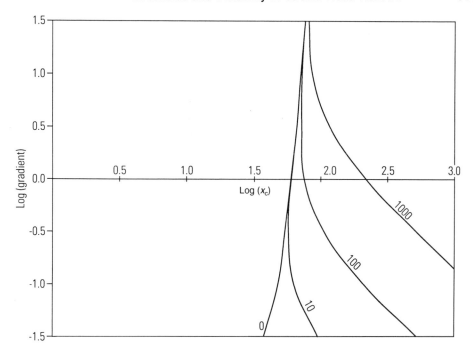

**Figure 14.3** Examples of the relationship between gradient and distance at the theoretical stream head. The curves are shown for conditions of appreciable saturation overland flow, and with a range of thresholds. The relationship is calculated on the assumption that profiles are in equilibrium with a constant rate of downcutting

This intersection gradient value level is necessarily greater than the critical gradient required for threshold behaviour to be dominant on the average, as defined by equation (26) above. Their ratio is $r_0/h \exp(h/2r_0)$, which has a minimum value of $e/2 \approx 1.36$. Thus we will always have the two regimes in conditions where threshold response is dominant. Where instability response is dominant overall, we may or may not have two regimes, depending on the value of the ratio $h/r_0$. For any particular case, the average stream head (ASH) position is as shown in Figure 14.2. The crossover point between regimes is defined by:

$$\frac{x_c}{u} = \frac{1}{\exp\left(\dfrac{h}{2r_0}\right) - \dfrac{h}{r_0}\left(\dfrac{ur_0\Lambda}{\Theta_c}\right)} \tag{28}$$

where the bracketed term $ur_0\Lambda/\Theta_c$ is the dimensionless ratio of gradient to threshold which was plotted in Figure 14.2.

Figure 14.6 sketches the possible relationships between regimes, using arithmetic scales, with a similar vertical axis to Figure 14.2. The horizontal line is taken as the stream head gradient, for a given threshold, below which threshold behaviour dominates on average. The curve shows the intersection gradient, referred to above,

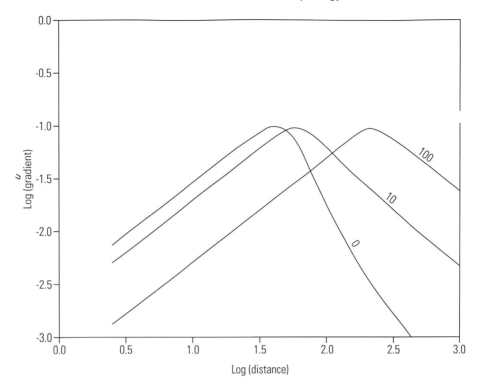

**Figure 14.4** Some constant downcutting profile forms for the examples shown in Figure 14.3. The curves are drawn for maximum gradients of 0.1 in each case, and for thresholds of 0, 10 and 100 m

and applies for each point on the slope profile. Since the stream head is usually close to the locus of maximum gradient (exactly so for the constant downcutting case), then almost all points on the slope plot below the horizontal line in cases where the integral behaviour is threshold dominated. It may be seen that there are three possible regimes for the valley head, as follows:

(i) Instability dominated integral behaviour for ASH:
    (a) storm behaviour is always instability controlled;
    (b) storm behaviour is threshold controlled in very large storms only.
(ii) Threshold dominated integral behaviour for ASH. The change to storm threshold control occurs upstream for the ASH.

At a downvalley site ($x_c$ larger than the value at P in Figure 14.5), instability as defined by equation (27) appears to set in before rillwash begins. However, the first term in equation (27) is zero because the flow power $(r - h)x\Lambda$ has not yet exceeded the traction threshold, $\Theta_c$. Thus the relevant criterion for hollow enlargement is the rillwash threshold. Once the threshold is reached, instability is already well developed. That is to say, the newly effective rill growth is already many times greater than the local rate of infilling by interrill processes. The field relationships associated

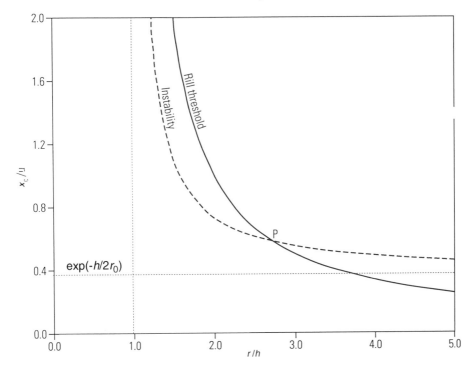

**Figure 14.5** Stable (unchannelled) valley length $x_c$ as a function of daily storm rainfall $r$ for an example set of parameter values, and for splash and wash erosion. In general the stability condition is critical at upvalley sites, corresponding to large critical storms; and the rill erosion threshold is critical at downstream sites, for smaller critical storms

with this regime are therefore thought to be well-defined channel banks, with sharp margins, and with relatively little sediment flowing across those margins to soften their morphology. Below the ASH position, this morphology will generally be more or less permanent; but above the ASH the sharp banks will be formed in major storms. For subsequent smaller storms there will be no rill transport, and the stream extensions will gradually degrade by splash infilling and bank collapse.

This sharp-bank morphology is usually clearly defined at downstream sites, but only at some stream heads. Generally it is very clear where there is a strong turf cover, as is the case for Dietrich's sites in northern California, and at the wide variety of other sites with clear headcuts. Plainly there are many cases where sharp stream heads are enhanced or created by subsurface piping. Piping is outside the scope of this discussion, but rapid subsurface flow can be seen as another factor which tends to increase the flow required to overcome traction thresholds. Both piping and strong turf covers have the effect of providing a higher threshold at the soil surface than at shallow depth, so that stream heads, and to a lesser extent banks, are additionally kept sharp by undermining and collapse.

At an upvalley site (i.e. $x_c$ less than the value at P in Figure 14.5), for which the erosion thresholds is reached before instability sets in, rillwash is actively eroding

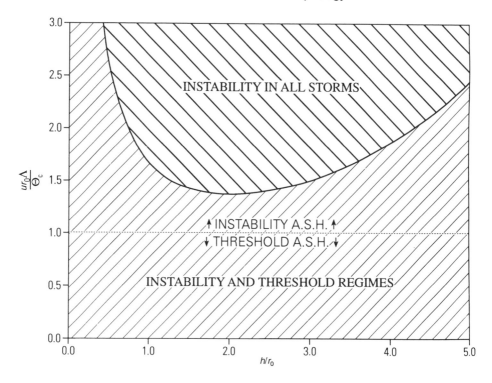

**Figure 14.6** Threshold and instability regimes. Broken line distinguishes regimes for average slope head position (ASH). Within instability ASH zone, heavy line outlines region where instability dominates in all storms from region with mixed response, dependent on storm intensity

and transporting material, but hollow enlargement is prevented in moderate storms by active splash infilling. For large enough storms, however, the rill flow becomes large enough to dominate, and rills can progressively begin to incise the surface faster than interrill processes can fill them in. In this regime, channel margins are receiving material over their banks as fast as it is carried away by sediment transport along the length of the channel. Channels therefore have less distinct banks. Out of the channels, at sites above the rill threshold but below the instability level, exceedance of the traction threshold also leads to effective transport of material, and some tendency to significant sorting and armouring of surface materials. The combination of this fuzzy-bank morphology and surface armouring is common, particularly in many semi-arid environments where there is an incomplete vegetation cover at the surface and appreciable stone content in the regolith. On the long unchannelled slopes described by Dunne & Aubry (1986) for Kenya, the area above permanent stream heads is associated with an increasing cross-slope roughness (T. Dunne, pers. comm.), which may also be diagnostic of this morphological regime.

The spatial expression of the three possible morphologies can be summarised as follows. Where the ASH is threshold dominated (Case (ii) above), then storm extensions above the ASH always show sharp headcuts, though these will degrade over time. Above the sharp headcut there may be further fuzzy extensions. Where the ASH is instability

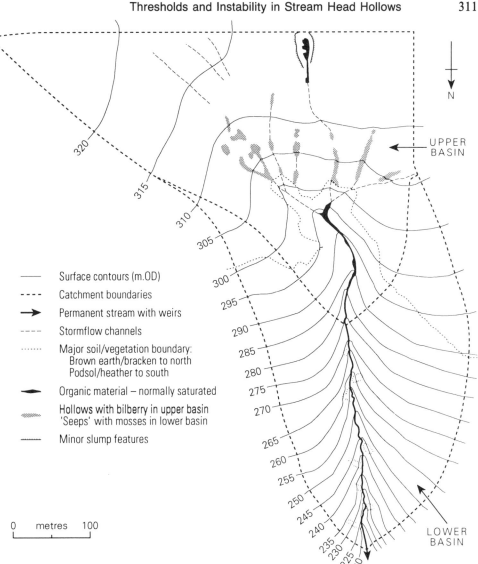

N

UPPER
BASIN

320

315

310

305

300

295

290

285

280

275

270

265

260

255

250

245

240

235
230
225
220

LOWER
BASIN

— Surface contours (m.OD)

- - - Catchment boundaries

→ Permanent stream with weirs

- - - - Stormflow channels

⋯⋯ Major soil/vegetation boundary:
    Brown earth/bracken to north
    Podsol/heather to south

➤ Organic material – normally saturated

▨ Hollows with bilberry in upper basin
  'Seeps' with mosses in lower basin

▬ Minor slump features

0    metres    100

**Figure 14.7**   East Twin Catchment, showing 'fuzzy' stream head extensions above a clearly defined headcut (from Weyman, 1974)

dominated, there will may or may not be a transition to sharply defined channel banks at some point downstream of the stream head (cases (i)a and (i)b above). Above the stream head, extensions will always be fuzzy. In all cases, the headward extensions are associated with increases in drainage density. We may therefore expect not only linear headward extensions, but also some fanning out into multiple stream heads. This is well shown in Figure 14.7 for the East Twin catchment (Weyman, 1974), and conforms with field descriptions of channel extensions after severe storms. The cumulative effect of the radiating channel extensions above the stream head is to erode the spoon-shaped

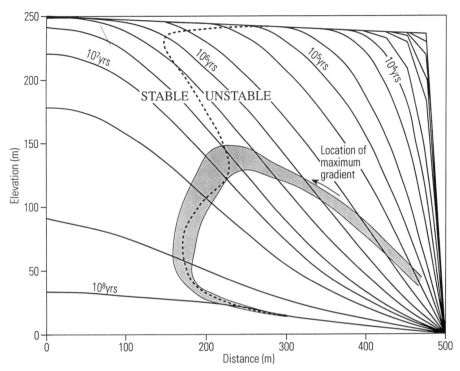

**Figure 14.8** Simulated evolution of a slope profile where subsurface flow is significant. Broken line distinguishes stable and unstable slope regions. Loops outline the locus of the maximum slope gradient as it evolves over time

are commonly found there. Where subsurface flow is important, there is a strong positive feedback between the formation of this bowl and the conditions for overland flow and channel extension, and this interaction is much weaker in semi-arid climates.

Figure 14.8 illustrates the complex behaviour which may occur where subsurface flow is important. This illustrates a slope profile evolving under the wash processes discussed here, and is for a one-dimensional profile without flow convergence in plan. No landslides have been included to degrade the initially steep slopes. To begin with, the profiles may be seen to depart very strongly from the condition of constant downcutting, and the boundary between stable and unstable zones, defining the ASH, is within the convex part of the profile. Once the slope profile reaches something like 'maturity', with no vestige of the initial plateau surface, the ASH corresponds more or less with the locus of maximum gradient. In interpreting the distance to the ASH, it should be remembered that the distance $u$ in equation (15) is also dependent on gradients through the average deficit, so that $u$ is tending to decrease as the slope gradients decline. Thus there are two periods in the slope evolution shown during which the distance to the ASH, $x_c$, increases as gradient falls, as is expected with threshold dominance. Between these two periods is a period of increasing $x_c$, in which the distance $u$ is falling rapidly in response to reduced saturation deficits, so that $x_c$ is falling, while the ratio $x_c/u$ continues to rise. Thus drainage density may not have a simple unidirectional history, even though the processes are idealised in a relatively simple way. Furthermore, there may be conditions, as in this example, where the relationship between gradient and $x_c$ appears to reverse, because of the influence of subsurface flow, even though threshold effects are dominant.

## CONCLUSIONS

It has been shown that the Smith & Bretherton instability criterion and the Horton/ Willgoose tractive stress criterion can be combined in a single comprehensive theory of stream head location. This theory forecasts a rather fixed drainage density and stream head location where instability is dominant, which is thought to occur typically in semi-arid environments with sparse vegetation, stony regolith and without strong subsurface piping. Where tractive thresholds are dominant, there is a strong inverse relationship between distance to stream head and gradient, but this can be partially masked where subsurface flow is strong. These conditions are best met in temperate conditions, with a strong turf cover.

A morphological classification is proposed on the basis of channel type. Channels may either have sharply defined or fuzzy banks. Banks are fuzzy when the sediment flux across them at the time of formation is similar to that along their length, so that there is no clean incision. The headmost storm extensions are always fuzzy. Where the average stream head is fuzzily defined, all its extensions are fuzzy, and there may be a transition to sharp banks downstream. Where the average stream head is sharp, it will have storm extensions which are fuzzy at their distal ends, and become sharp before reaching the average stream head. These sharp extensions will gradually degrade over time between major events. Above the average stream head position, the zone of storm extensions is typically associated with an increased channel density.

## REFERENCES

Beven, K. J. and Kirkby, M. J. (1979). A physically based, variable contributing area model of basin hydrology. *Hydrological Sciences Bulletin*, **24**, 43–69.

Carson, M. A. and Kirkby, M. J. (1972). *Hillslope Form and Process*. Cambridge University Press, 475 pp.

de Ploey, J., Kirkby, M. J. and Ahnert, F. (1991). Hillslope erosion by rainstorms – a magnitude-frequency analysis. *Earth Surface Processes and Landforms*, **16**, 399–409.

Dietrich, W. E. and Dunne, T. (1993) The channel head. In Beven, K. J. and Kirkby, M. J. (eds), *Channel Network Hydrology*, Wiley, Chichester, pp. 175–219.

Dietrich, W. E., Wilson, C. J. and Reneau, S. L. (1986). Hollows, colluvium and landslides in soil-mantled landscapes. In Abrahams, A. D. (ed.), *Hillslope Processes*, Allen and Unwin, pp. 361–388.

Dunne, T. and Aubry, B. F. (1986). Evaluation of Horton's theory of sheetwash and rill erosion on the basis of field experiments. In Abrahams, A. D. (ed.), *Hillslope Processes*, Allen and Unwin, pp. 31–53.

Horton, R. E. (1945). Erosional development of streams and their drainage basins: hydrophysical approach to quantitative morphology. *Geological Society of America Bulletin*, **56**, 275–370.

Melton, M. A. (1957). An analysis of the relations among elements of climate, surface properties, and geomorphology. *Technical Report 11, Project NR 389–042, Office of Naval Research*, Columbia University.

Smith, T. R. and Bretherton, F. P. (1972). Stability and the conservation of mass in drainage basin evolution. *Water Resources Research*, **8** (6), 1506–1529.

Weyman, D. R. (1974). Runoff processes, contributing area and streamflow in a small upland catchment. In Gregory, K. J. and Walling, D. E. (eds), *Fluvial Processes in Instrumented Watersheds,*, Institute of British Geographers, Special Publication 6, pp. 33–43.

Willgoose, G. R., Bras, R. L. and Rodriguez-Iturbe, I. (1991). A physically based coupled channel network growth and hillslope evolution model. *Water Resources Research*, **27** (7), 1671–96.

Wooding, R. A. (1965). A hydraulic model for the catchment stream problem. I: Kinematic wave theory. *Journal of Hydrology*, **3**, 254–276.

# 15 Theoretical Constraints on the Development of Surface Rills: Mode Shapes, Amplitude Limitations and Implications for Non-Linear Evolution

DEBORAH S. LOEWENHERZ-LAWRENCE
*Department of Geology and Geophysics, University of California, Berkeley, USA*

## ABSTRACT

The initial morphology and subsequent development of shallow surface incisions, such as rills and gullies, is assessed, based on the solutions for the associated linear stability problem and the derived scalings for those solutions. Of particular interest are the mode shape solutions, corresponding to the downslope longitudinal profiles of the incisions, and the hydrodynamic scaling for the surface water flux, which constrains the relative amplitudes attained by the various unstable modes within the linear regime. The mode shape solutions indicate that the linearly most unstable modes, which are the first-order modes of very short transverse wavelength, are spatially concentrated near the base of a hillslope and may never be apparent in the field. The more slowly growing unstable modes extend farther upslope and thus have a competitive advantage in subsequent modal interactions in that they have the first opportunity to capture and redirect surface water. The analysis also provides an amplitude limitation of the order of $1/k^2$, where $k = (2\pi)/\lambda$ is the transverse wavenumber of the surface disturbance. Consequently, the initially most unstable, short wavelength, modes rapidly saturate so that the marginally unstable modes, particularly the longer wavelength, first-order modes, ultimately determine the morphology of the drainage surface, despite their slow growth rates. Using the mode shapes, the modal growth rates and the amplitude constraints, a series of configurations is proposed which incorporate the combined effects of these factors on the development of a drainage surface.

## INTRODUCTION

In 1972, Smith and Bretherton proposed a model of drainage basin evolution which was rigorously developed from the fundamental mass balances for a hillslope system and also incorporated a very general transport law for the flux of surface sediments. Although their physical model was quite similar to other mathematical models for hillslope profile evolution appearing somewhat earlier and also concurrently (e.g. Culling, 1965; Ahnert, 1967, 1973; Kirkby, 1971), Smith & Bretherton (1972) additionally used a linear stability analysis to assess the tendency for the initiation of surface channels based on concepts derived from hydrodynamic stability theory. This latter component

*Process Models and Theoretical Geomorphology.* Edited by M. J. Kirkby
©1994 John Wiley & Sons Ltd

of their model represents an entirely unique and original contribution to theoretical geomorphology and those results are frequently cited in contemporary discussions of drainage network initiation and development (e.g. Dunne, 1980; Willgoose *et al.*, 1990; Dietrich and Dunne, 1993).

More recent work (e.g. Kirkby 1980, 1987, this volume, Chapter 14; Loewenherz 1990, 1991a,b) has reevaluated and extended the stability arguments posed by Smith (1971) and Smith & Bretherton (1972). It has thus been clearly established that, in mechanistic terms, instability leading to the initiation of shallow surface incisions on a previously unchanneled hillslope is associated with the dominance of advective over diffusive processes in the transport of surface sediments, as was suggested by Smith and Bretherton, though in somewhat different terms. Morphologically, this result implies that hillslope profiles which exhibit some concavity will be unstable to transverse perturbations (e.g. sinusoidal disturbances oriented perpendicular to the length of the slope). As a consequence of this physical instability, an unchanneled surface with a convexo-concave downslope profile will inevitably undergo a transition to a more stable configuration in which the surface is dissected by longitudinal channels, as illustrated in Figure 15.1. Beyond this fundamental result, however, two outstanding issues remain unresolved. Firstly, how is the incipient instability manifested on an unchanneled surface? In other words, what would one expect to observe in field and/or laboratory investigations which might be indicative of a transition to a surface configuration which is channelized? Secondly, how does the instability develop once initiated? Smith & Bretherton's (1972) work predicts that the fastest growing perturbations (and hence the first to be seen) have infinitesimally short transverse wavelength $\lambda = 2\pi/k \to 0$, where $k$ is the wavenumber.

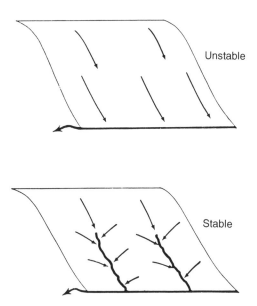

**Figure 15.1** Proposed unstable and stable configurations for a hypothetical drainage surface. The upslope boundary represents the drainage divide, while the downslope boundary consists of a stream or other 'sink' which is removing water and sediment from the bottom of the hillslope. The lateral (i.e. cross-slope) distance between the two channels corresponds to the transverse wavelength $\lambda = (2\pi)/k$ referred to in the stability analysis

Recent work (Loewenherz, 1990, 1991a,b) has shown that additional physical mechanisms, such as the hydrodynamics of the overland flow, which are not incorporated in Smith and Bretherton's model, are significant at short lengthscales and lead to a finite, though still quite short, transverse wavelength for the fastest growing modes. Furthermore, the linear stability analysis requires that the surface perturbation is quite small, such that those results must be expanded if one is to consider the development of stable surface incisions which are no longer infinitesimal. Both of these issues pertain to the morphologic characteristics of the surface instability and thus are dependent on the wavelength and on the mode shapes of the evolving surface disturbance, neither of which have been previously discussed. The purpose of this paper is therefore to consider the downslope morphology of initial surface incisions as given by the mode shape solutions for the stability problem and to also present the implications of these solutions and their scalings for the non-linear development of the drainage surface.

## THE THEORETICAL MODEL

A stability analysis may be used to assess the tendency for a transition from a simple equilibrium form to an alternative, preferred, configuration based on the dynamic response of a base state to an induced perturbation. The tendency for channel initiation can accordingly be evaluated by considering a hypothetical base state consisting of a smooth, unchanneled surface, from and over which water and sediment are transported subject to continuity constraints. By applying perturbation techniques to an appropriate mathematical representation of this base state, the dynamic response to an internal disturbance can be evaluated. If a particular configuration is stable, then the associated physical system possesses the capacity to internally damp disturbances, so that they are eliminated as the system evolves. If, however, the configuration is unstable, then the dynamics of the system will reinforce the perturbation such that its presence may ultimately dominate the behavior and morphology of the system. Based on field evidence (e.g. Melton, 1957; Montgomery & Dietrich, 1988, 1989, this volume, Chapter 11), we anticipate that the preferred form is one in which the hypothetical base state hillslope is dissected by surface channels which start at some distance from the divide and generally extend to the base of the slope. We also anticipate that these channels are characterized by a dominant wavelength or lateral spacing, which is roughly proportional to the inverse of the drainage density for first-order channels within the network.

To construct the mathematical model, we first pose a complete set of governing equations which characterize the movement of sediment through the model hillslope system. The fundamental equation for this system is a statement of continuity of the general form

$$T(x,y,t) - \nabla \cdot [mq_s] = h_{,t} \tag{1}$$

where $h$ is the elevation at a point on the surface, $q_s$ is the sediment flux, m is a unit vector denoting the local downslope direction, $T$ is the rate of tectonic uplift or stream downcutting which may be driving the system, and commas are used to denote differentiation with respect to the subscripted variable. The Cartesian $(x-y)$ coordinates lie in a horizontal plane and are respectively directed down and across the undissected

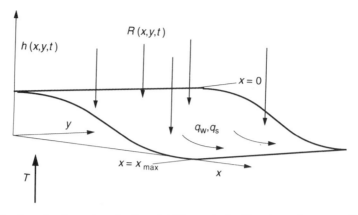

**Figure 15.2** Coordinate system for the stability analysis. The downslope coordinate is given by $x$ with $x=0$ at the divide and $x=x_{max}$ at the base of the slope, and the transverse coordinate is given by $y$, so that the elevation of the surface is $h(x,y,t)$, where $t$ is a geomorphic time variable. See text for additional details

slope (see Figure 15.2) and $t$ is a geomorphic (cf. hydrodynamic) time variable. In general, the sediment flux $q_s$ is at least partially dependent upon the surface water flux $q_w$, so that we require a separate mass balance equation which accounts for the surface water as given by

$$\nabla \cdot [mq_w] = R(x,y,t) \qquad (2)$$

where $R$ is the average effective rainfall after infiltration. Storage terms for surface water are negligible on a geomorphic timescale, such that the flow is effectively quasi-steady. The system of equations is completed by incorporating a general constitutive law for sediment flux,

$$q_s = F(S,q_w) \qquad (3)$$

where $S$ is the slope of the surface. Note that this constitutive function, $F$, encompasses both diffusive (slope-dependent) and advective (surface discharge-dependent) transport processes, but is given in a general rather than a specific form. If we assume that $R$ and $T$ are both steady and uniform, and that water and sediment at the base of a hillslope are removed by a stream channel, we obtain a solution for the hypothetical unchanneled state in which $q_w = Rx$ and $q_s = F(S_o, Rx) = Tx$, where $S_o = h_{o,x}$ and $h_o(x)$ describes the base state hill profile. Note that for a transport function of the specific form $S^n + S^p q_w^r$, with $n$ and $p$ greater than zero and $r$ greater than 1.0, the base state solution indicates a convexo-concave profile for a hillslope of sufficient length.

Based on the set of equations given by (1) and (2), together with the constitutive law (3) and the proposed base state solution, we can investigate the linear stability of the model system by introducing perturbed variables of the form $h = h_o(x) + \epsilon h'$, etc., where $\epsilon \ll 1$ and primes are used to denote the perturbation quantity. We retain terms to first order in $\epsilon$ and further represent the induced disturbance as a superposition of discrete Fourier modes such that $h' = \psi(x)\exp(iky + \sigma t)$, etc., where $k$ is the wavenumber and $\sigma$ is the growth rate of the perturbation. The wavenumber is given by $k = 2\pi/\lambda$, where $\lambda$ is the wavelength, and characterizes the spatial periodicity of the imposed disturbance in the transverse direction (i.e., perpendicular to the downslope length of a hillside). The growth rate of a disturbance is given by $\sigma$, which in the present problem always has real (cf. complex) values. If $\sigma > 0$, the system is unstable to disturbances; neutral stability occurs when $\sigma = 0$; and when $\sigma < 0$ the system is stable and will internally damp small disturbances. We are interested in examining the growth rates for all possible modes, so that the wavenumber of the most unstable mode (i.e., the mode for which $\sigma$ is largest) can be determined. The wavenumber of this most unstable mode corresponds to the transverse wavelength that should be present, and observable, in an unstable system in the absence of non-linear effects associated with the finite amplitude temporal evolution of the unstable modes. By introducing the perturbed, transformed variables into the governing equations and reducing the resulting system, we obtain the equation governing the stability of the model system as given by

$$\left(\frac{\sigma\psi}{f^{(2)}}\right)_{,x} = -\left(\frac{f^{(1)}S_{o,x}xk^2\psi}{f^{(2)}_{,x}S_o}\right)_{,x} + \left[\frac{(f^{(1)}\psi_{,x})_{,x}}{f^{(2)}_{,x}}\right]_{,x} + \frac{Rxk^2\psi}{S_o} \tag{4}$$

where $\psi(x)$ represents the surface perturbation or mode shape, $S_o$ is the base state surface slope, and $f^{(1)} \equiv \dfrac{\delta F}{\delta S}$ and $f^{(2)} \equiv \dfrac{\delta F}{\delta q_w}$ are both assumed positive.

Smith (1971) demonstrated, using fairly heuristic arguments based on a locally valid WKB analysis of the stability equation, that in the short wavelength limit (i.e. when $k \to \infty$), which represents the most unstable region of the system, the growth rate of a disturbance is given by

$$\sigma = -k^2 A(x) \tag{5}$$

where $x$ refers to a point on the base state hillslope and $A = (S_{o,x}xf^{(1)})/S_o$. Therefore, $A$ is positive on convex hillslope segments and negative on concave segments. This result suggests that convex segments of a base state profile are stable, while concave profile segments are unstable to lateral perturbations. However, using a more rigorous analysis (given in the Appendix to this paper), it is clear that stability is a global rather than a local property and it is shown that the fastest growing mode for large $k$ has a growth rate given by

$$\sigma = -k^2 A(x_{\max}) \tag{6}$$

where $x_{\max}$ is the total length of the base state slope. This latter solution gives the growth rate as a global property and implies that if a particular slope exceeds the length necessary for advective processes to become the dominant mode of sediment transport, then the entire slope system is unstable. More specifically, a base state hillslope is either stable or unstable, and it is not appropriate to evaluate the stability of convex and concave segments on the same hillslope, as illustrated in Figure 15.3. The fact that the unstable modes for large $k$ are apparent only in the concave section of the profile seems to lend credence to Smith and Bretherton's heuristic approach. However, for moderate values

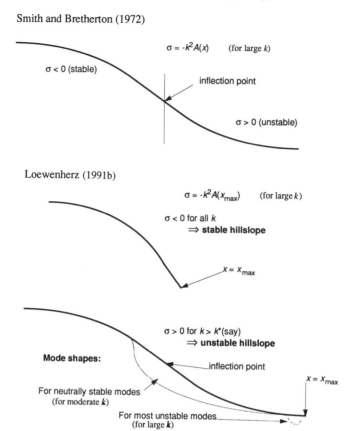

**Figure 15.3**  Comparison of the local (upper diagram) versus the global (lower two diagrams) stability problems. In the heuristic local analysis, stability is evaluated at any point $x$ along the base state hillslope, whereas in the global analysis, the maximum growth rate is determined by $x_{max}$. Representative first-order mode shape solutions for neutrally stable (moderate $k$) and very unstable modes (large $k$) are also indicated. The mode shapes represent the capacity for surface incision along the hillslope

of $k$, the unstable modes extend well above the inflection point and Smith and Bretherton's analysis is no longer applicable, as well be discussed later in this paper.

Smith's (1971) analysis is limited by the local character of the solutions he obtained, which are valid only in a piecewise rather than a uniform manner over the domain of interest. The solutions actually break down, in that they are undefined, at the inflection point between the convex and concave segments of the hillslope, whereas the global solution presented in the Appendix is uniformly valid throughout the length of the hillslope. Although the local analysis used by Smith (1971) does not alter the general validity of the $\sigma \propto k^2$ relationship he correctly identified, the more specific details obtained from his solutions are erroneous and incomplete. Discrepancies between the results obtained via Smith's approach versus a more rigorous global analysis become apparent when one considers (1) the significance of $A$, which serves as the constant of proportionality betweem $k^2$ and $\sigma$; (2) the behavior of the solution in the vicinity of the base state inflection point; and (3) the mode shapes associated with the unstable modes. Although the distinction between a local and a global analysis may initially appear to be a matter of hairsplitting only of interest to mathematicians, it is actually quite central to a range of issues which must be resolved before stability model results can be applied in field and laboratory investigations of drainage initiation. In particular, the morphologic properties, as characterized by the mean transverse spacing and the downslope profiles of initial incisions are quite dependent on the global character of the stability solution. Additionally, the weakly non-linear development of an unstable surface is largely determined by the profiles of the incisions as given by the global mode shape solutions in $x$, the downslope coordinate.

## THE MODE SHAPES AND AMPLITUDE LIMITATIONS

For each value of $k$, there may be a large family of solutions of the stability equation, each solution having a particular value of $\sigma$ and a characteristic mode shape. In the present problem, this mode shape is simply a solution for the surface perturbation, $\psi = \psi(x)$, and gives the downslope profile of the surface incision, with arbitrary (though necessarily small) amplitude. In practice, however, we are only interested in those modes which have a positive growth rate, $\sigma$, and are thus unstable. For each value of $k$, therefore, there may exist one or more unstable modes, each mode having a unique mode shape and an associated growth rate. The linear stability analysis indicates that the shortest wavelength, first-order modes are the most unstable (Loewenherz, 1991a,b). We must also recognize though that for larger values of $k$, higher-order unstable modes are also present, although their initial growth rates will be slower than that of the first-order mode for the same value of $k$. The asymptotic (large-$k$) solution procedure for obtaining mode shapes and growth rates is outlined in the Appendix and some computed mode shapes are shown in Figures 15.4a and 15.4b. The mode shapes are illustrated in non-dimensional form based on the characteristic lengthscale given by the distance from the divide to the inflection point of the base state profile. This scaling is used for both the longitudinal (downslope) and lateral (transverse) distances in this problem.

When $k$ is very large, the mode shapes are such that the first-order mode is spatially concentrated near the base of the slope, as in Figure 15.4a. In other words, the relative magnitude of the surface perturbation as a function of $x$ is very close to zero, except for a region which scales as $k^{-2/3}$ near $x_{\max}$. The first-order mode is identically zero only at the top of the hill and at the base of the slope, but the magnitude of the perturbation decays exponentially fast upslope from the deepest point of incision, hereafter referred to as the 'turning point' $x_{tp}$. At large $k$, there are also several unstable modes of higher order, with turning points located progressively upslope in small increments of order $k^{-2/3}$ as also shown in Figure 15.4a. Thus, the second-order mode, for example, is apparent in a region which is approximately twice as large as that for the first-order mode. The higher-order modes are also oscillatory in shape such that the $n$th-order mode has ($n$-1) zeroes between its turning point and the base of the slope. The growth rate for each mode is given approximately by

$$\sigma = -k^2 A(x_{tp}) \tag{8}$$

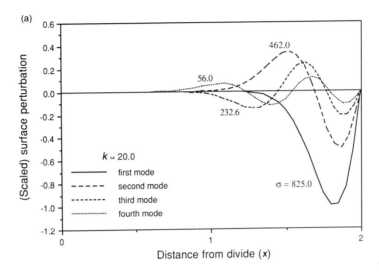

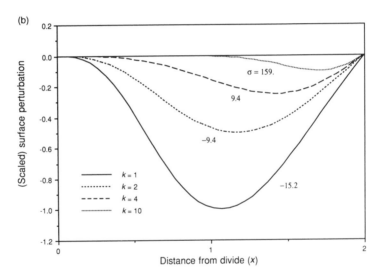

**Figure 15.4** (a) Mode shapes for first and higher-order modes at fixed wavenumber ($k = 20$). Numerical values give the growth rate ($\sigma$) associated with each mode. Positive values for the scaled surface perturbation correspond to a reduction in the net surface erosiion rate at that point, while negative values correspond to an increase. The base state inflection point occurs at $x = 1$ and the total length of the slope is given by $x_{max} = 2$. (b) Mode shapes for first-order modes of moderate wavelength. Negative values for the growth rate indicate a linearly stable mode, so that in this case a marginally unstable mode will have a mode shape which is intermediate between those given for $k = 2$ and $k = 4$

(cf. equations (5) and (6)), so that any mode whose turning point lies below the base state inflection point is unstable. The 'wavelength' of oscillation in the downslope direction is of order $2\pi/k$ except for the first few waves beyond the turning point which scale as $k^{-2/3}$. The length of the hill is scaled to be of order unity, so there are of order $k/\pi$ unstable modes for very large $k$.

The oscillatory character of the downslope incisions together with the sinusoidal waveform in the transverse direction, if realized in the field, might correspond to incisions running at an approximately constant angle to the downslope direction. An alternative possibility for the finite amplitude expression of these higher-order modes, which may be more consistent with field observations (e.g. Govers, 1987; Loewenherz, 1991b), is that the incision itself runs relatively straight down the hill, but that the depth of incision varies significantly along the course of a single incision or 'rill'. Downstream oscillation in the depth of incisions has also been observed in laboratory investigations of rill development (e.g. Bryan & Poesen, 1989). The lengthscale associated with this downslope oscillation should be of the same order as the lateral spacing of the incisions and, consequently, the incision may appear to be almost discontinuous in character.

For a fixed value of $k$, the least unstable mode (i.e. the mode having the smallest positive value of $\sigma$) will have a mode order given approximately for large $k$ by $(x_{max} - x_i)k/\pi$, and its turning point will be located immediately below the base state inflection point, $x_i$, which divides the convex and concave regions of the hillslope profile. These marginally unstable modes possess mode shapes which extend slightly above the inflection point, by an amount of order $k^{-2/3}$. Although these modes initially grow very slowly, they are significant in that they are apparent over a spatial interval extending up to and above the inflection point. Thus, these modes represent the 'first' surface perturbations encountered by infiltration-limited overland flow in its course down a hillslope. As such, they have the first opportunity to capture and channelize surface runoff and this may provide a significant advantage in subsequent competition between the unstable modes.

For a hillslope of length $x_{max} > x_i$, there are two types of modes which are marginally unstable: (1) higher-order, large wavenumber (i.e. short wavelength) modes; and (2) longer wavelength modes of lower to first order. The short wavelength modes will generally be highly oscillatory and will extend only a short distance above the hillslope inflection point. The longer wavelength modes, however, may extend a considerable distance above the turning point or inflection point (see Figure 15.4b). In particular, one would expect the first-order mode to extend farthest up the hillslope. Thus, the least unstable mode, which is the first mode with a turning point located just below the base state inflection point, has a competitive advantage over all other modes in that it represents the surface perturbation which is evident nearest the divide. Surface water may thus be captured and concentrated in these depressions in the upper regions of the hillslope so that this mode is reinforced further down the slope, possibly in preference to other modes which are more unstable in the linear case. Even though this mode initially grows very slowly, it may actually become dominant as the system evolves.

A limited degree of upstream propagation is always possible in a physical system which is in part diffusive in character. In this system, the initial incisions have the capacity to incise above the turning point a distance which scales as $k^{-2/3}$ for large $k$ and which is of order unity for $k$ of order unity. As mentioned above, the stability or instability of different modes is determined by the location of their turning point above or below the inflection point in the base state profile. Marginally unstable modes have a turning point just below the inflection point and so grow very slowly. For $k$ of order unity, these modes may actually extend a significant distance upslope from that point. This upslope extension not only represents the diffusive capacity of the system, but, more importantly, it reflects the global rather than local nature of the stability solution.

In other words, a local analysis which treats a hillslope as consisting of discrete convex and concave segments would necessarily neglect this tendency for upslope extension.

In addition to their differing mode shapes, the various unstable modes also encounter non-linear effects at differing relative amplitudes. A constraint on the linear regime can be rigorously derived by incorporating the hydrodynamics of the surface water flux into the governing equations and also then using a more general constitutive model for the surface sediment flux of the form

$$q_s = F_D(S, q_w)m + F_A(S, q_w)r \tag{9}$$

where $F_D$ and $F_A$ refer to diffusive and advective sediment transport, respectively, and r is the direction of the local surface water flux as determined by the hydrodynamics of the surface water (as discussed in Loewenherz, 1991b). In this revised constitutive model, diffusive sediment transport occurs in response to a local elevational gradient, and is thus oriented in the local downslope direction, as given by m. Although we expect the diffusive component of sediment transport to be largely dependent upon the surface slope, $S$, it is conceivable that the presence of surface water increases $F_D$, for example through buoyancy effects, so that $q_w$ is included as a variable in that term of the function. In contrast, the advected component of the sediment flux, which we expect to be dependent on both $S$ and $q_w$, moves in the same direction as the surface water, as given by r, which is not necessarily coincident with the local ground surface gradient. This revised constitutive model thereby not only distinguishes the relative magnitudes of diffusive versus advective components of the surface sediment flux at points along the hillslope, but also accounts for differing local directions for each of these fluxes.

By incorporating the revised constitutive function given by equation (9) into the stability model, it can be shown that the amplitude of the perturbation in surface elevation scales with $k^{-2}$, for large $k$ (see Loewenherz, 1991b, for additional details). Furthermore, for large $k$, and in the linear regime of the developing instability, the perturbation in the surface elevation is therefore on the same scale as the perturbation in the surface water depth. This is intuitively plausible, as we would expect that very short wavelength perturbations of the order of the depth of the surface water have a strong effect on the direction and redistribution of the water flux. In particular, when the perturbation in elevation becomes greater than the depth of the surface water, segments of the perturbed surface may effectively 'dry out', preventing the possibility of further concentration of water and limiting the rapid exponential growth in amplitude which is achieved by these small incisions. In contrast, if we consider long wavelength perturbations in surface elevation, we would not expect to find noticeable changes in the depth and direction of the surface water flux until the amplitude of the perturbation was a significant fraction of the total elevation of the surface. We would therefore expect that very long wavelength disturbances could continue to grow and concentrate water until their amplitude is a significant fraction of the total height of the base state hillslope.

A limitation on the attainable amplitude for a perturbation of a given wavelength is common in physical stability problems and in simpler problems can be rigorously derived by applying a weakly non-linear analysis in the vicinity of neutral stability. Of particular significance in the present problem is that the physical limitations which constrain the attainable amplitude arise very quickly for the short wavelength perturbations and much more slowly for the longer wavelength perturbations. This constraint can be summarized by expressing the maximum amplitude $B(k)$ as a function of $k$, which for large $k$ takes the form $B(k) \sim k^{-2}$. A precise expression for $B(k)$ cannot be readily obtained nor does this function necessarily correspond directly to any particular feature that might be observed in the field, due to the presence of modal interactions associated with the non-linear development of the incisions. However, the expression given is a mathematically rigorous scaling law which is valid for large $k$ and which strongly suggests that the rapid exponential growth of short wavelength features ceases before they reach large amplitude. We thus have a well-grounded conjecture that there is an effective amplitude limitation which preferentially and progressively retards the growth of the shorter wavelengths. This implies further that if there is large amplitude development of the surface over a substantial period of time, then this development must be associated with moderate wavelengths. Consequently, and somewhat ironically, it may well be the case that the longest wavelength, most slowly growing modes ultimately dominate the developed drainage surface.

There are potentially several other constraints on the finite amplitude development of channel incisions which have not been incorporated into this model but which may arise in the field. For example, removal rates may eventually be limited by the rate of substrate weathering so that the system is no longer transport limited in character. In this case, the susceptibility of the channel bed to erosion may be considerably less than that of the surrounding hillslopes, thereby limiting the amplitude that the surface incision can achieve. The simulation model for drainage surface development recently presented by Howard (1990) utilizes this type of amplitude limitation. Similarly, there may be sedimentation in the downslope reaches of the developing incisions which could significantly retard downcutting and would also modify the morphology of the finite amplitude form.

## THE FINITE AMPLITUDE PROBLEM

Based on the presence of multiple competing modes and using the proposed amplitude limitation as a constraint, we can pose a conceptual model for the development of a hypothetical hillslope from an initial equilibrium state in which there is no channelized drainage to that of a more developed drainage surface possessing equally spaced channels. The model is most conveniently posed as a series of three distinct configuraions from an initially perturbed surface to that of a steady state drainage system consisting of first-order channels as will be illustrated below. The proposed configurations are necessarily idealized and are used to illustrate the implications which the mode shapes and amplitude constraints have for the early development of a drainage surface. The timescale associated with each configuration is much longer than for previous configurations, so that the sequence presented is logarithmic, rather than being linear, in its progression.

The linear stability theory indicates that initially the most rapidly growing incisions are of very short wavelength and are deepest near the base of the slope. We would thus expect these surface perturbations to take the form of closely spaced microincisions restricted to a region very near the bottom of a hillslope. Although these modes are mathematically significant, however, such incisions may never be apparent in the field due to their infinitesimal character. Growing simultaneously, but somewhat more slowly than the fastest growing mode(s), are two other sets of incisions. The first consists of incisions which are qualitatively similar to the fastest growing mode, but of slightly longer wavelength. When such incisions appear, they may lead to a beating phenomenon in which the amplitude of the disturbance is modulated in the transverse direction. It does not seem though that such a phenomenon could be directly related to the appearance of longer wavelength features, since the disturbance would still be confined to the very base of the hillslope. Thus, the contribution of these other wavelengths is merely to introduce a range of wavelengths comparable to that of the fastest growing mode. The second set of incisions corresponds to the higher-order, more slowly growing modes of similar wavelength to the fastest growing mode, which have a deepest point of incision higher up the hillslope. These modes are entirely independent of the fastest growing mode, so that although the wavelength is the same, the relative phase is not fixed. Thus, the lateral location of the 'head' of an incision may not correspond to the position of the incisions formed further downslope.

As a consequence of the amplitude constraint, we expect that the fastest growing modes will quickly saturate, reaching a fairly stable state in which their depth is of the order of the maximum surface water depth (i.e. they are quite shallow). Higher modes will also begin to saturate, although they grow somewhat more slowly, so that the saturation will also proceed more slowly. As a consequence of the progressive saturation of the modes, the surface incisions may appear to propagate or incise upslope in an erratic manner. However, this does not necessarily correspond to a headward growth mechanism for the upslope extension of these incisions in that the location of the highest point of incision has been predetermined by the upslope location of the turning point for the higher-order modes. In fact, the location of the head of the incision at any given time reflects the relationship $\sigma = -k^2 A(x_{tp})$ for a mode with a turning point located at $x_{tp}$. At any location on the hillslope, the time taken for an incision to appear is therefore of order $1/\sigma$. The lateral location of the incision head will be determined by the phase of the corresponding mode.

The higher modes are oscillatory in nature, which in a linear system could potentially produce incisions at an angle to the downslope direction. However, we would expect that in the non-linear development of the system, there would be a propensity for water to converge into pre-existing incisions. This might explain the erratic planform often exhibited by individual incisions and also the apparent capture (or cross-grading and micropiracy as suggested by Horton, 1945) of one incision by another. Alternatively, the finite amplitude incisions may be relatively straight, but show oscillations in their depth. Despite the irregularity in the planform of the incisions, however, one would anticipate some consistency in the lateral spacing since that is determined by the fastest-growing wavelength as predicted by the stability theory. This initial wavelength is strongly dependent upon the material properties of the surface material, as suggested by recent field studies of rill 'density' (e.g. Parsons, 1987; Govindaraju & Kavvas, 1990; Loewenherz, 1991b). Note also that in this first configuration the erratic planform of the incisions is inherent in the process by which they are formed. In other words, it is not necessary to have lateral migration of incisions after formation in order for the incisions to merge and/or exhibit irregular downslope flow paths. However, it is possible that once formed, the incisions would meander, or that new incisions would continually develop and modify the planform. Within field contexts, one would also anticipate that macroscale surface roughness associated with coarse debris or vegetation would locally divert the surface water, thus constraining the planform. We therefore expect that the

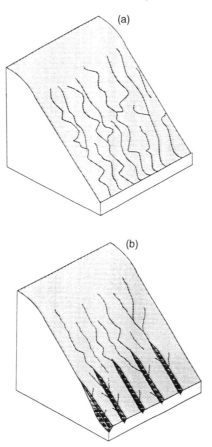

**Figure 15.5**  (a) Spatial pattern of surface incisions associated with the presence of multiple modes early in the developing instability. (b) Surface topography associated with the weakly non-linear development of the instability, as inferred from the mode shapes, modal growth rates and amplitude limitations for a superposition of short and longer wavelengths

hillslope will tend towards a state in which shallow, closely spaced incisions extend in an erratic path from a point just below the inflection point in the base state profile to the base of the slope, as illustrated in Figure 15.5a. This state is not necessarily a stable configuration, but in the transient development of a hillslope, it may well dominant the early stages.

Simultaneous with the development of the first configuration, but on a much longer timescale, there are disturbances growing at longer wavelengths. These long wavelength disturbances correspond to lower-order mode shapes with turning points located upslope towards the inflection point. We would in general expect the development of these much longer wavelengths to be insensitive to the existence of the fully developed shorter wavelength incisions, although the inverse would not be true. In other words, the short wavelength (shallow) incisions should not directly affect the development of the longer lengthscale features; however, as the amplitude of the longer wavelength features becomes finite (given that they will grow to be quite large before they saturate), these deeper incisions may systematically modify the shorter wavelength, shallower incisions so that they are diverted laterally into the large amplitude features. During this secondary phase in the development of the surface, one would expect to actually see finite amplitude incisions of moderately long wavelength which extend a substantial distance toward the inflection point, coexisting with a modified pattern of the initial short wavelength incisions developed in the earlier configurations (see Figure 15.5b and also Photo 1 in Kashiwaya, 1978). We would accordingly expect the short wavelength incisions to exhibit convergence in the valleys and divergence on the noses of the larger scale features, although the process by which the planform of the smallest incisions changes is not given by the present theory. If the planform is in fact dynamic in time, then the change in orientation could occur by elimination of the original incisions and the development of new incisions which conform to the larger scale features. Lateral migration of the incisions may also contribute to a reorientation of the short wavelength incisions as finite amplitude development proceeds. At present, however, there is insufficient field or laboratory evidence to suggest the manner in which this reorientation might occur and its significance.

The slowest growing of the unstable modes are the first modes with turning points just below the base state inflection point. These modes are, however, also capable of attaining the largest amplitude and have the first opportunity to collect surface water in its course downhill. One would therefore expect such modes to dominate in any 'steady state' drainage network realized over a substantial time period (as suggested in Figure 15.1). For these very large modes, one would also expect a significant deviation in the transverse cross-section from the simple sine wave imposed in the linear stability theory. In order to fully characterize the morphologic structure of this system, the stability model would have to be modified to account for fully developed channel flow as well as the differential erodibility of the bed of the channel, if bedrock is exposed or if downstream sedimentation is significant. In this third configuration, the sides of the noses of the large-scale features can themselves be considered as shorter hillslopes which might be susceptible to channel initiation and which could accordingly exhibit their own characteristic channel spacing, as first discussed by Horton (1945).

The field parameters which can be used to describe the morphologic structure of the drainage surface at a given point in time include the mean spacing of surface incisions, the depth or amplitude of the incisions and the point along the base state hillslope at which the incisions are initiated, i.e. the head of the incisions. At any given time $t$ (say), the highest point on the hill at which dissection is apparent, $x_{head}(t)$ (say), will correspond roughly to the turning point of the shortest wavelength mode whose growth rate corresponds to $1/t$ i.e., $-A(x_{head}) \sim 1/(k^2t)$. The head of the incisions $x_{head}(t)$ thereby appears to propagate up the hill as time increases. Similarly, at any given time $t$ and any given location on the hillslope for which $x > x_{head}$, the wavenumber which is most apparent, $k_a(x,t)$ (say), will be the wavenumber characterized by the largest amplitude. This should correspond to the longest wavelength mode which has saturated, i.e. the most recently saturated mode as given by

$$k_a(x,t) \sim [-A(x)t]^{-1/2} \text{ with amplitude } B(x,t) \sim k^{-2} \sim -A(x)t \qquad (10)$$

Therefore, at a given point on the hillslope, the amplitude of the deepest incision grows approximately linearly with time, while the effective wavelength of the deepest incisions grows as the square root of time. Similarly, at any fixed time the wavelength of the deepest apparent incisions grows down the slope from $x_{\text{head}}$ according to the square root of $-A(x)$, while the amplitude of these incisions grows linearly as $-A(x)$. As discussed previously, the slowest growing mode corresponds to the first marginally unstable mode and has a turning point located at the base state inflection point. This mode has the longest possible wavelength and can grow to the largest amplitude, so that once it appears and attains a maximum depth, the time-dependent development of the drainage structure would effectively cease, until the system is (perhaps) subjected to an externally imposed disturbance of sufficient magnitude.

## SUMMARY

In many physical systems which exhibit instability, the first disturbance to actually appear in the system corresponds to the fastest growing unstable mode as given by a linear stability analysis. It is indeed often the case that there may be a new stable state which is somewhat more complex that the original base state, but which is qualitatively similar in appearance to the initially most unstable mode as determined by a linear analysis. However, the instability associated with surface drainage initiation does not exhibit this behavior. The linear analysis indicates that the most unstable modes are of very short wavelength and are spatially concentrated near the base of a hillslope. On the basis of field evidence though we expect that the less unstable, longer wavelength modes eventually dominate at large amplitude in that surface channels exhibit a finite lateral spacing as represented by the network drainage density and the requisite contributing area associated with channel initiation. This incongruity between field observations and linear stability analysis results seems to be a consequence of two primary factors:

1.  The base state solution depends on $x$, the downslope direction. In conjunction with this downslope variability, the physical system is also 'advective' in that information can be passed upstream only a very short 'diffusive' distance, while information passing downstream is propagated much further via the movement of surface water. Therefore, the unstable modes which extend furthest upslope have the best opportunity to capture and channelize water and will eventually prevail in the competition for dominance in the morphologic structure of the drainage surface, independent of their initial growth rates.
2.  The amplitudes of the unstable modes are strictly limited due to saturation. Even though the shorter wavelength modes are initially the fastest growing, they also achieve their maximum amplitude very quickly, so that the slowest growing modes can ultimately determine the behavior and morphologic configuration of the system.

These factors impose constraints on the modal interactions which occur as drainage development proceeds and they can be used, together with the stability model results, to construct a hypothetical series of surface configurations associated with early drainage development as has been outlined herein. It should also be recognized though that field realizations of the physical system tend to be fairly well developed, i.e. they do not usually represent a marginally unstable surface. Hence, the drainage networks often observed in the field (including rilled surfaces) may be well beyond the critical values of parameters associated with possible bifurcations and secondary instabilities such that we cannot reasonably expect them to resemble the results of a linear stability theory or even a weakly non-linear analysis near criticality. This latter factor strongly suggests the need for more extensive laboratory investigations (such as those previously undertaken by Bryan and Poesen, 1989) specifically designed to clarify the morphologic structure of initial and developing microchannels.

## ACKNOWLEDGEMENT

Partial support for this work was provided by the American Geophysical Union 1990 Horton Research Grant.

## REFERENCES

Ahnert, F. (1967). The role of the equilibrium concept in the interpretation of landforms of fluvial erosion and deposition. In Macar, P. (ed.), *L'Evolution des Versants*, University of Liege, pp. 23–41.

Ahnert, F. (1973). COSLOP2 – a comprehensive model program for simulating slope profile development. *Geocom Bulletin*, **6**, 99–122.

Bryan, R. B. and Poesen, J. (1989). Laboratory experiments on the influence of slope length on runoff, percolation and rill development. *Earth Surface Processes and Landforms*, **14**, 211–231.

Culling, W. E. H. (1965). Theory of erosion on soil-covered slopes. *Journal of Geology*, **73**, 230–255.

Dietrich, W. E. and Dunne, T. (1993). The channel head. In Kirkby, M. J. and Beven, K. J. (eds), *Channel Network Hydrology*, Wiley, Chichester, pp. 175–219.

Dunne, T. (1980). Formation and controls of channel networks. *Progress in Physical Geography*, **4**, 211–239.

Govers, G. (1987). Spatial and temporal variability in rill development processes at the Huldenberg experimental site. In Bryan, R. B. (ed.), *Rill Erosion*, Catena Verlag, Cremlinger, West Germany, pp. 17–34.

Govindaraju, R. and Kavvas, M. L. (1990). Influence of rills on erosion over steep hillslopes. *EOS, Transactions of the American Geophysical Union*, **71**, 1345.

Horton, R. E. (1945). Erosional development of streams and their drainage basins; Hydrophysical approach to quantitative morphology. *Geological Society of America Bulletin*, **56**, 275–370.

Howard, A. D. (1990). Parsimonious simulation model for drainage basin evolution. *EOS, Transactions of the American Geophysical Union*, **71**, 576.

Kashiwaya, K. (1978). On the rill net in a slope system. *Bulletin of the Disaster Prevention Research Institute of Kyoto University.*, **23**, 69–93.

Kirkby, M. J. (1971). Hillslope process-response models based on the continuity equation. In *Slopes: Form and Process, Special Publication 3* (ed D. Brunsdon), Institute of British Geographers, London, pp. 15–30.

Kirkby, M. J. (1980). The streamhead as a significant geomorphic threshold. In Coates, D. R. and Vitek, J. D. (eds), *Thresholds in Geomorphology*, Allen and Unwin, Boston, pp. 53–73.

Kirkby, M. J. (1987). Modelling some influences of soil erosion, landslides and valley gradient on drainage density and hollow development. In Ahnert, F. (ed.), *Geomorphological Models*, Catena Verlag, Cremlinger, West Germany, pp. 1–4.

Loewenherz, D. S. (1990). Lengthscale-dependent material transport and the initiation of surface drainage: A modified linear stability theory. *EOS, Transactions of the American Geophysical Union*, **71**, 515–516.

Loewenherz, D. S. (1991a). Stability and the initiation of channelized surface drainage: A reassessment of the short wavelength limit. *Journal of Geophysical Research*, **B96**, 8453–8464.

Loewenherz, D. S. (1991b). Stability and the role of lengthscale-dependent material transport in the initiation and development of surface drainage networks. Unpublished Ph.D. Thesis, University of California, Berkeley, 206 pp.

Melton, M. A. (1957). *An Analysis of the Relations Among Elements of Climate, Surface Properties and Geomorphology*. Office of Naval Research Geography Branch, Project NR 389–042, Technical Report 11. Columbia University, 102 pp.

Montgomery, D. R. and Dietrich, W. E. (1988). Where do channels begin? *Nature*, **336**, 232–234.

Montgomery, D. R. and Dietrich, W. E. (1989). Source areas, drainage density and channel initiation. *Water Resources Research*, **25**, 1907–1918.

Parsons, A. J. (1987). The role of slope and sediment characteristics in the initiation and development of rills. In *Processes et Mesure de l' Erosion*, C.N.R.S., pp. 211–220.

Smith, T. R. (1971). Conservation principles and stability in the evolution of drainage systems. Unpublished Ph.D. Thesis, Johns Hopkins University, Baltimore, 77 pp.

Smith, T. R. and Bretherton, F. P. (1972). Stability and the conservation of mass in drainage basin evolution. *Water Resources Research*, **8**, 1506–1529.

Willgoose, G., Bras, R. and Rodriguez-Iturbe, I. (1990). A model of river basin evolution. *EOS, Transactions of the American Geophysical Union*, **71**, 1806–1807.

## APPENDIX: SOLUTION FOR THE STABILITY EQUATION

The linear stability equation is given by

$$-k^2\left(\frac{A\psi}{B}\right)_{,x} + \left[\frac{(D\psi_{,x})_{,x}}{B}\right]_{,x} + \frac{Rk^2x\psi}{S_0} = \left[\left[\left(\frac{\sigma\psi}{f^{(2)}}\right)_{,x}\right]_{,x}\right] = \left\{\left\{\left(\frac{\sigma\psi}{B}\right)_{,x}\right\}\right\} \tag{A.1}$$

where $A = (S_{0,x}xf^{(1)})/S_0$, $B = f^{(2)}_{,x}$ and $D = f^{(1)} = \dfrac{\partial F}{\partial S}$ is the diffusion coefficient. For large $k$, this has a slowly varying solution of the form

$$\psi = \left[\frac{B}{\mathcal{A}}\right]\exp\left\{\int\frac{RxB}{AS_0}dx\right\} \tag{A.2}$$

where $\mathcal{A} = A + \sigma/k^2$.

This solution breaks down when $A = -\sigma/k^2 = (S_{0,x}xf^{(1)})/S_0$. Therefore, for a given $\sigma$ and $k^2$ there may exist a point $x = x_{tp}$ (referred to as the 'turning point') such that

$$\left[\frac{S_{0,x}xf^{(1)}}{S_0}\right]\Bigg|_{x=x_{tp}} = -\frac{\sigma}{k^2} \tag{A.3}$$

Furthermore, we know that $S_{0,x}$ and $f^{(1)}$ are always positive, and that $S_{0,x}$ is positive in the convex segment of the profile, negative in the concave segment, and zero at the inflection point, $x_i$. Therefore if $\sigma = 0$, then $x_{tp} = x_i$, and if $\sigma > 0$, then $x_{tp} > x_i$. Thus, for positive values of $\sigma$, the turning point of the differential equation is always located downslope of the inflection point of the base state profile. Similarly, if $\sigma < 0$, then $x_{tp} < x_i$.

There are also WKB-type solutions (in addition to the slowly varying solution given by equation (A.2)) which are valid when $k^2$ is large and which break down at the turning point. Therefore, if we introduce

$$\psi = p(x)\exp[kr(x)] \tag{A.4}$$

into equation (A.1) we can obtain the solutions

$$\psi = \frac{1}{(D\mathscr{A})^{1/4}}\exp\left[ -\int\frac{RxB}{S_o\mathscr{A}}dx \pm k\int\left(\frac{\mathscr{A}}{D}\right)^{1/2}dx\right] \tag{A.5}$$

In the vicinity of the turning point, we can obtain an inner equation which is valid within that region by simply defining $a \equiv -\mathscr{A}'(x_{tp}) = -A'(x_{tp})$ and introducing an inner coordinate for $x$ of the form $\xi = (x - x_{tp})k^{\nu}$. If we substitute these into the governing equation, we find that the terms balance if $\nu$ is taken to be 2/3. The inner equation is thus given by

$$\left(\frac{D}{a}\right)\psi''' + \xi\psi' + b\psi = 0 \tag{A.6}$$

where

$$b = \left[1 + \frac{RBx}{aS_o}\right]\Bigg|_{x=x_{tp}} \tag{A.7}$$

In the case of neutral stability (i.e when $\sigma = 0$) $b = 2$. For non-zero values of $\sigma$, we have

$$b = \left[1 - \frac{Rf^{(2)}{}'x}{[(S_o'xf^{(1)})/S_o]'S_o}\right]\Bigg|_{x=x_{tp}} \tag{A.8}$$

If we use the second derivative of the constitutive relationship, i.e. $Rf^{(2)}{}' = -(f^{(1)}S_o')'$ then we can write

$$b = 1 + \frac{[x(f^{(1)}S_o')']/S_o}{[(xf^{(1)}S_o')/S_o]'} \tag{A.9}$$

Note that in general $R$, $B$, $x$, $a$, and $S_o$ are positive, so that $b$ is always greater than 1.

Now for fixed values of $k$ and $x_{max}$, the turning point may be located almost anywhere on the slope, above or below the inflection point, and thus there exists a range of modes associated with different values of $\sigma$. As $k \to \infty$, this range becomes a continuum. However, we are interested in the largest value of $\sigma$, which corresponds to the largest negative value of $A$. Provided that $A(x)$ is monotonic, this occurs for the largest value of $x_{tp}$ which is consistent with the downslope boundary condition. We therefore obtain

$$x_{tp} = x_{max} - \xi_1(b)k^{-2/3}\left[\frac{D}{a}\right]^{-1/3} \tag{A.10}$$

where $b$, $D$ and $a$ are evaluated at $x = x_{tp}$ and $\xi_1(b)$ is the location of the first zero of $\psi_b(\xi)$ which satisfies

$$\psi''' + \xi\psi' + b\psi = 0 \tag{A.11}$$

Numerical solutions for equation (A.11) are shown in Figure 15.A1 for $b = 1.0$, $b = 1.5$ and $b = 2.0$. The solutions exhibit exponential decay upslope from the turning point and oscillation in the downslope region. Note also that with increasing $b$, the location of the first zero (corresponding to the most unstable mode of the system) moves upslope. The largest value of $\sigma$ is given by the first mode of the system, i.e.

$$\sigma = -k^2 A(x_{tp}) \tag{A.12}$$

or, substituting in equation (A.10),

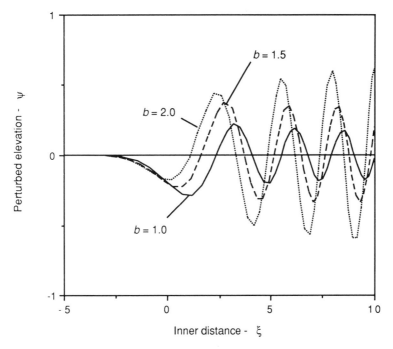

**Figure 15.A1**   Inner solutions for the perturbed surface elevation, $\psi$, as a function of the inner variable, $\xi$. See Appendix for details

$$\sigma \approx -k^2 \left[ A(x_{\max}) + \xi_1(b)k^{-2/3} \left[ \frac{D}{a} \right]^{-1/3} a(x_{\max}) \right] \tag{A.13}$$

$$\sigma \approx -k^2 A(x_{\max}) - \xi_1(b)k^{-4/3} \left[ \frac{D}{a} \right]^{-1/3} a(x_{\max}) + O(k^{2/3}) \tag{A.14}$$

Thus, we confirm explicitly Smith and Bretherton's assertion that in the short wavelength limit $\sigma$ behaves like $k^2$. However, this short wavelength instability is a global rather than a local phenomenon. Smith and Bretherton's analysis and discussion suggests that the growth rate $\sigma$ is given locally by $\sigma = -k^2 A(x)$ such that above the base state inflection point the system is stable and below the inflection point the system is unstable. Furthermore, Smith's (1971) set of solutions implies a WKB-type oscillatory mode shape below the inflection point. We have alternatively demonstrated that a rigorous analysis of the stability equation gives $\sigma = -k^2 A(x_{\max})$. This is a necessarily global result which implies that if $\sigma$ is less than zero for all values of $k$, then the hillslope is stable, and that if $\sigma$ is greater than zero for a particular value of $k$, then the hillslope is unstable. We have also here shown that at neutral stability the deepest point of incision is located at the base state inflection point. However, the shorter wavelength, initially more unstable modes are spatially isolated further downslope, due to the migration of the turning point below the inflection point with increasing values of $\sigma$.

# Part 4

## APPLICATIONS

# 16 Variation in Runoff on Steep, Unstable Loess Slopes near Lanzhou, China: Initial Results Using Rainfall Simulation

**TATIANA MUXART, ARMELLE BILLARD**
*Laboratoire de Géographie Physique, CNRS, Meudon, France*

**EDWARD DERBYSHIRE**
*Department of Geography, Royal Holloway, University of London, UK and Geological Hazards Research Institute, Gansu Academy of Sciences, China*

and

**JINGTAI WANG**
*Geological Hazards Research Institute, Gansu Academy of Sciences, China*

## ABSTRACT

The Lanzhou region, China, lies at the western margins of the Loess Plateau. The terrain consists of steep bedrock mountains and hills with a thick drape of loess. The climate is sub-humid to semi-arid, the mainly summer rainfall occurring as high intensity events, over half the annual total frequently falling in a few days in July or August. In natural conditions, such a rainfall regime results in overland flow which is followed by rapid infiltration into the fissure systems and sub-surface pipes of the widespread 'loess karst'. This provides efficient infiltration routes and, in turn, discrete zones of saturation between loess units of different permeability or at the loess–bedrock contact, leading to collapse (hydroconsolidation) and slope failure.

A series of rainfall simulation experiments was undertaken on a steep loess slope near Lanzhou city as a first attempt to assess the order of magnitude of percentage overland flow in this environment within the framework of the two principal land use types (pasture and cultivation of hand-built terraces). It was shown that for similar slopes of similar gradient and rainfall intensities, runoff is clearly greater on recently ploughed plots than on a plot under pasture. On the cultivated field, the runoff coefficients during the simulation experiments increased with rainfall intensity, reaching 95% at a rainfall rate of 87 mm/hr. Also, runoff coefficients remained the same or increased with repeated simulations on the same plot, at similar or even lower intensities. These results are a product of modification of the soil surface by splash and sheetflow which leads to destruction of clods and aggregates, sealing of pores, and the formation of a surface crust of low permeability. On sealed soil after natural heavy rain (the situation at the field site one month before the experiment), the overland flow was found to be reduced by infiltration into the desiccation features which had developed after the last rainfall event. A relationship between rainfall intensity, sealing of the soil surface during rainfall events, the development of overland flow, infiltration of water into the sub-surface fissures and pipe systems making up the loess karst, and subsidence and sliding on loess slopes is postulated.

*Process Models and Theoretical Geomorphology.* Edited by M. J. Kirkby
©1994 John Wiley & Sons Ltd

# INTRODUCTION

The Loess Plateau of north-central China occupies about three-quarters of that huge country's loess-covered terrain (estimated at 440 000 km$^2$: Liu *et al.*, 1985), by which is meant loess at least 10 m in thickness (Figure 16.1). However, the loess is very much thicker over enormous areas, the thickest-known sequences being on the western margins of the Plateau around the city of Lanzhou. Lying to the north and east of the Tibetan Plateau, this region suffers frequent earthquakes associated with the continuing rapid uplift of the Qinghai–Xizang (Tibet) Plateau. This has resulted in the notable incision of the Yellow River (Huang He) and its tributaries to produce a generally mountainous landscape of elongated steep hills cut into the argillites, sandstones and conglomerates of Mesozoic and Cenozoic age. This impressive terrain is covered by loess, the thickness of which varies considerably in relation to the buried bedrock, from less than 20 m to more than 300 m. The eastern Gansu region thus contrasts strongly with the more regular *yuan* (plateau) and *liang* (ridge) landforms of the Loess Plateau proper to the east in the provinces of Shaanxi and Shanxi.

The mountainous loess landscape of eastern Gansu lies across the climatic transition from sub-humid to semi-arid. Mean annual precipitation ranges from about 500 mm in the south to less than 300 mm in the north, with mean annual potential evapotranspiration exceeding 1000 mm. As much as 70% of the mean annual precipitation (300–400 mm) may fall in July, August and September in any one year, and 40% of this annual total may sometimes be concentrated in a single storm (Derbyshire *et al.*, 1991). Rainfall intensities are very high, rates in excess of 50 mm/hr having been recorded quite frequently. Eastern Gansu thus suffers both water stress and extreme rainfall events, variability in rainfall one year to the next at Lanzhou often being three-fold.

# SURFACE MATERIALS AND SLOPES: BACKGROUND

The region is characterized by large numbers of landslides which are triggered by earthquakes and heavy monsoonal rains. The susceptibility of this seismically active and climatically extreme landscape to such geomorphological hazards is enhanced by (a) the physical, chemical and mineralogical characteristics of the loess, together with its behavioural properties; and (b) the steeply sloping terrain. These are discussed below.

(A) Loess consists of sub-angular, platy particles mainly of quartz in the silt range (20–60 $\mu$m) with (in the Lanzhou region) rather small percentages of clay-sized particles which include quartz and feldspars as well as clay minerals. Salts and carbonates are also present, the latter occurring as both clastic components and interstitial cements. Loess is a sensitive material (Smalley, 1971; Cabrera & Smalley, 1973). When dry and unfractured, loess is capable of sustaining considerable static loads. However, it may fail in the dry state in response to earthquake shock with consequent fluidization and associated powder flows. In moist conditions, on the other hand, loess may fail instantaneously at given values of overburden stress as moisture contents approach saturation. This is termed hydroconsolidation. The highest values of collapse coefficient in the Chinese loess come from the Upper Pleistocene formations of the Lanzhou region

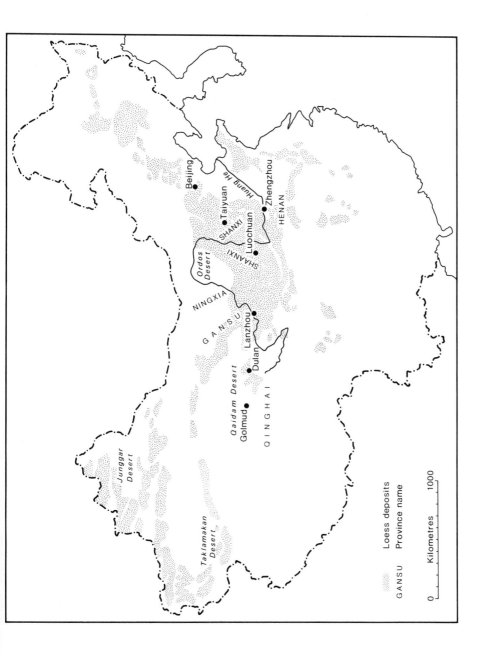

**Figure 16.1**  Distribution of thick loess in China, and location of Lanzhou, Gansu Province and Shaanxi Province

(Derbyshire & Mellors, 1988). Coefficients of vertical permeability in this younger loess average around $40 \times 10^{-5}$ cm/sec for intact specimens (Derbyshire *et al.*, 1991). These different characteristics play an important role in the landsliding process, as will be seen later.

(B) Deep incision of the Yellow River and its tributaries has produced a landscape characterized by long and steep slopes, with gradients predominantly in the range 30–40°, on which gravity-induced mass movements are widespread. Moreover, failure of these slopes is greatly influenced by the common presence of joint systems and slope-parallel planar surfaces produced by both dilation and shearing. Joints, enlarged partly by suffosion and subsequent collapse, become the locus of sub-vertical sinkholes which commonly occur as aligned series within shallow surface gully networks and which lead down to large sub-surface pipe systems, the pipes being several metres in diameter. These fissures, sinkholes and pipes are collectively known as 'loess karst' in China (Figure 16.2). The presence of loess karst is important because field permeabilities are consequently much higher than those measured in the laboratory as quoted above. It is also important because, in combination with the ubiquitous joint systems, loess karst has been shown by engineering excavation to reach landslide shear zones not only at the loess–bedrock contact (beneath relatively thin loess) but also at zones of reduced permeability within the thicker loess mantles, including some of the many buried palaeosols to be seen in this region.

Loess karst plays a critical role in the degradation of the loess landscape around Lanzhou. In general, much of the loess is too thick for the whole mass to become saturated in such a climate: rather, surface waters enter the loess and infiltrate to considerable depths by way of the joint and pipe systems. Thus, the loess reaches saturation in two discrete and widely separated zones, namely the surface and zones of low permeability at depth fed by the loess karst and joint system (Derbyshire *et al.*, 1991). It is along these critical zones of subsurface water circulation, both at certain horizons within the body of the loess and at the loess–bedrock contact, that liquefaction occurs, leading to mass failure of the loess slopes (Derbyshire *et al.*, 1991; Billard *et al.*, 1992; Billard *et al.*, 1993). This succession of processes, culminating in slope failure, is controlled in the first instance by rate and volume of overland flow.

With this background, and given that traces of surface flow (both as sheets and as rills and shallow gullies) have been observed on crusted soil surfaces following rainfall events, and that landslides, both large and small, are demonstrably triggered by heavy rains, it is clearly important to study overland flow on the Lanzhou loess terrain if the landslide initiation mechanisms are to be better understood. In particular, it is important to attempt to determine how much water, within a range of rainfall intensities, flows over the surface and is therefore available to enter the loess karst.

To this end, a series of rainfall simulation experiments was undertaken in the summer of 1990 on a loess-covered mountain slope overlooking Lanzhou city. The specific objective was a study of the surface hydrology, not the organization of the rills or the measurement of the surficial erosion on the slopes. As overland flow depends, amongst other things, on the vegetation cover, the experimental plots were located on surfaces under the two principal land use types found in the Lanzhou area, namely pasture and the ploughed surfaces of sloping (3–23°) hand-built terraces which continue to be constructed in response to rural population pressures.

**Figure 16.2** Destruction of the only access road to Gaolan Shan summit by extension of huge pipe system (foreground) and secondary landsliding (middle distance) following the record rainstorm intensity in August 1990

## THE RAINFALL SIMULATION: EXPERIMENTAL CONDITIONS

The rainfall simulation experiments described here are the first to be attempted on the mountainous loess terrain of the Lanzhou region. The field conditions are difficult, not only because of the steep gradients but because the isolation of the site meant that the experiments were dependent on water piped under gravity from oil drums brought in lorries up the 1500 m high mountain (Gaolan Shan) from Lanzhou city. These difficult conditions dictated the choice of rainfall simulator. The *'mini simulateur de pluie'* designed by Asseline & Valentin (1978), Valentin (1978) and Asseline (1981, 1984) is readily portable, requires only a modest water supply, and is designed for use with small plots on steep, isolated slopes. It was known that this equipment had already been used with success on very steep (40–60%) slopes of marl in the south of France. Furthermore, the use of small plots is perfectly adequate for the study of overland flow in such terrain. As pointed out by Collinet (1985), it has been demonstrated that in cases where the surface hydrology is controlled by the formation of a crusted surface (as is the case in silt-dominated soils), the use of small plots (1 m²) for hydrodynamic experiments leads to the same conclusions as studies based on much larger plots (200 m²).

Using this particular piece of equipment, water is pumped to the top of a 3.5 m high steel frame at a pressure of 0.5 bar where a spray of water droplets is emitted from a jet-head which is driven from side to side in a pendular motion by electrical battery power. The rainfall intensity is adjusted by altering the angle of swing of the water jet. This jet is centred over a rectangular metal frame, half-inserted into the soil and so delimiting the experimental 1 m² microplot (length 1.25 m aligned down the local slope, and width 0.8 m). Using a constant pressure to produce raindrops, their diameter depends on their sorting which is controlled by the swing of the jet-head. Consequently, the mean drop size varies slightly with the rainfall intensity ($I$) in the range 0.84 mm ($I = 35$ mm/hr) to 1.35 mm ($I = 120$ mm/hr) according to the work of Asseline and Valentin (1978). Two gutters, each with an area of 0.1 m² are set up adjacent to the two longer sides of the metal frame in order to provide check readings of the rainfall amounts during the experiment. The rainfall intensity is determined prior to any exposure of the soil in the plot by placing over the soil a rectangular metal cover exactly the same area as the microplot. Rainfall intensity to be used in the succeeding experiment is then determined by measuring, every 2 or 5 minutes, the volume of water which flows out from the metal cover and the two gutters. After removal of the cover, the rainfall intensity continues to be monitored by repeated measurements of the amount of water issuing from the two gutters. Overland flow is determined by measuring the volume of water flowing out of the microplot by way of the exit tube on the lower side of the microplot frame.

The selection of rainfall intensities used in these pilot experiments was guided by the known range of variation in precipitation intensities in the Lanzhou region, as derived from over half a century of records held by the Meteorological Bureau. Fortuitously, also, the historically greatest rainfall rates were recorded at Lanzhou only one month before these experiments began, 93 mm of rain being measured in 1 hour and 40 minutes. This remarkable event began with a fall of 62 mm in the first 15 minutes (an hourly rate of 248 mm/hr), the fall after 25 minutes reaching 80 mm, and the first hour seeing the total reach 90 mm. Strong overland flow was observed on the bare surfaces of the hand-built terraces, the resultant crust, rills and shallow gullies still being visible in September 1990 (Figures 16.3 to 16.5).

Rainfall intensities were selected in the range 30–90 mm/hr, i.e. a range representative of the historical values known from this area. However, because of the lack of abundant water, practical constraints meant that the same values were not strictly reproduced for each experiment, so that the actual range of intensities achieved was 29 to 87 mm/hr (Table 16.1). The duration of the simulation runs (see Figures 16.6 to 16.9) varied from 40 to 150 minutes, depending on the rainfall intensity in use at the time.

The microplots were set up on slopes of similar gradient in the following different land use situations:

(A) on a pasture site with a gradient of 38% and a surface covered (up to 80%) by grass and small bushes, this selected slope being rather gentler than most of the pastures on the hills and mountains around Lanzhou city;

(B) on a field ploughed only 5 days previously, its gradient being 29–31% (Figure 16.3); and

(C) on a field with a gradient of 30%, and not ploughed since the heavy rainfall of August 1990, whose surface still showed some rills and a widespread crust cut by desiccation cracks (Figure 16.4). This plot was located on the upper part of the field, on the sealed surface and so avoiding most of the rills.

Samples of soil were collected from various depths around the plots before and after each experiment in order to determine the soil moisture content by weighing. The roughness of the ploughed soils was estimated by calculating the mean value of the true

**Table 16.1** Rainfall simulation test data

| Land use | Plot | Experiment | Slope gradient (%) | Rain intensity (mm/hr) | Volume of rain per 5 min. (cm³) | Volume runoff per 5 min. (cm³) | Runoff coefficient (%) | Starting time of runoff (min.sec) | Time for reaching stabilized runoff (min) |
|---|---|---|---|---|---|---|---|---|---|
| A: Pasture | 1 { | P1 | 38 | 70 | 5833 | 2160 | 37.0 | 1.30 | ~35 |
| | | P1bis | 38 | 56 | 4666 | 2920 | 62.6 | 0.30 | ~20 |
| B: Surface of field | 2 { | P2 | 29 | 87 | 7250 | 6920 | 95.4 | 1.30 | ~70 |
| | | P2bis | 29 | 29 | 2416 | 2200 | 91.1 | 3.45 | ? |
| ploughed | 3 | P3 | 31 | 70 | 5833 | 4600 | 78.9 | 3.00 | ~90 |
| 5 days before | 4 | P4 | 30.5 | 33 (120) | 2750 | 1250 | 45.4 | ~65.00 | ~90 |
| experiment | | | | 34 | 2833 | 1560 | 55.1 | | |
| C: Sealed surface of field | 5 | P5 | 30 | 35 | 2916 | 1610 | 55.2 | 7.45 | ~85 |

**Figure 16.3** Rainfall simulation experiment on a cultivated terrace above Lanzhou city. Simulator is on plot P3: the arrow indicates plot P5

**Figure 16.4**   Crust and rills produced during severe rainfall events of August 1990 still visible in fields not yet ploughed in September. The terrace slope is towards the upper right, i.e. rills are aligned down the local slope

surface lengths measured along five different traverses of a plot, both before (initial value) and after (final value) each experiment. The measurements were made by laying a fine metal chain carefully across the irregular surface of the soil. The chain was marked where it touched the two edges of the metal frame and then straightened, this length being measured so as to compare it with the width of the plot (metal frame: 0.80 m) and so derive a simple roughness ratio.

## RESULTS

This section consists of a description of the variation in runoff on the five plots (Figures 16.3 and 16.5) under different land use and the changes in soil roughness which occurred during the experiments.

### Variation of Runoff in Relation to Land Use

*On Pasture*

Two experiments, denoted as P1 and P1bis (Figure 16.6 and Table 16.1), were undertaken, 24 hours apart, on the same plot. Rainfall rates used were 70 mm/hr in the first experiment and 56 mm/hr in the second. Clearly, the soil moisture content of the near-surface samples (5–20 cm depth: Figure 16.7) was higher before the beginning

**Figure 16.5**  Terrace cultivation of steep slopes above Lanzhou. View from plot P5 showing pasture plot P1 (arrow right) and P2, P3 and P4 all located on the same field (arrow left)

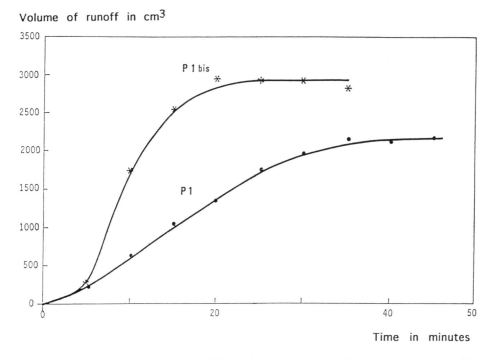

**Figure 16.6**  Runoff in relation to rainfall simulation on pasture. For data on slope gradient and rainfall intensities see Table 16.1

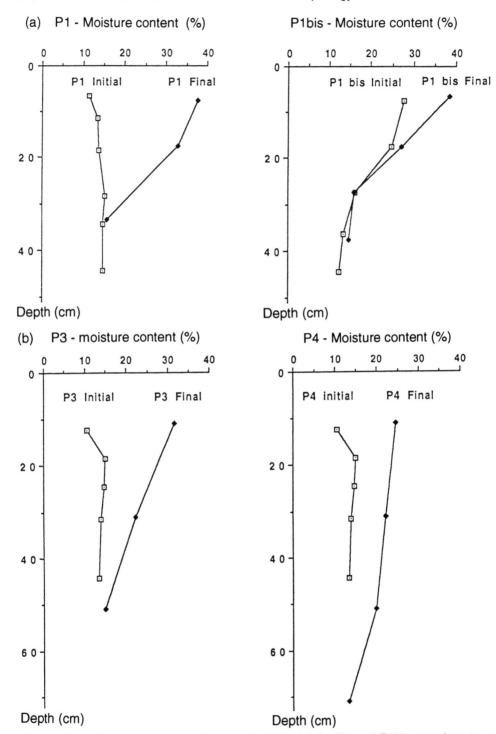

**Figure 16.7**  Variations in soil moisture content for (a) the P1 and P1bis experiments on pasture; and (b) the P3 and P4 experiments on a recently ploughed terrace field

of the P1bis experiment (27% at 5 cm) than before the P1 (10% at 5 cm). In the case of the P1 experiment, the simulated rainfall lasted for 45 minutes (Figure 16.6). Runoff began less than 2 minutes after the beginning of the rain, and the overland flow amount increased for the first 35 minutes to 37% of the total rainfall entering the plot (so reaching, at that stage, a stabilized runoff coefficient). A crust developed on the 20% of the plot surface covered by bare soil. During the P1bis experiment, the runoff started almost immediately after the beginning of the rain. Moreover, the runoff stabilized at a higher value (62.6%) than for P1, even at lower rainfall intensities. In this case, the increased amount of runoff may have been caused by both the development of a crust and the higher moisture content of the surficial soil layers.

### On a Recently Ploughed Field

Four experiments, denoted P2, P2bis, P3 and P4 (Table 16.1 and Figures 16.8 and 16.9) were set up on the lumpy soil surface of a field which had been ploughed 5 days earlier (Figure 16.10).

This situation represents that of maximum surface roughness and maximum soil porosity prior to the action of any rain. The moisture contents of the samples taken from the profile before the experiment, and especially those from the near-surface layers (Figure 16.7), were very similar to those taken from the pasture plot, so that the results of the runs on the different land use plots may be fairly compared.

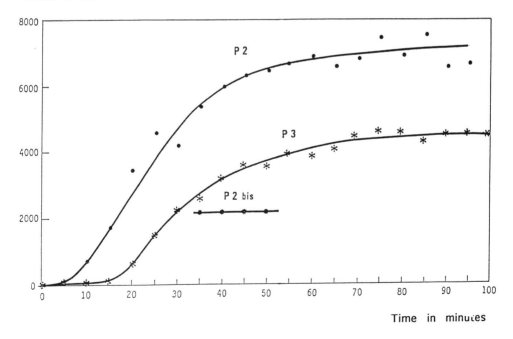

Volume of runoff in cm³

**Figure 16.8**  Runoff in relation to rainfall simulation on a recently ploughed field. For data on slope gradient and rainfall intensities see Table 16.1

Volume of runoff in cm$^3$

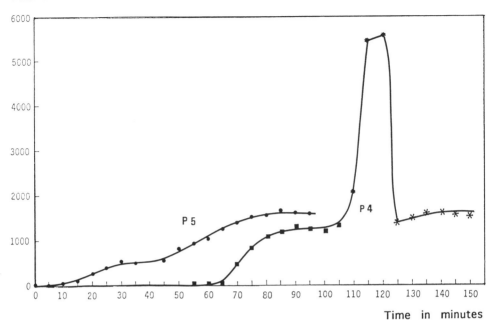

Time in minutes

**Figure 16.9** Runoff in relation to rainfall simulation on a recently ploughed field (P4) and on the sealed surface of the ploughed field (P5). For data on slope gradient and rainfall intensities see Table 16.1

**Figure 16.10** Lumpy soil surface of recently ploughed field: example of plot P4 before the experiment

The most striking characteristic of the results obtained on the ploughed field is the marked runoff coefficient rise with rises in the rainfall intensity values: runoff reached 95.4% at a rainfall rate of 87 mm/hr in experiment P2. These results clearly underline the well-known effect of rainsplash and sheetflow on bare soil surfaces.

The soil surface of the ploughed field became sealed progressively quickly as rainfall rates rose (Figure 16.11). This explains the variations in the runoff coefficients, $K_r$, for the P2bis and P4 experiments, even though the rainfall intensities were rather similar (29 and 33 mm/hr, respectively). The value of $K_r$ was substantially higher for experiment P2bis (91.1%) than in the case of P4 (45.4%) owing to the fact that P2bis had been preceded (the day before) by the heavy simulated rainfall of the P2 experiment (87 mm/hr) which produced a long-lasting crust and some rills.

The same effect was observed during successive rainfall events simulated during the same day on the P4 plot (Figure 16.9). With an initial rainfall intensity of 33 mm/hr, the runoff coefficient was about 45%. However, after a succeeding short and heavy shower, starting at the 105th minute (120 mm in less than 10 minutes) $K_r$ increased to 55.1% despite the fact that the rainfall intensity was reduced to 34 mm/hr after 125 minutes. Compared with the data obtained from the pasture plot, the results derived from the ploughed field clearly underline the effect of land use on overland flow. For a similar rainfall intensity (70 mm/hr), the runoff coefficient was twice as high on the ploughed field (P3: 78.9%) as on the pasture (P1: 37%).

### On the Sealed Surface of a Ploughed Field

The P5 experiment (Table 16.1, Figure 16.9), using the low rainfall rate of 35 mm/hr, was sited on a ploughed field where clods and aggregates had been destroyed by the heavy monsoonal rain in August 1990: this had resulted in the formation of a continuous and rather thick crust (several millimetres). As a result of the dry period which followed this severe rainfall event, the crust became very hard and was cut by desiccation fissures 1 to 3 mm wide. The moisture contents in the soil profile measured before the run (Figure 16.12) are quite similar to those of the previous plots. During the first part of the P5 experiment (Figure 16.9), in which a low rainfall intensity was simulated (35 mm/hr), part of the overland flow was collected by these natural desiccation fissures, so that water was lost within the plot after very short distance flow. Runoff stabilized after 30 minutes once the fissures situated in the lower part of the plot became infilled with fine particles. As a direct result of this local sedimentation, the runoff coefficient increased again some 45 minutes after the start of the experiment until it stabilized a second time at about 75 minutes. Even at that stage other fissures, mostly situated in the upper part of the plot, had remained open (Figure 16.13) and continued to intercept part of the runoff which would otherwise have reached values higher than 55.2%. It was at this stage that exhaustion of the portable water supply forced the termination of the experiment.

### Variations of Soil Roughness on Ploughed Fields

Measurement of roughness on bare soils (Table 16.2) showed that the simulated rainfalls produced changes which depend on the state (more or less crusty or lumpy) of the surface before the beginning of the experiments. In the case of the recently ploughed field, they also relate to the rainfall intensities.

**Figure 16.11**   Sealing of the surface after rainfall simulation: example of plot P2 at the end of the experiment

On the sealed soil, the mean roughness decreased slightly after the P5 experiment as the overland flow partly smoothed out the inter-rill rough features across the plot. The ratio of mean final true length ($L_f$) versus mean initial true length ($L_i$) was 0.98 in this situation. The slightly lower $L_f/L_i$ ratios (0.96 and 0.95, respectively) found on the recently ploughed field after the P3 and P4 experiments express, in this case, a decrease of roughness due to the destruction of clods and aggregates on the lumpy surfaces. In contrast, the plot used for the P2 and P2bis rainfall simulations undertaken on the same field as P3 and P4, showed an increased roughness at the end of the experiment. This is a result of the formation of rills initiated by the peculiar P2 heavy rain (87 mm/hr) and to their development during the P2bis experiment.

## DISCUSSION AND CONCLUSIONS

The pilot study reported here consists of a series of observations of the response to simulated rainfall of some steep slopes in thick loess above the city of Lanzhou.

The aim was to characterize the hydrodynamic behaviour and, specifically, to assess the ratio of infiltration to overland flow on this highly porous soil under different land use and a range of rainfall intensities. An understanding of this behaviour is essential to determination of the relative importance of infiltration as against overland flow, and so to establish the dominant water route in the coarse silty loess which mantles so much of the western margins of the Loess Plateau. Slow infiltration through the micropores leads to saturation of the upper layers. This process contrasts with the interception of

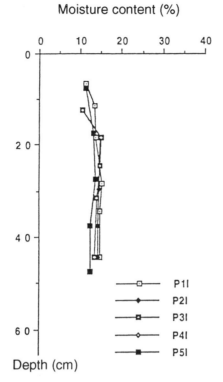

**Figure 16.12**  Variations in soil moisture content for the initial experiment (P1I to P5I) on the five plots

overland flow by fissures and sinkholes which conduct water through the loess karst until it is impeded by zones of lower permeability both within the loess and at the loess–bedrock contact. In this situation, overland flow may play an indirect role (by way of several mechanisms) in the triggering of mass movements.

The number and duration of the experiments reported here were constrained by certain practical difficulties, notably water supply and accessibility to suitable sites. Because of this, only slopes with gradients of about 30% were tested. Despite these constraints, the results clearly show the formation of crusting under the action of rainsplash and sheetflow on the bare surfaces of both recently ploughed fields and pasture land. Such crusting leads to a sealing of the soil which drastically reduces the rate of infiltration and so enhances the rates of overland flow.

Soil surface crusting, development of a surface seal, overland flow, and the relation of these to soil erosion have been widely studied since the first permeability measurements of soil crusts by McIntyre (1958a, b), the subject having been reviewed by Mualem *et al.* (1990a–c). Specifically, the effects of slope steepness and rainfall intensity on the development of a soil crust have been discussed by Römkens *et al.* (1986) and Poesen (1986), respectively. Cai *et al.* (1986) and Luk & Cai (1990) subjected Chinese loess to rainfall simulation in laboratory conditions and reported a cyclic process made up of crust formation followed by disruption during non-equilibrium stages of seal development. The experiments reported here are in line with the general view that

**Figure 16.13**   Sealed surface of plot P5 with fissures still open at the end of the experiment

**Table 16.2**   Changes in roughness on the surface of the experimental plots

| Plot (125 cm × 80 cm) | Experiment | Rain intensity (mm/hr) | Mean initial true length $L_i$ (cm) | Mean initial roughness $R_i$ ($L_i/80$ cm) | Mean final true length $L_f$ (cm) | Mean final roughness $R_f$ ($L_f/80$ cm) | $L_f / L_i$ |
|---|---|---|---|---|---|---|---|
| 2 | P2 | 87 | 88.8 | 1.11 | 89.6 | 1.12 | 1.01 |
|   | P2bis | 29 | 89.6 | 1.12 | 92.3 | 1.15 | 1.03 |
| 3 | P3 | 70 | 93.3 | 1.17 | 89.9 | 1.12 | 0.96 |
| 4 | P4 | 33 (120) 34 | 92.5 | 1.16 | 87.6 | 1.09 | 0.95 |
| 5 | P5 | 35 | 85.1 | 1.06 | 83.6 | 1.04 | 0.98 |

overland flow on bare loess soils, together with rainsplash, destroys clods and aggregates, sealing the surface to produce a long-lasting structural and sedimentary soil crust.

On the steep slopes above Lanzhou, the action or rainfall in the formation of a crust is shown to be all the more rapid because of the rise in the rainfall intensities. Also, in the case of the initial experiments carried out on each plot in the recently ploughed field, the rapidity of initiation and the relatively high volumes of the overland flow matched the rises in the rainfall intensities. The results also clearly show the very high values of runoff coefficient in response to rainfall rates commonly experienced under natural conditions in this region: runoff coefficients as high as 95.4% were measured with a simulated rainfall intensity of 87 mm/hr on this highly porous loess (first experiment on plot P2). Moreover, as a result of the formation of the crust, a severe rain shower on the ploughed loess surface greatly increases the percentage of over-land flow in any rainfall event occurring soon afterwards, even given a lower intensity. However, in the case of a long interval between two rainfall events, the complete drying up of the sealed surface following rainfall leads to its hardening and to the formation of rather deep (up to 5 cm) and narrow (1–3 mm) desiccation cracks. During subsequent rainfall experiment P5, the interception of part of the overland flow by these fissures, and its infiltration into the loess by this route, serves to reduce the runoff until such time as the cracks become clogged with redeposited silt. This process takes some time, of course, because the strength of the crust is such that both soil erodibility and soil detachment and transportation are reduced. This interception by fissures is an interesting phenomenon which, although demonstrated here only at the centimetre scale, is essentially the same at the scale of the loess slope and loess karst system.

It is interesting to note that such crusting also plays a part in enhancing runoff percentages on land under pasture because of the presence of bare patches in this semi-arid environment: all of the runoff coefficients calculated for pasture on the Lanzhou loess are thus relatively high although, of course, lower than for the recently ploughed field. The initial conditions of soil moisture content on both ploughed and pasture plots were found to be very similar (Figure 16.12). In contrast, the differences between the relative values of the runoff coefficient are very great and clearly demonstrate the influence of land use upon the proportion of water movement by overland flow. The strikingly high runoff coefficients demonstrated for seasonally bare surfaces of cultivated loess must be a matter of concern in this very poor region where the rural population is still expanding and cultivation continues to extend on to ever steeper slopes of loess.

Evidently, further work is now required involving a wider range of slope gradients, more varied and different successions of rainfall intensities, and the use of plots on cultivated land with different crop types and on pastures with different percentages of vegetation cover. Nevertheless, the results of this first series of experiments confirm field observations and the hypotheses formulated in the introduction.

It appears that, whatever the field situation and the experimental conditions, the overland flow organizes itself in response to the simulated rainfall. It occurs on pasture with its bare patches as well as on cultivated fields. It does not give rise, however, to a sheet of water flowing to the base of the slope and escaping infiltration through the loess. On the contrary, as observed during heavy natural rainfall events, the overland flow moves across a terrace surface and then down the front face, or riser, which frequently coincides with a natural fissure because these features are used by the farmers as an aid in the construction of the hand-built terraces. Water becomes concentrated in the rills and small gullies, ultimately being engulfed by the fissure networks and the pseudo-karst pipes. It is the efficiency with which the overland flow is thus organized which controls the speed with which the water infiltrates and enters the sub-surface drainage networks in the loess. There is a direct relationship between that proportion of the rainfall running off as overland flow and that which enters into the loess karst system. An important result is that, during periods of heavy and prolonged rain, large volumes of water may find their way down the large pipe systems to zones of much lower permeability within or at the base of the loess. The role of this by-pass drainage varies according to local circumstances. Many systems are highly integrated and drain freely for long periods. Others, however, suffer roof falls, a process which explains the coincident distribution of sinkholes and the shallow but dry gully networks on the loess slopes. Some sub-surface distributary pipes develop and lead down into reduced-diameter pipes or even dead ends. In such situations, water pressure potentials may rise dramatically, producing local saturation in the loess, creating temporary perched water tables, and commonly leading to liquefaction followed by collapse of the superjacent loess mass. Such discrete zones of saturation, most notably associated with buried soils and the loess–bedrock contact, together with the well-known tendency of the young (Upper Pleistocene) loess to hydroconsolidation, is highly conducive to small but frequently very numerous landslides during summer in eastern Gansu. Thus it is that overland flow plays a critical part in a combination of processes involved in the activation and reactivation of slab-failure type landslides in loess as well as in the extension of the pipe systems of the loess karst. There is some evidence, also, that large landslides (up to an area of 7 km$^2$) in very thin loess resting on steeply-sloping argillite bedrock, such as the 1989 slides at Tala and Hong Quan (west of Lanzhou), coincide with the overland flow induced by heavy and prolonged rainfall events.

Given the hydrodynamic properties of loess as derived from the authors' field observations over several years, taken together with the initial experiments described here, it seems clear that sealing of the soil and overland flow play an important role in an erosion–transport system characterized much more by tunnelling, collapse and mass-sliding than by rilling and gullying. On the basis of the experiments reported here, further experiments are being planned which will take account of a greater number of variables in order to quantify the processes and to specify more precisely the nature of the links between rainfall-induced overland flow, infiltration (including water movement down fissures and other discontinuities in the loess), and landsliding.

## ACKNOWLEDGEMENTS

This is contribution No. 7 of CEC Contract CI.1.0109. UK(H). We are pleased to acknowledge the financial support provided by the Council of the European Communities and the Gansu Academy of Sciences. We are grateful to E. Roose, J. Asseline, A. Gioda and P. Ribstein of the Institut Français de Recherche Scientifique pour le Développement en Coopération (ORSTOM) for providing the plans of the simulator frame (built in Lanzhou) and for their invaluable scientific support as well as practical guidance on the selection of a rainfall simulator and its use in difficult field conditions. We acknowledge the team of the Geological Hazards Research Institute in Lanzhou, especially Wang Xu Quan and Ma Jing Hui, for a preparatory survey of water routes within the loess and for collection of the meteorological data, together with Wang Dekai, Zhou Zi Qiang, Li Bao Xing, Cheng Jing, Wang Nian Qin, Qiao Li Min and Wei Yong Ming for their enthusiastic participation in laboratory work and long, hot days in the field.

# REFERENCES

Asseline, J. (1981). Notice technique – Construction d'un infiltromètre à aspersion. *Rapport interne, ORSTOM*, Abidjan, 26 pp.

Asseline, J. (1984). Notice technique – Particularités du nouvel infiltromètre à aspersion construit à Gabès (mini simulateur de pluie). *Rapport interne, ORSTOM*, 16 pp.

Asseline, J. and Valentin, C. (1978). Construction et mise au point d'un infiltromètre à aspersion. *Cahiers ORSTOM, Série Hydrologie*, **15** (4), 321–349.

Billard, A., Muxart, T., Derbyshire, E., Egels, Y., Kasser, M., and Wang, J. T. (1992). Les glissements de terrain induits par les pluies dans les loess de la province de Gansou, Chine. *Annales de Géographie*, **566**, 495–515.

Billard, A., Muxart, T., Derbyshire, E., Wang, J. T. and Dijkstra, T. A. (1993). Landsliding and land use in the loess of Gansu Province, China. *Zeitschrift für Geomorphologie*, Supplement Band 87, 117–131.

Cabrera, J. C. and Smalley, I. J. (1973). Quickclays as products of glacial action: a new approach to their nature, geology, distribution and geotechnical properties. *Engineering Geology*, **7**, 115–133.

Cai, Q. G., Luk, S. H., Chen, H. and Chen, Y. Z. (1986). Effect of surface crusting on water erosion: laboratory experiments on loess soils, China. *Assessment of Soil Surface Sealing and Crusting. Proceedings of the Symposium, Ghent, Belgium, 1985*, pp. 99–105.

Collinet, J. (1985). Présentation générale du 'programme simulateur' et des premières techniques utilisées a l'ORSTOM (simulateur de type Swanson). *Journées Hydrologiques de l'ORSTOM à Montpellier*, 17–18 Septembre, pp. 5–9.

Derbyshire, E. and Mellors, T. W. (1988). Geological and geotechnical characteristics of some loess and loessic soils from China and Britain: a comparison. *Engineering Geology*, **25**, 135–175.

Derbyshire, E., Wang, J., Jin, Z., Billard, A., Egels, Y., Kasser, M., Jones, D. K. C., Muxart, T. and Owen, L. (1991). Landslides in the Gansu loess of China. *Catena Supplement*, **20**, 119–145.

Liu, T. S. *et al.* (1985). *Loess and the Environment*, China Ocean Press, Beijing, 251 pp.

Luk, S. H. and Cai, Q. G. (1990). Laboratory experiments on crust development and rainsplash erosion of loess soils, China. *Catena*, **17**, 261–276.

McIntyre, D. S. (1985a). Permeability measurements of soil crusts formed by raindrop impacts. *Soil Science*, **85**, 185–189.

McIntyre, D. S. (1958b). Soil splash and the formation of surface crusts by raindrop impacts. *Soil Science*, **85**, 261–266.

Mualem, Y., Assouline, S. and Rohdenburg, H. (1990a). Rainfall induced soil seal. (A) A critical review of observations and models. *Catena*, **17**, 185–203.

Mualem, Y., Assouline, S. and Rohdenburg, H. (1990b). Rainfall induced soil seal. (B) Application of a new model to saturated soils. *Catena*, **17**, 205–218.

Mualem, Y., Assouline, S. and Rohdenburg, H. (1990c). Rainfall induced soil seal. (C) A dynamic model with kinetic energy instead of cumulative rainfall as independent variable. *Catena*, **17**, 289–303.

Poesen, J. (1986). Surface sealing as influenced by slope angle and position of simulated stones in the top layer of loose sediments. *Earth Surface Processes and Landforms*, **11**, 1–10.

Römkens, M. J. M., Baumhardt, R. I., Parlange, J. Y., Whisler, F. D., Parlange, M. D. and Prasad, S. N. (1986). Effect of rainfall characteristics on seal hydraulic conductance. *Assessment of Soil Surface Sealing and Crusting. Proceedings of the Symposium, Ghent, Belgium, 1985*, pp. 228–235.

Smalley, I. J. (1971). 'In situ' theories of loess formation and the significance of the calcium carbonate content of loess. *Earth Science Reviews*, **7**, 67–85.

Valentin, C. (1978). Problèmes méthodologiques de la simulation de pluie. *Rapport interne, ORSTOM*, 11 pp.

# 17 Modelling Hydrogeomorphic Processes to Assess the Stability of Rehabilitated Landforms, Ranger Uranium Mine, Northern Territory, Australia – A Research Strategy

**S. J. RILEY**
*Alligator Rivers Region Research Institute, Jabiru, Australia*

## ABSTRACT

Design and assessment of rehabilitation of the Ranger Uranium Mine, Northern Territory, Australia, will require the development of hydrologic and geomorphic models to predict whether the engineered landforms are stable. Non-stationarity of the modelled system, largely because of soil formation and topographic changes, complicates the modelling and requires a strategy different from that adopted in modelling soil erosion of agricultural lands. Preliminary research suggests that gullying will be the critical erosion process. However, modelling will be further complicated by the fact that erosion will probably be supply limited in the short term and weathering limited in the long term.

The strategy adopted to develop geomorphic models and calibrate them involves a two-fold approach:

1. Simulation experiments on the existing waste rock dump of Ranger Uranium Mine and on natural areas, designed to identify the important processes and provide a data set to calibrate and test existing models.
2. Study of natural processes of weathering and soil formation in areas with rock types and landforms similar to those of the engineered landforms, designed to develop temporal models of soil formation, ecosystem succession, erodibility and hydrology.

It is expected that geomorphic models will be modular, allowing the incorporation of new models and refinements in sub-models of the important hydrogeomorphic processes.

## INTRODUCTION

Within the World Heritage Listed Area of Kakadu National Park, Northern Territory, Australia, is the Ranger Uranium Mine (Figure 17.1). At the completion of the project the mill tailings will be returned to the pits (the 'below-grade' option) and capped with waste rock unless the regulatory authorities allow *in situ* rehabilitation of the tailings dam (ER29: Clause 29, Ranger Environmental Requirements, Commonwealth of

*Process Models and Theoretical Geomorphology*. Edited by M. J. Kirkby
©1992 Commonwealth of Australia. Published 1994 by John Wiley & Sons Ltd

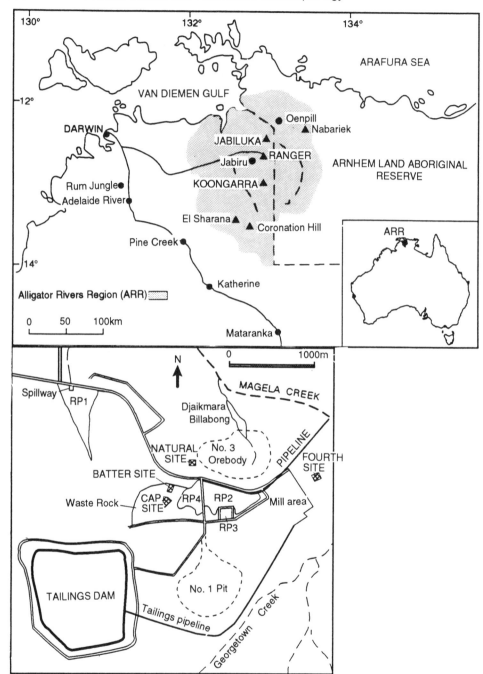

**Figure 17.1** Location of Ranger Uranium Mine, Northern Territory, Australia. Kakadu National Park, with an area of 19 757 km², is the largest terrestrial National Park in Australia, and one of the largest in the world, and incorporates the drainage basins of the South, West and East Alligator Rivers. There are four retention ponds (RP) in the lease. Geomorphic research related to modelling is concentrated on four areas; two on the waste rock (Cap and Batter sites) and two on adjacent relatively undisturbed areas (Natural and Fourth sites)

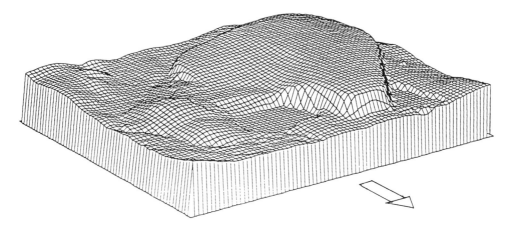

**Figure 17.2**   Digital elevation model of the 'above-grade' option for the rehabilitation of Ranger Uranium Mine, Northern Territory. The view is from the northeast looking towards the southwest. The tailings dam would be contained underneath the highest area in the southwest. Pit 3 would act as a sediment trap and a large portion of retention pond 1 (RP1) would be retained

Australia, 1979). The latter option ('above-grade') involves the containment of tailings in the existing 1 km$^2$ 'turkey nest' dam which occupies a topographic high within the Ranger Lease area (Figure 17.2). Both options would involve the containment of low-grade ore and the waste and discards of the processing plant when dismantled at the completion of mining. For the 'above-grade' option erosion of the containment structure would lead to dispersal of tailings over a substantial area of the lower Magela catchment (Riley and Waggitt, 1992). The dispersal of tailings from 'below grade' as a consequence of erosion is much less likely within the next 10 000 to 100 000 years.

Kakadu National Park is ecologically and archaeologically significant. The diversity, refuge value and pristine nature of the ecosystems of the Park are well documented (Braithwaite & Werner, 1987; Braithwaite & Woinarski, 1990). The oldest known (dated) human occupancy site within Australia is found within the area (Roberts *et al.*, 1990). Preservation of the ecosystem and the values of the Traditional Owners is essential (Fox, 1977).

Fluvial erosion of waste rock dumps and contained contaminated material is a potential hazard to Kakadu National Park and its users. Thus, a primary objective in rehabilitation is to ensure that the products of weathering and erosion do not degrade the Park ecosystems or put at risk the culture of the Traditional Owners.

The mining company, Ranger Uranium Mines Pty Ltd (RUM), prefers the mill tailings to be rehabilitated 'above-grade' (Unger *et al.*, 1989). However, it is necessary to show that the environment is 'no less well protected' (ER29) by this option. As one of the major problems will be the geomorphic stability of the structure over a long period a geomorphic model is required to predict the rate of erosion and the time for the tailings to possibly be exposed.

This paper presents an analysis of the problems of modelling the long-term behaviour of rehabilitated (engineered) landforms at Ranger Uranium Mine and the approach to developing models that can be used in the design and assessment process.

## BACKGROUND

The two main potential sources of contaminants from a rehabilitated mine are the mill tailings and the waste rock dumps, which include the very low-grade ore stockpiles (Pidgeon, 1982; Ellis, 1989).

The mill tailings are the processed waste of uranium oxide extraction. Their containment is essential because of the potential radiological and toxicity (e.g. uranium) risks. Problems of containment are increased by the fine-grained texture of the tailings (approximately 50% silt and 50% fine sand; Pickup *et al.*, 1987; Wasson, 1991) which gives them a high potential mobility. The intense oxidation of the extraction process (sulphuric acid digestion) removes the more active minerals but the tailings are deposited as a slurry. The soluble component of the tailings water has a high ionic strength (e.g. the $SO_4$ ion concentration is approximately 23 000 ppm). Release in groundwater of soluble toxicants contained in the tailings water could be a serious problem, although the issue is not addressed in this paper.

Apart from the tailings there are also environmental problems related to the release of containment materials (waste rock and low-grade ore in the case of Ranger Uranium Mine). Weathering and erosion of the waste rock may release toxicants whose environmental impact could be as great as that of the tailings. The waste rock has a variety of lithologies (Table 17.1). Some of the sulphide-rich rocks release sulphuric acid during weathering (Figure 17.3). Leaching of the waste rock and low-grade ore stockpiles, particularly with acidic waters, can release uranium and other heavy metals. The potential for acid-sulphate soils to develop, and the problems associated with them, has been recognised (Fitzpatrick & Milnes, 1988).

Increases in the suspended and solute loads of natural streams resulting from erosion of the engineered landforms may cause significant environmental damage (Riley & Waggitt, 1991). For rehabilitation to be judged successful the off-site impacts of erosion and weathering of waste rock, low-grade ore and tailings must be minimal with the ecosystems showing no sign of degradation. Several research programmes are directed towards examining these potential impacts (OSS, 1991). The tolerance of the receptive ecosystems to the influx of eroded material (soluble, coloidal and particulate) needs to be determined to provide the acceptable limits of off-site contamination (Riley & Waggitt, 1991).

**Table 17.1**  Lithology of material of the waste rock dump (modified from Milnes *et al.*, 1986)

| Rock | Percentage of total |
|---|---|
| Footwall schists | 1.3 |
| Chert | 20.5 |
| Basic dykes | 10.0 |
| Pegmatites | 22.4 |
| Degradable pegmatites | 2.1 |
| Upper mine schists | 14.7 |
| Carbonates | 14.4 |
| Porphyroblastic schist | 1.8 |
| Others | 12.9 |

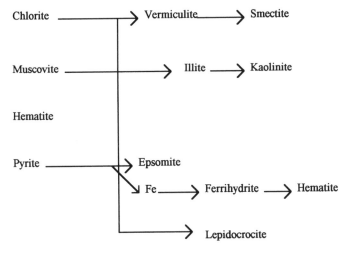

**Figure 17.3** The result of weathering of waste rock boulders exposed at the surface for 3 to 5 years. Weathering leads to the rapid leaching of Mg, Fe and SO₄ ions and the formation of smectitic clays. Salt crystallisation probably contributes to the disintegration. The model for the weathering reactions is modified from Milnes *et al.* (1986). Note the platy shape of the smaller gravels

The geomorphic issues related to the dispersal of erosion products are not the subject of this paper. These have been discussed by Riley & Waggitt (1992) and Waggitt & Riley (1992). Geomorphic modelling of rehabilitation structures will need to predict the likely levels of contaminants (loads and concentrations) flowing off-site, information that can be used in environmental impact models. The primary research thrust of the modelling at this time is to predict the on-site weathering and erosion processes operating within geomorphic time frames. Once a suitable model is developed it will be possible to examine the off-site impacts of discharged material. The successful geomorphic model must be a process-response one because of the need to predict not only the total erosion but also the rate and temporal and spatial pattern of erosion.

The geomorphic problems of containment are related to the stability of structures and sites to weathering and erosion over long periods (Schumm et al., 1982a,b; Nelson et al., 1983, 1986; East, 1986). The geomorphic problems link to the ecological problems through the discharge of potentially toxic materials from the site (Ellis, 1989) and the impact of on-site instability (erosion and release of toxic materials) on restoration of the site. There are feedback links between the geomorphic and the ecological problems simply because erosion is a biophysical phenomenon (Hole, 1981; Thornes, 1985, 1990).

The issues of containment are not simply biophysical. Some of the non-biophysical issues are:

- Economic: primarily related to the cost of rehabilitation, environmental repair and maintenance.
- Cultural: which is particularly important in the case of Ranger where the Traditional Owners continue aspects of their culture that depend on a high quality of the environment.
- National: expressed by the Australian people in terms of the values they attribute to the maintenance, conservation and preservation of the environment of Kakadu National Park.

These issues will also be considered in an evaluation of the rehabilitation but clearly are not the subject of this paper.

## Period of Containment

Fry (1982) suggests that the life of the containment structure is the critical factor in its design and assessment. The Guidelines of the Code of Practice on Management of Radioactive Waste from the Mining and Milling of Radioactive Ores (Commonwealth of Australia, 1987, p. 26) recognise two periods for containment structures, namely the design life and the structural life.

> Design life is the period after completion of an engineered structure, during which the structure and all its components are expected to perform in accordance with design objectives . . . a more conservative design methodology needs to be adopted, providing for a design life of at least 200 years without supportive maintenance or renovation.
>
> In this context 'structural life' is defined as the period over which a structure is expected by the designer to continue to perform its basic functions, even if at a reduced level. Structural life is therefore a measure of the useful life of rehabilitation structures in terms of their performance, and is obviously capable of being extended by renovation and maintenance activity. . . . 'Structural life' in these terms could reasonably be expected to extend to thousands of years.

It is obviously necessary to predict the long-term erosional and hydrologic behaviour of the rehabilitated structures, the nature of material released by weathering and erosion processes, and the probabilities of failure of containment structures. It is imperative for design and subsequent assessment by regulatory authorities that erosional and hydrologic models are available to make these predictions. The time scales of the predictions (centuries to millennia) force the modelling into a geomorphic time frame where the evolution of landscape (topography, soils, vegetation and hydrology) is a critical consideration in developing reliable predictions.

A long-term monitoring program (> 10 years) is impractical as part of the rehabilitation design process because design needs to be finalised and construction commenced long before completion of mining – the quantity of waste rock constrains double-handling on economic grounds. Furthermore, it is unlikely that monitoring will record the very large and infrequent erosion events whose probability of occurrence is high during the long period that the rehabilitated landforms must survive. Clearly, numerical modelling is the only viable approach to predicting whether erosional and hydrologic stability will be maintained for several millennia. Modelling must be used if the impact on the environment over long periods is to be predicted because, as indicated previously, it is necessary to forecast the loads and concentrations of potential contaminants that will be discharged from the rehabilitated sites (Jorgensen, 1983; Khanbilvardi et al., 1983; Nelson et al., 1986; Riley & East, 1990; Ringrose et al., 1991).

## GEOMORPHIC PROBLEMS OF REHABILITATION DESIGN AT RANGER

The engineered landforms of Ranger Uranium Mine will be largely constructed from schist waste rock, which weathers in the seasonally Wet Tropical environment of Kakadu to produce a fine-grained material rich in clay (Figure 17.3). The waste rock is the majority of the material extracted from the open pits and dumped at the surface. The weathering environment is intense, with daytime temperatures commonly in excess of 35°C and most of the yearly rainfall (1500 mm average annual) falling in 4 to 5 months of the year (McAlpine, 1976; Armstrong & Reid, 1989). Weathering releases a large quantity of soluble material, particularly Mg and $SO_4$, and average solute concentrations of runoff from individual storms in the early phases of the Wet season are often higher than $200 \, mg \, L^{-1}$ (Uren, 1991). The solute loads from the recently deposited waste rock decline dramatically during the Wet season (Table 17.2). Solute loads and weathering rates of the waste rock are several orders of magnitude higher than those on surrounding undisturbed areas.

**Table 17.2** Variation of mean solute concentration (mg $L^{-1}$) through 1988/89 Wet season Batter plot 3 (covered with run-of-mine waste rock), Ranger Uranium Mine* (modified from Uren, 1991)

| Ion | Early Wet season (1 Nov 1988) | Mid-Wet season (26 Jan 1989) | Late Wet season (23 March 1989) |
|---|---|---|---|
| Cl | 1.5 | 0.9 | 0.9 |
| Na | 3.8 | 0.8 | 2.3 |
| K | 6.6 | 0.5 | 2.9 |
| Ca | 18 | 1.0 | 1.1 |
| Fe | 9.3 | 5.6 | 4.2 |
| $HCO_3$ | 13 | 7.6 | 15 |
| Mg | 60 | 4.6 | 5.5 |
| $SO_4$ | 223 | 6.6 | 3.6 |

*This experiment involved a plot approximately 80 m long and 20 m wide on a concave batter slope with an average gradient of 20%. Samples were frequently collected during runoff from each of the storms.

If the 'above-grade' option is approved, the final engineered landform will cover an area of approximately 5 km² and rise an average of 15 m above the surrounding lowland. Some areas on the engineered landform will be higher than 20 m (Figure 17.2). The rehabilitated landforms of the 'below-grade' option will occupy a smaller area, approximately 3 km², but will still rise some 15 m above the lowland (Unger *et al.*, 1989). In both cases the waste rock will form the cover over the tailings (i.e. the tailings cap) and the majority of the volume of the engineered landforms. Except for the cap over the tailings there are presently no plans for engineering the surface of the rehabilitated structure other than ripping and shaping it with a bulldozer. The engineered landforms as presently proposed will have batter slopes with gradients of 10 to 20%, some ten times steeper than the natural slopes, and will have a much higher erosion potential as a result of their topography (Bryan, 1979; Singer & Blackard, 1982; Poesen, 1984) than the surrounding lowlands of Koolpinyah surface.

The Koolpinyah surface (Williams, 1969a) on which the mine is located has a very low denudation rate, less than 0.1 mm y$^{-1}$ (Duggan, 1988; Roberts, 1991). The receiving waters (Magela Creek) have very low suspended sediment concentrations (average non-filterable residue concentrations $<13$ mg L$^{-1}$; Hart *et al.*, 1986; Roberts, 1991) and equally low solute concentrations (average concentration $<10$ mg L$^{-1}$; Hart *et al.*, 1986; Roberts, 1991). Suspended sediment and solute concentrations in undisturbed tributaries draining the Koolpinyah surface near the mine are not much higher than those for Magela Creek (Nanson *et al.*, 1990, p. 33). A possible erosion standard for the rehabilitation of Ranger is that the sediment solute concentration of runoff is similar to that of the natural environment (Waggitt & Riley, 1992). Such a standard would result in very low acceptable levels of erosion and weathering and place considerable constraints on engineering design.

## THE MODELLING PROBLEM

The geomorphic model needs to address the two aspects of the geomorphic assessment of the stability and success of rehabilitation, namely, the stability of the engineered structures and the discharge of materials into the surrounding area. For the assessment of stability it is sufficient that the changes in topography be modelled. Information on topographic changes will provide data on the containment of materials and the primary sites of erosion. The discharge of materials has to be considered in terms of both total loads and peak loads. The total load estimates are required for sedimentation models. The peak load (concentration) predictions are needed to determine whether environmental thresholds are crossed beyond which degradation occurs. Both the erosion and the sediment discharge components of the geomorphic model will require a reliable hydrologic model. The hydrologic model will be used to predict the surface and subsurface losses of water.

A further demand made of a geomorphic model will be prediction of the soil water regime. Such information is invaluable in the prediction of the type of ecosystems likely to develop on the rehabilitated waste rock landforms. Part of the model predictions will also relate to the development of soil depth and possibly texture. Again, these data are relevant in the prediction of ecosystem development. Because of the feedback between ecosystems and erosion it is likely that a fully developed geomorphic model will incorporate modules that model soil moisture and development.

Thus the successful geomorphic model must provide the following information:

1. Digital terrain models of the topography at each time increment of the simulation for monitoring of the changes in topography and drainage.
2. Sedigraphs, solu-graphs and hydrographs of discharge to receptor systems (billabongs and streams) for assessment of the environmental impact of discharge over short periods.
3. Annual sediment and solute loads discharged to receptor systems for assessment of impacts over long periods.
4. Groundwater losses for assessment of changes in groundwater and potential impact on contained materials.
5. Soil moisture regimes to enable the prediction of hydrologic stress on vegetation.

The model must give predictions with acceptable errors that allow comparisons between different design options. Models in which the confidence limits are poorly defined will be of little use. The model must not show chaotic behaviour within a geomorphic time frame (0–10 000 years).

## RESEARCH STRATEGY

Clearly, it is necessary to identify the principal processes that will determine the geomorphic stability of engineered landforms in the long-term and the model that will accurately predict these processes. There is little known about the response of engineered landforms in a geomorphic time frame for the seasonally Wet Tropics. The majority of research on mine rehabilitation has been undertaken in the temperate and semi-arid areas of Europe and the United States of America (e.g. McCarter, 1985; Toy *et al.*, 1987), and this is particularly true of research on uranium mill tailings containment structures (Nelson *et al.*, 1983, 1986). Research on the rehabilitation of mine spoil and tailings has concentrated on short-term problems of erosion and hydrology (Beedlow *et al.*, 1982; Hartley *et al.*, 1982; Toy *et al.*, 1987). As indicated previously, currently used erosion models do not accommodate feedback resulting from erosional modification of the landforms and cannot be applied to long-term prediction. Thus, there is a two-fold research problem in developing a predictive capability based on geomorphic modelling. Firstly, it is necessary to identify and understand the major processes and secondly it is necessary to adopt or develop a suitable model.

The geomorphic strategy adopted to develop and assess geomorphic models (Figure 17.4) uses a combination of short-term monitoring and simulation experiments (Riley & East, 1990) and long-term studies of environments that may resemble the 'stable' ecosystem of the rehabilitated landforms.

Monitoring and simulation experiments are designed to indicate the significant hydrologic and erosion processes and evaluate the parameters of the models that describe these processes. Kirkby (1980, p. 53) indicates that it is necessary to neglect many minor processes and concentrate on the dominant ones. However, care must be taken with this philosophy as the minor processes, operating within a geomorphic time frame (graded time, Schumm, 1977) may be significant. At this stage it appears that the most important erosion process is gullying (Abt *et al.*, 1986; Riley & Williams, 1991), as it will impact

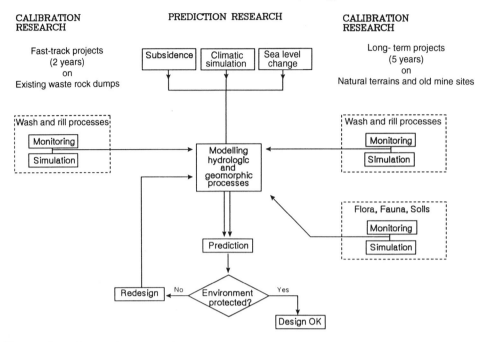

**Figure 17.4**   Research strategy, illustrating the short and long-term research studies, oriented towards developing and testing geomorphic models. The major objective of the research programme is the assessment of the long-term stability of engineered landforms and the prediction of impacts on the adjacent environment (Kakadu National Park in the case of Ranger Uranium Mine)

on the stability of the engineered landforms to a greater extent and more quickly than any other erosion process.

Monitoring experiments provide a data set suitable for model testing and validation as well as presenting the opportunity to assess whether simulation will reproduce the significant processes. There is a high probability that significant, low-frequency events will not be monitored because the duration of monitoring is limited by resources and the need to obtain results in as short a period as possible.

Simulation experiments allow low-frequency events to be studied. Simulation of low-frequency events is particularly important for identifying significant thresholds of erosion that only occur in the more intense storms. The simulation experiments combine rainfall simulation and *in situ* concentrated flow (flume) simulation tests. The rainfall simulator is capable of rainfall intensities up to $300 \, \text{mm h}^{-1}$ and can cover an area between 500 and $1000 \, \text{m}^2$ (Riley *et al.*, 1991), while the flume is 13 m long and 500 mm wide with the capacity for a $40 \, \text{L s}^{-1}$ discharge (Figure 17.5).

Monitoring and simulation experiments on the waste rock dump of Ranger Uranium Mine are limited by available surfaces. These surfaces are only 5 to 8 years old and are not mature in terms of total ecosystem interaction in the weathering and erosion processes; they do not resemble those likely to be present after several centuries. Thus, the research strategy incorporates an 'analogue' approach, a study of landforms with geologic and geomorphic properties similar to those likely to develop on the rehabilitated landforms. For Ranger the most likely analogue areas are schist hills with hillslope gradients between 5% and 20%. In these analogue areas simulation experiments will provide a further data set which will be used to develop and calibrate geomorphic models and compare the erosion and hydrology of old (natural) and new (engineered) landforms. An analogue area has been identified to the west of Ranger at Tin Camp Creek (Uren, 1992).

It is highly unlikely that field sites can be found that encompass (as analogues) a continuum of landforms ranging from the active newly developed rehabilitated landforms to the 'equilibrium' conditions of an old landscape. Landforms in the Alligator Rivers Region generally developed well before the Holocene (Williams, 1969a). Limited linear studies of the development of erosion and hydrologic processes are possible on the waste rock dumps of abandoned mine sites in the Alligator Rivers Region. These are small dumps and they are less than 100 years old. No suitable landforms appear to fill the time gap between 100 and 1000 years, although field investigations are still under way.

One research task will be to identify whether it is the long or short-term conditions that are critical for the discharge of sediment and solutes. This task is part of the assessment of off-site impacts. Models of temporal trends in weathering, soil development and topographic change are unlikely to be precise. Hence, the geomorphic model using the sub-models of temporal trends will have a degree of uncertainty in its prediction resulting from the difficulties of confirming the temporal trends. 'High' (high erosion) and 'low' (low erosion) case scenarios for erosional stability will be developed by examining conditions likely to exist at the beginning of the post-rehabilitation phase and after a long period has elapsed.

A potential problem of the long-term 'analogue' approach is the assumption that engineered landforms will survive long enough for steady state conditions to develop. It is possible that erosion will proceed at such a pace that stability is not reached until the landforms are significantly degraded. Rapid degradation would place the containment structure at risk in the case of the above-grade option. Modelling of the engineered landforms using the conditions that will exist at the beginning of the post-rehabilitation phase should indicate inherent instability.

Attempts to model the temporal trends in erosion and hydrology are worthwhile because the costs of overemphasising either the 'low' or 'high' cases of erosion are high. If the 'low' case is emphasised then the predicted rates of erosion of the engineered landforms may give a false sense of security of containment and lesser impacts on the environment than may actually occur. If the 'high' case is emphasised it may be impossible, at one extreme, to engineer any rehabilitated landforms and containment structures. At the very least, an overemphasis on the 'high' (environmentally conservative) case scenario may impose a substantial cost on the mine operation. However, paramount in all considerations is the protection of the environment.

'First approximation' answers to questions about the geomorphic stability of engineered landforms are required within 2 years, which provides insufficient time to do detailed work on models of soil development. The strategic approach involves a two-pronged research programme. The first is a 'fast-track' one in which models will be tested and applied with the data that are at hand or that can be collected rapidly. The second is a 'slow-track' one in which models will be modified and refined as the results of more detailed studies become available. It is entirely possible that the 'fast-track'

368

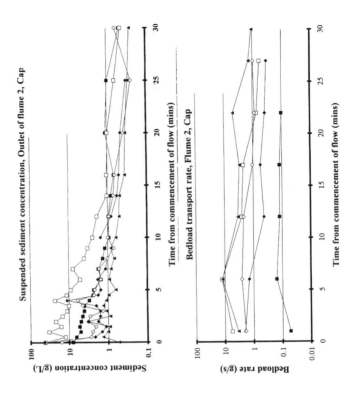

Suspended sediment concentration, Outlet of flume 2, Cap

Bedload transport rate, Flume 2, Cap

(a)

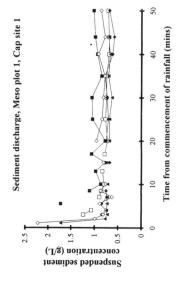

**Sediment discharge, Meso plot 1, Cap site 1**

**Figure 17.5** Concentrated flow (flume) simulator (a) and large-scale rainfall simulator (b); with examples of bedload and suspended sediment load discharge from a concentrated flow experiment on the Cap site. Each run, of 30 minutes duration, has a higher discharge than the preceding run. The bedload is sampled at the outlet of the flume using a US hand-held Arhnem-type bedload sampler. Suspended load is sampled by hand with wide-mouth bottles. Note that the graphs show different threshold responses for the suspended and bedloads, with obvious depletion of sediment over time and between successive runs (—■—, run 1; —□—, run 2; —◆—, run 3; —◇—, run 4; —▲—, run 5)

approach will indicate aspects of the long-term programme not worth pursuing although field observations suggest otherwise.

## THE MODELS

The preceding review suggests that a geomorphic model must meet several criteria and incorporate the several processes that are distinctly related to landscape evolution of rehabilitated mines. These processes are discussed below.

There are many models which describe components of landscape evolution, such as Kirkby's various models of slope development and weathering (Kirkby, 1980, 1985a, b, 1989) or Montgomery & Dietrich's (1992) model of channel initiation. Some distributed hydrologic models have been applied to erosion prediction (Moore & Clarke, 1983; Moore *et al.*, 1988) but there are few that incorporate a variety of processes and allow for the feedback between erosion, hydrology and landform development. Two-dimensional models of erosion (Lane *et al.*, 1988), while conceptually useful, will be inadequate for the three-dimensional problem of landform evolution. Clearly the models of the most use will be process-response models that are distributed, i.e. they model the four dimensions of geomorphic processes.

Amongst the 'complete' models are those of Ahnert (1987) and Willgoose *et al.* (1989, 1991). Both have attempted to model landscape development using a process-response approach. All the models need to be calibrated and for this the data derived from the simulation and monitoring experiments is essential. It is desirable that the models are modular to allow for developments in sub-models, such as sediment transport, entrainment and hydrology, to be incorporated with a minimum of programming. Continued research at the Alligator Rivers Region Research Institute and elsewhere will inevitably lead to improvements in the understanding of processes and consequent developments of new models or modifications to existing models of the hydrogeomorphic processes of landscape development.

Geomorphic modelling of the engineered landforms must accommodate the non-stationary nature of the system. To be relevant a geomorphic model must take account of the following:

1. That soils will develop that do not resemble the initial waste rock dump surface materials.
2. That erodibility and infiltration will change as a result of soil development, itself promoted by a variety of faunal, floral, and mineral weathering processes.
3. That developing drainage networks could be disrupted by settlement of the waste rock as it weathers and consolidates.
4. That landform shape will change with each erosive storm, possibly leading to the development of different erosion processes (e.g. slope wash dominated to gully dominated). These changes must be incorporated into digital terrain models for each successive simulated storm.
5. That the bed-level of the nearby Magela Creek may fluctuate because of climatic and sea level changes.
6. That erosivity may change as a result of changing vegetation cover.

In the remainder of this paper particular aspects of modelling the environment over time will be examined.

## SOIL DEVELOPMENT

The present weathered surface of the waste rock dump is dominated by materials finer than coarse gravels (Figures 17.3 and 17.6; Fitzpatrick and Milnes, 1988). In the immediate future the surface materials, or 'mine soils' will be largely composed of fine gravels and mixtures of clays and sands (Figure 17.6). Over the long-term, resistant quartz within the schist will weather-out and accumulate on the surface to form a gravel and

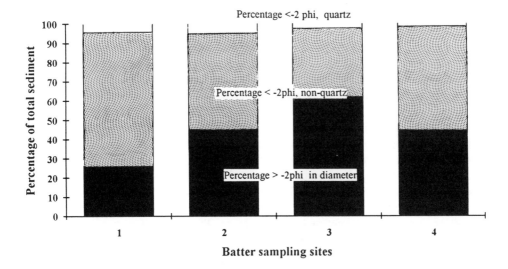

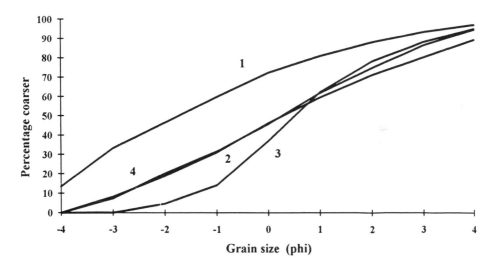

**Figure 17.6**   Grain size analysis of surface material from 5-year-old waste rock dumps. The sampling is biased towards the finer grained material by the collection method, which involved scraping material from the top 5 cm of the weathered surface

cobble lag. Whether the lag will protect the surface from erosion is uncertain at this stage. Preliminary research at the Tin Camp Creek analogue area suggests that the lag could be easily removed in concentrated flows with depths and velocities in excess of 50 mm and $1 \, \text{m s}^{-1}$, respectively.

The present gravel surface of the waste rock dump tends to accumulate for two reasons:

1. Some of the fine-grained materials are transported vertically down through the profile where they fill voids and begin the process of sealing the surface from the remainder of the waste rock pile.
2. Some of the fine-grained materials are transported off the waste rock dumps by wash and rill flow.

The gravels are dominated by schist fragments, platy in shape, and largely 1 to 2 cm in diameter (Figure 17.3). In the older waste rock dumps (up to 6 years old) the gravels form a surface layer 1 to 5 cm thick. This gravel layer is very mobile and transient, as indicated in rainfall simulation experiments. Observations suggest that large quantities are removed during the Wet season only to be replaced when the exposed rock weathers during the late Wet and Dry seasons (Gardiner and Riley, in press).

The gravels are underlain by a clay-rich, highly impermeable layer that can be over a metre deep. This layer has developed as a result of both mechanical action on the surface during the formation of the dump and subsequent weathering. The clays fill the voids between the less weatherable boulders. A tentative model for the development of this layer is presented in Figure 17.7. The mechanical action of heavy machinery and weathering lead to a compact near-surface layer. The upper 10 mm of the weathered surface 'soils' of the waste rock dump, immediately below the loose surface-gravel layer, frequently show the vesicular structure typical of crusting (Moss, 1991). The infiltration rates of these 'mine soils' are low; infiltration loss rates for individual storms are seldom more than 3 to 5 mm (Williams *et al.*, 1990) and at depths of 500 mm soils seldom reach saturation during the Wet season (Gardiner *et al.*, 1990). Below the compact layer the waste rock is porous, but the layer effectively seals (in a hydrologic sense) the upper 1 m from the underlying waste rock.

Soils in areas surrounding the waste rock dump (Wells, 1979) have developed on materials that are not similar to the waste rock. The Koolpinyah surface which surrounds the mine is covered with colluvial and alluvial deposits and soils on these cannot be used as analogues for soils likely to form in the long-term on the waste rock. Analogue areas at Tin Camp Creek with outcrops of schist, whose lithology is similar to that of the waste rock, are the sites for comparisons.

Preliminary investigations of soils in schist terrain elsewhere in the Alligator Rivers Region show that their development is strongly influenced by a combination of bioturbation (Gardiner, 1992; see general review by Lobry de Bruyn & Conacher, 1990) and subsequent wash erosion of silts and clays brought to the surface by termites and ants (Williams, 1978). The significance of bioturbation is not quantified at this stage, but it must be significant because all soils show significant bioturbation in the profile and at the surface. The waste rock dumps of Ranger will be invaded by meso-fauna (termites and ants) and it is highly likely that soils will develop with characteristics similar to those on schist terrain. The schist soils are either gradational or duplex (Story *et al.*, 1976; Northcote, 1979), with quartz lag gravels covering up to 80% of the surface (Figure 17.8).

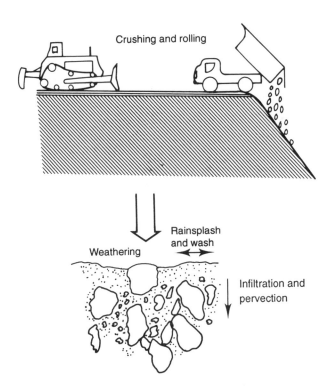

**Figure 17.7**    Model for development of soils on the waste rock dump of Ranger Uranium Mine. The photograph shows the relatively impermeable cap in a section cut through one surface

**Figure 17.8** Examples of soil typical of the schist terrain of Tin Camp Creek. Note the lag gravel at the surface, dominantly quartz

The soils on schists in natural areas do not resemble in any way the 'soils' that have developed to this time on the waste rock. Considerable modification of the surface material of the waste rock by weathering and erosion is expected. Consequently soils will develop on the waste rock that bear little resemblence to the present waste rock dump surface materials in terms of erodibility, hydrology or texture. For example, the mine soils have little or no organic matter content at present. This will increase probably reducing in erodibility (Bryan, 1976). There is a need to model the changes in erosion and hydrologic responses of the surface if predictions of landform changes and discharges of contaminants are to be realistic.

The natural schist terrain soils have infiltration capacities one to two orders of magnitude higher than those that are typical of the mine soils (5 to 10 mm h$^{-1}$; Gardiner et al., 1990). Thus, the permeability of the upper 400 mm of the developing mine soils probably will increase over time, but the quantity of water retained in the soil is unlikely to increase by the same amount. The marked vertical contrast in permeability obvious in the profiles of schist soils will probably develop in the waste rock dump soils. This vertical contrast in hydraulic conductivity may encourage throughflow/interflow, although there is no evidence of piping in schist soils. The main hydrologic consequence of soil development on the waste rock dump is likely to be an increase in the lag time and recession coefficient of run-off. There may be an increase in deep percolation losses, but this is unknown at present.

Natural schist terrain soil surfaces are partly covered with a cryptogam that appears to retard infiltration (similar to that described by Mucher et al., 1988). However, it is not a continuous cover and probably its greatest significance is in reducing wash erosion, reinforcing the protection offered by the quartz lag gravel. The influence of cryptogam covers on infiltration and the effect of fire on them are matters of contention (Green et al., 1990).

The role of vegetation cover in the erosion and hydrologic processes of the area is uncertain. Previous work (Williams, 1969b; Duggan, 1988) suggests that vegetation cover protects the surface. However, this protection may only be relevant in the case of splash-dominated erosion and may offer minimal protection against wash and concentrated flow-dominated erosion. Preliminary studies have shown that rainfall erosivity is less underneath shrub and grass covers (Table 17.3). On the other hand, simulation experiments on natural schist slopes where the vegetation cover was removed by intense fires suggests that the bare surface is resistant to erosion. It is possible that vegetation is important in promoting erosion stability not because of its cover but because of its direct effect on soils (mechanical or chemical) and indirect effects through the fauna it attracts. The frequency of fire in the Alligator Rivers Region and the obvious stability of many natural slopes suggests that cover is not the only critical factor in erosion stability.

The rates of soil formation are not known and unlikely to be known for some time. Erosion rates and solute loads give an indication of the rates. Comparison of soil development on waste rock dumps of ages up to approximately 100 years (the oldest in the area) may give a clue to the rates of soil formation. The geochemical balance method of Wakatsuki & Rasyidin (1992) may also provide a method of determining weathering and soil formation rates.

**Table 17.3**   (a) Rainsplash (g). Mean and standard deviation for the four vegetation covers

| Vegetation cover | Mean | Minimum value | Maximum value | Standard deviation (SD) |
|---|---|---|---|---|
| Bare | 1.2 | 0.0 | 13.6 | 2.3 |
| Grasses | 0.7 | 0.0 | 10.8 | 1.5 |
| Shrubs | 0.6 | 0.0 | 14.9 | 1.5 |
| Trees | 1.3 | 0.0 | 18.6 | 2.9 |

(b) One-way analysis of variance of rainsplash beneath vegetation covers

| SOURCE | DF | SS | MS | F | p |
|---|---|---|---|---|---|
|  | 3 | 12.529 | 4.176 | 8.60 | 0.000 |
| ERROR | 701 | 340.439 | 0.486 |  |  |
| TOTAL | 704 | 352.968 |  |  |  |

| LEVEL | $N$ | MEAN | SD | INDIVIDUAL 95 PCT CI'S FOR MEAN BASED ON POOLED SD |
|---|---|---|---|---|
|  |  |  |  | ---------+---------+---------+------- |
| Bare | 179 | −0.4462 | 0.7510 | (------*------) |
| Grasses | 181 | −0.6364 | 0.6752 | (------*------) |
| Shrubs | 171 | −0.7758 | 0.6391 | (------*------) |
| Trees | 174 | −0.4677 | 0.7154 | (------*------) |
| POOLED SD = 0.6969 |  |  |  | ---------+---------+---------+-------<br>−0.75    −0.60    −0.45 |

The little known so far about the evolution of soils on the rehabilitated surfaces of Ranger suggests the following hypotheses:

1. Some metres of surface lowering will be required before a quartz-rich gravel layer develops that will protect the surface from wash erosion.
2. The infiltrability of the surface will increase but there may not be a commensurate increase in the water-holding capacity of the soil.
3. The erodibility of the surface material will decrease because of profile development and an increase in the organic matter content.
4. Meso-fauna activity will probably determine the rate of development of 'mature' soil profiles.

These hypotheses are being tested in the Alligator Rivers Region. The research should lead to models that predict the temporal trends in soil development.

## DOMINANT EROSION PROCESSES AND THEIR TEMPORAL TRENDS

Both simulation and monitoring studies of gully and inter-rill erosion on the waste rock dump of Ranger Uranium Mine are under way. The simulation experiments involve the use of a large-scale rainfall simulator for wash, rainwash, splash and micro-rill studies, and an *in situ* flume for studying concentrated flow (Figure 17.5). Monitoring involves a number of catchment-size experiments designed to study a range of processes (Riley and East, 1990), and includes bedload, wash, and splash traps, flumes for measuring overland and rill flow, piezometers and tensiometers, erosion pins, painted stone lines and raingauges (Figure 17.9). The monitoring provides a means of assessing the validity

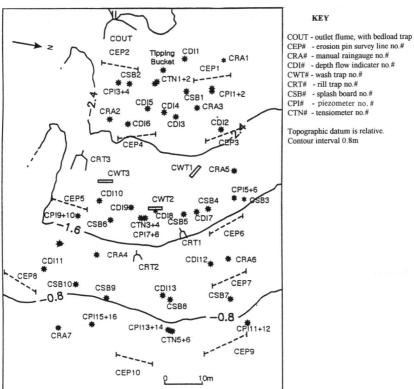

**Figure 17.9**  Map and picture of monitoring experiment, waste rock dump of Ranger Uranium Mine. The photograph is taken from the southwest of the area, looking towards the east. The vehicle is parked near the tipping-bucket raingauge. Steel posts indicate the ends of lines of erosion pins as well as providing support for cabling between dataloggers and sensors

of the results from simulation experiments, focusing simulator experiments on the critical processes, and extending the scale of the experiments beyond that possible with the simulators.

Initial results indicate a significant depletion of particulate and solute loads during storms. Monitoring also indicates depletion through the Wet season (Riley *et al.*, 1991; Uren, 1991). Suspended sediment concentrations rapidly decline in the early phase of monitored storms and simulation experiments (Figure 17.5), attributed to depletion of available material. As the smaller particles are eroded from the surface the imbedded rocks are exposed. Sometimes these exposed rocks are so intensively weathered that they break down under the hydraulic stresses, but more commonly they remain intact for some time after exposure. This depletion is likely to be a feature of the rehabilitated landforms.

During the Dry season the waste rock dump surface rocks are 'shattered' by weathering that is aided by the accumulation of salt brought to the surface by evaporation and by dew (wetting and drying) that sometimes accumulates on the surface during the mornings. Hydration of clays and thermal expansion and contraction may also contribute to the breakdown. Bioturbation also brings sands and clays to the surface although to a minor extent at this time because the meso-fauna population on the waste rock dumps is small by comparison with nearby natural areas. By the end of the Dry season the partly stripped surface of the previous Wet season is again covered with gravels and smaller particles. It is also possible that the wetting and drying cycles of the early Wet season, when storms are separated by several days, contribute to weathering and heaving and cracking of the more clay-rich mine soils.

Solute and particulate concentrations of runoff from the waste rock are highest in the early part of the Wet season and low by the middle of the Wet season (Table 17.2). However, the concentrations are not as low as they are in the natural streams where solute concentrations are almost as low as those in rainfall (Roberts, 1991). These trends of decreasing sediment and solute from the waste rock suggest rapid depletion of available sediment and a combination of flushing-out of soluble materials and a change in the chemistry of the weathering reactions during the Wet season (Uren, 1991).

Geomorphic models must incorporate modules that allow for depletion of loose surface material and its subsequent replacement by the following Dry season weathering (Figure 17.5). A weathering module needs to predict the quantity and texture of erodible material made available by weathering in each successive time increment. Ahnert (1987) includes a weathering module in the SLOP3D model. Kirkby (1985a, b, 1989) details a more elaborate weathering model that incorporates several processes involved in weathering and soil formation. At this stage insufficient work has been completed to decide on which approach to adopt to modelling weathering and the availability and depletion of sediment.

There are thresholds of erosion to wash erosion on the surface of the waste rock dump (Riley, 1992; Figure 17.10). However, these thresholds are so low that they are not significant – the hydraulic conditions of the thresholds are commonly exceeded. There may, however, be a series of thresholds, related to the different materials. Some concentrated flow simulation experiments indicate multiple thresholds (Figure 17.5).

There are clearly defined thresholds of gully erosion in natural schist terrains (Riley & Williams, 1991; Figure 17.10). It is highly likely that similar gully thresholds will be evident on the waste rock dumps as the drainage system develops. The gully thresholds will be incorporated into models of landscape evolution (e.g. Willgoose et al., 1991).

Gully erosion is a critical erosion process as recognised in a number of studies of other rehabilitation structures (Nelson et al., 1983; Soulliere & Toy, 1986; Toy et al., 1987). The low permeability of the waste rock dump surface (Gardiner et al., 1990; Williams et al., 1990) causes very high rates of runoff, particularly in this monsoonal environment where five-minute storm intensities commonly exceed $50\,\mathrm{mm\,h^{-1}}$. Rill networks can cover more than 20% of the surface at the peak of the storm runoff (Figure 17.11). In the schist-dominated terrains of the Alligator Rivers Region, which are being used as analogues of the long-term development of landforms on the schist-dominated engineered landforms, the potential for gully erosion is indicated by the high drainage density and the small catchment area thresholds (Riley & Williams, 1991).

Models must incorporate gullying and include the complexities of sediment depletion by erosion and replacement of erodible material by weathering. Nelson et al. (1983) and Abt et al. (1986) present models for predicting gully incision on reclaimed slopes. Abt et al. (1986) incorporate an exponential function for the growth and decay of a gully and the model is related to slope length, gradient and stable slope geometry. Both models require calibration by reference to slopes in various stages of gullying. More sophisticated hydraulic and sediment transport modelling of the type suggested by Bhallamundi & Chaudhry (1991) in which the Saint-Venant equations of unsteady flow are solved, may not be warranted, although aspects of this type of approach are contained in the Willgoose et al. (1989) model.

Jones (1987) lists several mechanisms for the initiation of drainage networks that are relevant. Recent work by Montgomery & Dietrich (1992) using a combination of distributed hydrological modelling and estimates of gully thresholds suggests a method for predicting gully formation. It is possible that Hortonian processes will be dominant in the early stages of the rehabilitated Ranger Uranium Mine and saturated overland flow and dynamic area flow in the more 'mature' engineered landform. Piping is unlikely to be an important mechanism for channel initiation as there is no evidence of it in a variety of terrains in the Alligator Rivers Region.

# A VARIABLE TOPOGRAPHY

Topographic changes as the engineered landforms erode and settle need to be incorporated into the modelling. The modification of topography by gullying will change drainage density, slopes, drainage vectors, and the quantity of material available for erosion. Several erosion models (e.g. ANSWERS, CREAMS, WEPP; see reviews by Rose, 1985 and Pickup, 1988) do not contain facilities for varying the topography during the simulation period and thus are unsuitable for erosion and hydrologic modelling in a geomorphic time frame. Models by Ahnert (1987) and Willgoose et al. (1991) provide the opportunity for the interactive feedback between erosion, topographic change, and process response.

A further complication in modelling erosion and hydrology is the surface deformation caused by unavoidable settlement. While the gross characteristics of settlement can be predicted (for Ranger Uranium Mine, of the order of 1 m; Richards, 1987), the magnitude

380

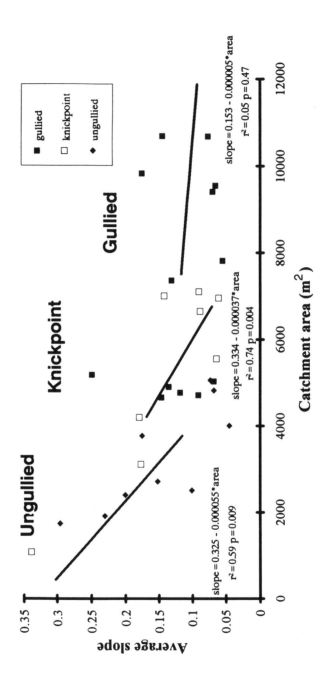

381

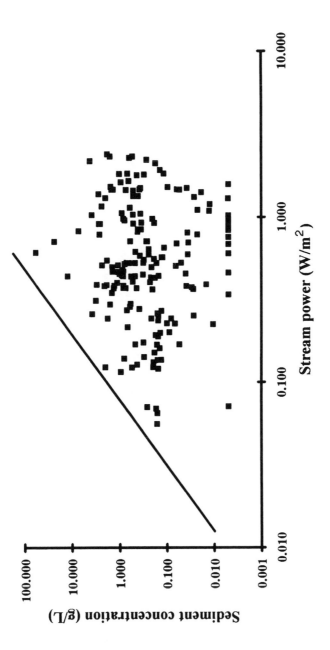

**Figure 17.10** Thresholds of erosion are not evident in small flume (Riley and Gore, 1990) experiments on wash erosion on the waste rock dump. Even at the low stream power the sediment concentration of runoff was high relative to that experienced in nearby natural slopes. There is, however, clear evidence of thresholds of erosion for gullies in the Tin Camp Creek area, although the threshold is on the basis of catchment area, rather than around a regression line

**Figure 17.11**  View of catchment on waste rock dump after a storm, showing the dense network of rills

and rate of settlement at particular locations cannot. A stochastic approach to settlement and its effect on topography must be adopted, with the significance of the process being determined by sensitivity analysis.

## CLIMATIC SERIES AND CHANGE

Climatic change is noticeable in the stratigraphic record of northern Australia. Lees (1992) suggests that the climate since 5500 BP has been increasingly variable with possibly four wetter periods since then. The most significant impact of climatic change on the stability of Magela Creek near the mine will be related to sea level change, particularly a fall in sea level (Nanson *et al.*, 1990; Wasson, 1991). The available sedimentary record for Magela Creek suggests that catchment response is insensitive to climatic fluctuations. There is no evidence of catastrophic flooding in the Magela. Analysis of storm frequency, duration, magnitude and intensity suggests that the variation in climate may only be expressed by a change in storm frequency (Riley, 1991). Wetter or drier climatic regimes, while they may be important in modifying the rates of erosion of the rehabilitated structure, are less likely to have a significant impact on the elevation of the bed of Magela Creek (Roberts, 1991). Erosion models will need to be flexible enough to incorporate scenarios of base-level changes imposed by external factors, although the triggering of these changes in simulated time may be random.

Simulated climatic time series based on real climatic series are required. However, for Jabiru (Figure 17.1) the pluviometer records are only available for the period 1971–1990 and the longest daily raingauge record for the region (80 years) is for Gunbalanya (Oenpelli), some 40 km north of the mine. Therefore, there will be problems in defining the present climate and the nature of the climatic time series appropriate for other climatic regimes that may be typical for Jabiru when climate changes. The north–south shifts of climate, the likely consequence of climatic change, will dictate the choice of climate stations for alternative climatic series.

## CONCLUSIONS

The requirements of geomorphic modelling of engineered landforms are distinctly different from those of a soil erosion modelling problem. The differences arise from the time frame and the intentions of the models. Erosion models are primarily extension tools, conventionally for agricultural use, predicting the immediate erosional behaviour of small parcels of land under different land-use practices. Geomorphic models were originally designed to understand landform and landscape development, although there is an increasing practical application for them. Their time frame is several orders of magnitude larger than soil erosion models and they tend to incorporate 'averaging' procedures.

It is unlikely that the existing geomorphic models will suffice for the problems of predicting landform stability and off-site impacts. A marriage of convenience is required between the soil erosion and geomorphic models. A tandem approach suggests itself, allowing a switch from the detailed analysis provided by the soil erosion models to the analysis in graded time provided by the geomorphic models. Detailed analysis of catchment hydrology and runoff is unlikely to be required for every runoff event, but assessment of design requires this detail at key times.

This paper has outlined the demands that will be placed on geomorphic modelling of engineered landforms in the Alligator Rivers Region of the Northern Territory of Australia. The research strategy to calibrate and assess the models uses both short- and long-term approaches. The short-term approach concentrates on the assessment of models and evaluation of landform stability to determine whether there is an immediate threat to the environment arising from the rehabilitation design and whether models need modification. The long-term approach develops models that account for temporal changes in soil and hydrologic properties of the landforms.

It appears that no single geomorphic or erosion model is suited to the task of rehabilitation assessment. For Ranger Uranium Mine a requirement is the ability to predict processes of erosion and sediment transport related to gully development.

## ACKNOWLEDGEMENTS

This paper has arisen from research conducted by the Office of the Supervising Scientist. Several staff and consultants have contributed to the work. In particular I appreciate the assistance of co-workers in the Alligator Rivers Region Research Institute and especially the help given by Mr P. Waggitt, Mr C. Uren and Mr B. Gardiner.

## REFERENCES

Abt, S. R., Falk, J. A., Nelson, J. D. and Johnson, T. L. (1986). Gully incision prediction on reclaimed slopes. In *Water Forum '86 World Water Issues in Evolution'*, *Proceedings American Society Civil Engineers*, pp. 412–419.

Ahnert, F. (1987). Approaches to dynamic equilibrium in theoretical simulations of slope development. *Earth Surface Processes and Landforms*, **12**, 3–15.

Armstrong, A. and Reid, A. (1989). Environmental auditing at Ranger. *Proceedings North Australian Mine Rehabilitation Workshop*, No. 11, 95–100.

Beedlow, P. A., McShane, M. C. and Cadwell, L. L. (1982) Revegetation/rock cover for stabilisation of inactive uranium-mill tailings disposal sites. *US Dept Energy UMT-0210 PNL-4238*.

Bhallamundi, S. M. and Chaudhry, M. H. (1991). Numerical modeling of aggradation and degradation in alluvial channels. *Journal of Hydraulic Engineering*, **117**, 1145–1164.

Braithwaite, R. W. and Werner, P. A. (1987). The biological value of Kakadu National Park. *Search*, **18** (6), 296–301.

Braithwaite, R. W. and Woinarski, J. C. Z. (1990). Coronation Hill Stage III – assessing the conservation value. *Australian Biologist*, **3** (1), 3–13.

Bryan, R. B. (1976). Considerations on soil erodibility indices and sheetwash. *Catena*, **3**, 99–111.

Bryan, R. B. (1979). The influence of slope angle on soil entrainment by sheetwash and rainsplash. *Earth Surface Processes*, **4**, 43–58.

Commonwealth of Australia (1979). *Atomic Energy Act 1953, Authority under s.41, The Ranger Authority, Schedule 2, Appendix A, Ranger Environmental Requirements*.

Commonwealth of Australia. Department of the Arts, Sport, The Environment, Tourism and Territories (1987). *Code of Practice on the Management of Radioactive Waste from the Mining of Radioactive Ores 1982. Guidelines*. Australian Government Publishing Service, Canberra.

Duggan, K. (1988). Mining and Erosion in the Alligator Rivers Region of Northern Australia. Ph.D. Thesis, School of Earth Sciences, Macquarie University.

East, T. J. (1986). Geomorphological assessment of sites and impoundments for long-term containment of uranium mill tailings in the Alligator Rivers Region. *Australian Geographer*, **17**, 16–21.

Ellis, D. (1989). *Environments at Risk: Histories of Impact Assessment*. Springer-Verlag, Berlin.

Fitzpatrick, R. W. and Milnes, A. R. (1988). Characteristics of soil forming on waste-rock dumps in the Ranger Project Area, Jabiru, NT. *Focus Report No. 6 to Ranger Uranium Mine Pty Ltd*. CSIRO Division of Soils, Adelaide.

Fox, R. W. (1977). *Second Report, Ranger Environmental Inquiry*. Australian Government Publishing Service, Canberra.

Fry, R. M. (1982). Criteria for the long-term management of uranium mill tailings. In *Management of Wastes from Uranium Mining and Milling*. International Energy Agency, Vienna, pp. 71–83.

Gardiner, B. (1992). The role of bioturbation in the rehabilitation of mine landforms in the Alligator Rivers Region: literature review and research proposal. *Commonwealth of Australia, Office of the Supervising Scientist for the Alligator Rivers Region Internal Report* IR 63.

Gardiner, B. and Riley, S. J. 1989–91 Wet season monitoring results, Waste Rock Dump, Ranger Uranium Mine, Commonwealth of Australia, Office of the Supervising Scientist for the Alligator Rivers Research Region Internal Report (in press).

Gardiner, B., Riley, S. J. and Williams, D. (1990). Infiltration characteristics of the surface of the waste rock dump, Ranger Uranium Mine, Northern Territory, under natural rainfall. *Proceedings 7th Australian Hydrographic Workshop, Darwin, 1990*, Northern Territory Power and Water Authority, Darwin, 2 vols.

Green, R. S. B., Chartres, C. J. and Hodkinson, K. C. (1990). The effects of fire on the soil in a degraded semi-arid woodland. I. Cryptogam cover and physical and micromorphological properties. *Australian Journal of Soil Research*, **28**, 755–777.

Hart, B. T., Ottaway, E. M. and Noller, B. N. (1986). Nutrient and trace metal fluxes in the Magela Creek system, Northern Australia. *Ecological Modelling*, **31**, 249–265.

Hartley, J. N., Gee, G. W., Freeman, H. D., Cline, J. F., Beedlow, P. A., Buelt, J. L., Relyea, J. R. and Tamura, T. (1982). Uranium mill tailings remedial action project (UMTRAP) – cover and liner technology development project. In *Management of Wastes from Uranium Mining and Milling*, International Atomic Energy Agency, Vienna, pp. 429–448.

Hole, F. D. (1981). Effects of animals on soil. *Geoderma*, **25**, 75–112.

Jones, J. A. A. (1987). The initiation of natural drainage networks. *Progress in Physical Geography*, **11** (2), 207–245.

Jorgensen, S. E. and Mitsch, D. W. (1983). *Applications of Ecological Modelling in Environmental Management*, Elsevier, Amsterdam.

Khanbilvardi, R. M., Rogowski, A. S. and Miller, A. C. (1983). Predicting erosion and deposition on a strip mined and reclaimed area. *Water Resources Bulletin*, **19**, 585–593.

Kirkby, M. J. (1980). The stream head as a significant geomorphic threshold. In Coates, D. R. and Vitek, J. D. (eds), *Thresholds in Geomorphology*. Allen & Unwin, London, pp. 53–74.

Kirkby, M. J. (1985a). A basis for soil profile modelling in a geomorphic context. *Journal of Soil Science*, **36**, 97–121.

Kirkby, M. J. (1985b). A model for the evolution of regolith-mantled slopes. In Woldenberg, M. J. (ed), *Models in Geomorphology*. Binghamton Symposium in Geomorphology, International Series No. 14, Allen & Unwin, pp. 213–237.

Kirkby, M. J. (1989). A model to estimate the impact of climatic change on hillslope and regolith form. *Catena*, **16** (4/5), 321–339.

Lane, L. J., Shirley, E. D. and Singh, V. P. (1988). Modelling erosion on hillslopes. In Anderson, M. G. (ed.), *Modelling Geomorphological Systems*, Wiley, London, pp. 287–308.

Lees, B. G. (1992). Geomorphological evidence for late Holocene climatic change in Northern Australia. *Australian Geographer*, **23** (1), 1–10.

Lobry de Bruyn, L. A. and Conacher, A. J. (1990). The role of termites and ants in soil modification: a review. *Australian Journal of Soil Research*, **28** (1), 55–94.

McAlpine, J. R. (1976). Climate and water balance. In *Lands of the Adelaide–Alligator Area, Northern Territory*, CSIRO Land Research Series No. 38, pp. 35–49.

McCarter, M. K. (ed.) (1985). *Design of Non-Impounding Mine Waste Dumps*, American Institute of Mining, Metallurgical and Petroleum Engineers, New York.

Milnes, R. R., Riley, G. G. and Raven, M. D. (1986). Rock weathering, landscape development and the fate of uranium in waste-rock dumps and the low grade ore stockpile. In *Rehabilitation of Waste Rock Dumps, Ranger No. 1 Mine, NT*, CSIRO Division of Soils, Adelaide, Chapter 1.

Montgomery, D. R. and Dietrich, W. E. (1992). Channel initiation and the problem of landscape scale. *Science*, **225**, 826–830.

Moore, I. D., O'Loughlin, E. M. and Burch, G. J. (1988). A contour-based topographic model for hydrological and ecological applications. *Earth Surface Processes and Landforms*, **13**, 305–320.

Moore, R. J. and Clarke, R. T. (1983). A distribution function approach to modelling basin sediment yield. *Journal of Hydrology*, **65**, 239–257.

Moss, A. J. (1991). Rain impact soil crust. *Australian Journal of Soil Research*, **29**, 271–337.

Mucher, H. J., Chartres, C. J., Tongway, D. J. and Green, R. S. B. (1988). Micromorphology and significance of the surface crusts of soils in rangelands near Cobar, Australia. *Geoderma*, **42**, 227–244.

Nanson, G. C., East, T. J., Roberts, R. G., Clark, R. L. and Murray, A. S. (1990). Quaternary evolution and landform stability of Magela Creek catchment near the Ranger Uranium Mine, Northern Australia. *Commonwealth of Australia, Office of the Supervising Scientist for the Alligator Rivers Region Open File Record* OFT 63, 119 pp.

Nelson, J. D., Volpe, R. L., Wardwell, R. E., Schumm, S. A. and Straub, W. P. (1983). Design considerations for long-term stabilization of uranium mill tailings impoundments. *US Nuclear Regulatory Commission* NUREC/CR-3397, ORNL-5979.

Nelson, J. D., Abt, S. R., Volpe, R. L., van Zyl, D., Hinkle, N. E. and Straub, W. P. (1986). Methodologies for evaluating long-term stabilisation designs of uranium mill tailings impoundments. *Division of Waste Management, Office of Nuclear Material Safety and Safeguards, US Nuclear Regulatory Commission*, NRC FIN B0279.

Northcote, K. H. (1979). *A Factual Key for the Recognition of Australian Soils*, 4th edition, Rellim Technical Publication, South Australia.

OSS (Office of the Supervising Scientist) (1991). Alligator Rivers Region Research Institute 1991–92 Research Strategy. *Commonwealth of Australia, Office of the Supervising Scientist for the Alligator Rivers Region Internal Report No. 40*.

Pickup, G. (1988). Hydrology and sediment models. In Anderson, M. G. (ed.), *Modelling Geomorphological Systems*, London, pp. 153–215, Wiley.

Pickup, G., Wasson, R. J., Warner, R. F., Tongway, D. and Clark, R. L. (1987). A feasibility study of geomorphic research for the long term management of uranium mill tailings. *CSIRO Division of Water Resources Research Divisional Report* 87/2, 65 pp.

Pidgeon, R. T. (1982). Review of non-radiological contaminants in the long-term management of uranium mine and mill wastes. In *Management of Wastes from Uranium Mining and Milling*. International Atomic Energy Agency, Vienna, pp. 263–284.

Poesen, J. (1984). The influence of slope angle on infiltration rate and Hortonian overland flow volume. *Zeitschrift für Creomorphologie, Supplement*, **49**, 117–131.

Richards, B. (1987). Settlement under rock dumps. *Focus Report No. 1 to Ranger Uranium Mines Pty Ltd*, CSIRO Division of Soils, Adelaide.

Riley, S. J. (1991). Analysis of pluviometer records, Darwin and Jabiru. *Commonwealth of Australia Office of the Supervising Scientist for the Alligator Rivers Region. Internal Report IR-48*.

Riley, S. J. and East, T. J. (1990). Investigation of the erosional stability of waste rock dumps under simulated rainfall: a proposal. *Commonwealth of Australia, Office of the Supervising Scientist for the Alligator Rivers Region Technial Memorandum* 31, 48 pp.

Riley, S. J. and Gardiner, B. (1991). Characteristics of slope wash erosion on the waste rock dump, Ranger Uranium Mine, Northern Territory. *International Hydrology and Water Resources Symposium, Perth, 2–4 October 1991, Institution of Engineers Australia National Conference Publication* 91/22, pp. 295–300.

Riley, S. J., Gardiner, B. G. and Hancock, F. (1991). A large scale rainfall simulation facility for studying the erosion and hydrology of mine sites (abstract). *Proceedings, Second International and Sixteenth Annual Environmental Workshop, Australian Mining Industry Council, Perth*, 7–11 October 1991.

Riley, S. J. and Gore, D. (1990). Aspects of the design and calibration of a portable flume. *Soil Technology*, **1**, 297–312.

Riley, S. J. and Waggitt, P. W. (1991). Discussion paper on issues on the acceptable design life of structures to contain mill tailings. *Commonwealth of Australia, Office of the Supervising Scientist for the Alligator Rivers Region Internal Report No. 58.*

Riley, S. J. and Waggitt, P. W. (1992). The potential fate of particulate contaminants from the rehabilitated Ranger Uranium Mine. *Water Forum '92, 'Saving a Threatened Resource – In Search of Solutions', Proceedings Water Resources Sessions, American Society of Civil Engineers*, pp. 884–889.

Riley, S. J. and Williams, D. (1991). Some geomorphic thresholds related to gullying Tin Camp Creek, Arnhem Land, Northern Territory, Australia. *Malaysian Journal of Tropical Geography*, **22** (2), 133–143.

Ringrose, P., Bonne, A., Peaudecerf, P., Fourniguet, J., Klussin, F., Patyn, J. and Wilmot, R. (1991). Geoforecasting: assessing the long-term evolution of geological confinement systems. In Cecille, L. (ed.), *Radioactive Waste Management and Disposal*, Elsevier, London, pp. 472–487.

Roberts, R. G. (1991). Sediment budgets and Quaternary history of the Magela Creek catchment, tropical Northern Australia. Ph.D. Thesis, University of Wollongong, 565 pp.

Roberts, R. G., Jones, R. and Smith, M. A. (1990). Thermoluminescence dating of a 50,000 year-old human habitation site in Northern Australia. *Nature*, **345**, 153–156.

Rose, C. W. (1985). Developments in soil erosion and deposition models. *Advances in Soil Science*, **2**, 1–61.

Schumm, S. A. (1977). *The Fluvial System*, Wiley, Toronto.

Schumm, S. A., Costa, J. E., Toy, T. J., Knox, J., Warner, R. and Scott, J. (1982a). Geomorphic assessment of uranium mill tailings disposal sites. Summary report of the workshop by the panel of geomorphologists. In *'Uranium Mill Tailings Management', Proceedings of Nuclear Energy Agency Workshops on Geomorpological Evaluation of the Long-Term Stability of Uranium Mill Tailings Disposal Sites, Colorado State University, Fort Collins, 28–30 October 1981.* OECD NEA and USDOE, pp. 69–79.

Schumm, S. A., Costa, J. E., Toy, T. J., Knox, J. C. and Warner, R. F. (1982b). Geomorphic hazards and uranium-tailings disposal. In *Management of Wastes from Uranium Mining and Milling.* International Atomic Energy Agency, Vienna, pp. 111–124.

Singer, M. J. and Blackard, J. (1982). Slope-angle – interrill soil loss relationship. *Transactions of the American Society of Agricultural Engineers.*

Soulliere, E. J. and Toy, T. J. (1986). Rilling of hillslopes reclaimed before 1977 surface mining law, Dave Johnston Mine, Wyoming, *Earth Surface Processes and Landforms*, **11**, 293–305.

Story, R., Galloway, R. W., McAlpine, J. R, Aldrick, J. M. and Williams, M. A. J. (1976). Lands of the Alligator Rivers Area, Northern Territory. *CSIRO Land Research Series 38.*

Thornes, J. B. (1985). The ecology of erosion. *Geography*, **70**, 222–235.

Thornes, J. B. (ed.) (1990). *Vegetation and Erosion*, Wiley, London.

Toy, T. J., Hadley, R. F. and Frederick, R. (1987). *Geomorphology and Reclamation of Disturbed Lands*, Academic Press, London.

Unger, C., Armstrong, A., McQuade, C., Sinclair, G., Bywater, J. and Koperski, G. (1989). Planning for rehabilitation of the tailings dam at Ranger Uranium Mines. *Proceedings North Australian Mine Rehabilitation Workshop No. 11*, pp. 153–165.

Uren, C. (1991). The application of geomorphic variables for improving the erosional stability of artificial hill slopes at the Ranger Uranium Mine. *Commonwealth of Australia, Office of the Supervising Scientist for the Alligator Rivers Region, Open File Record* OFR 79, 167 pp.

Uren, C. J. (1992). An investigation of surface geology in the Alligator Rivers Region for possible analogues of uranium mine rehabilitation structures. *Commonwealth of Australia, Office of the Supervising Scientist for the Alligator Rivers Region Internal Report* IR 56.

Waggitt, P. W. and Riley, S. J. (1992). Development of erosion standards for use in the rehabilitation of uranium mines in Northern Australia. Younos, T., Diplas, P., Mostaghimi, S. (eds). Land reclamation advances in research and technology. American Society of Agricultural Engineers, Nashville Conference, 204–212.

Wakatsuki, T. and Rasyidin, A. (1992) Rates of weathering and soil formation. *Geoderma*, **52**, 251–263.

Wasson, R. (ed.) (1991). Modern sedimentation and late Quaternary evolution of the Magela Creek Plain. *Commonwealth of Austtralia, Office of the Supervising Scientist for the Alligator Rivers Region, Open File Record* OFR 88.

Wells, M. R. (1979). *Soil Studies in the Magela Creek Catchment, 1978. Part 1.* Land Conservation Unit, Territory Parks and Wildlife Commission, Darwin, 101 pp.

Willgoose, G., Bras, R. L. and Rodriguez-Iturbe, I. (1989). Modelling of the erosional impacts of landuse change: a new approach using a physical based catchment evolution model. *Proceedings Hydrology and Water Resources Symposium 1989, Christchurch, NZ.* Institution of Engineers Australia, pp. 325–329.

Willgoose, G., Bras, R. L. and Rodriguez-Iturbe, I. (1991). Results from a new model of river basin evolution. *Earth Surface Processes and Landforms*, **16**, 237–254.

Williams, D., Riley, S. J. and Gardiner, B. (1990). Infiltration characteristics of the surface of the waste rock dump, Ranger Uranium Mine, Northern Territory, under simulated rainfall. *Proceedings 7th Australian Hydrographic Workshop, Darwin, 1990*, Northern Territory Power and Water Authority, Darwin, 2 vols.

Williams, M. A. J. (1969a). Geomorphology of the Adelaide–Alligator area. In Lands of the Adelaide–Alligator Area, Northern Territory. *CSIRO Land Research Series* No. 25 pp. 71–94.

Williams, M. A. J. (1969b). The predictability of rainsplash erosion in the seasonally wet tropics. *Nature*, **222**, 763–765.

Williams, M. A. J. (1978). Termites, soils and landscape equilibrium in the Northern Territory of Australia. In Davies, J. L. D. and Williams, M. A. J. (eds), *Landform Evolution in Australasia*, Australian National University Press, Canberra, pp. 128–141.

# 18 Applications of a Numerical Model for Shore-normal Sediment Size Variation

DIANE P. HORN
*King's College London, UK*

## ABSTRACT

This paper describes the performance and applications of a numerical model which is designed to predict and explain shore-normal sediment size variation. The model simulates wave shoaling, wave height attenuation due to frictional losses, and breaking. Peak horizontal orbital velocities at the bed are calculated from Stokes' second order wave theory. The peak onshore and offshore celocities are used with the threshold expression of Komar & Miller (1975) to predict both the spatial variation in threshold sediment size across a beach profile, and the grain size distribution at any particular position on the beach profile. The problems that have been observed with the present model, which has attempted to go from known waves to known sediments, are used to identify some of the possible errors implicit in estimates of palaeowave conditions, and to assess the extent to which individual parameters may influence the environmental interpretation. Examples are also given of potential applications for coastal management and predictions of beach response to sea level change.

## INTRODUCTION

The concept of beach equilibrium is a central theme in coastal research, both with respect to equilibrium morphology and equilibrium sediment dynamics. The seabed slope and grain size gradient have been generally considered to represent an equilibrium response to hydrodynamic conditions in the nearshore environment, where the beach morphology and sediment grain size distribution across the profile are in equilibrium with the wave-generated hydraulic regime. Tanner (1958) provided a concise definition of the equilibrium beach as 'one whose curvature in plan view and profile is adjusted in such a way that the waves impinging on the shore provide precisely the energy required to transport the load of sediments supplied to the beach'.

Equilibrium beach morphology has received more attention from coastal researchers than has its sedimentological analogue, the equilibrium sediment size gradient. Confusion can arise from the fact that some investigators refer to morphology in speaking of a profile of equilibrium, whereas other investigators have referred to the nature of the bottom sediments. The problem of explaining and predicting shore-normal sediment size variation is part of the larger problem of predicting the rates and quantities of shore-normal sediment transport. Although in recent years coastal scientists have recognised

*Process Models and Theoretical Geomorphology.* Edited by M. J. Kirkby
©1994 John Wiley & Sons Ltd

the interrelationship between beach profile morphology and wave and current processes, present shore-normal sediment transport models have not yet been expanded to include sediment characteristics such as grain size variation and sediment sorting processes.

## MODELLING SHORE-NORMAL SEDIMENT SIZE

Despite the range of geological conditions and wave and tidal environments, beaches everywhere are remarkably similar. Most observations indicate that grain size decreases with increasing depth in the nearshore and offshore zones, and that sediments are coarser on the intertidal profile than on the offshore and nearshore profiles (e.g. Inman, 1953; Fox *et al.*, 1966; Swift *et al.*, 1971; Greenwood & Davidson-Arnott, 1972; Shipp, 1984; Niedoroda *et al.*, 1985). The seaward-fining grain size gradient is usually attributed to progressive sediment sorting, or size grading, by which grains of a particular size reach an equilibrium position. The degree of similarity that has been observed in shore-normal sediment characteristics implies that similar processes may be acting on the beach profile. Two models have been suggested to predict and explain the processes responsible for this selective shore-normal sorting of grain sizes: the null-point hypothesis, and the hypothesis of asymmetrical thresholds under waves. Although these hypotheses do not represent our present understanding of nearshore sediment transport, particularly in their treatment of the hydrodynamics involved in the sediment transport processes, they still continue to be discussed, even in such recent works as Carter (1988) and Wright *et al.* (1991), and there is little alternative but to include these simple hypotheses in any consideration of the equilibrium sediment size gradient.

### The Null-point Hypothesis

The null-point hypothesis was first proposed by Cornaglia in 1889, emphasising the concept of the balance of forces controlling the stability of single particles on nearshore slopes. The null-point hypothesis was expressed in mathematical terms by Ippen & Eagleson (1955), Eagleson *et al.* (1958, 1961, 1963), Miller & Zeigler (1958, 1964) and Eagleson & Dean (1959, 1961), who tested their equations with laboratory experiments and field measurements. This model combines the forces due to flow asymmetry with the downslope component of the gravitational force, suggesting that both onshore wave drift and an offshore gravitational component produced by bottom slope affect sediment grains on the sea bed. The null-point model suggests that nearshore slopes as well as grain size are controlled by this mechanism. The model presumes that, for every grain size, there exists a unique depth at which the wave-induced onshore flows exactly balance the offshore force of gravity. The position of zero net transport is called the null-point. Large grains will have a null-point further onshore than smaller ones, and the distribution of null-points will vary from fine to coarse sizes in the onshore direction as long as threshold velocities are exceeded. This null-point is an unstable equilibrium, however, as in slightly deeper water the equilibrium grain size will move offshore, and in slightly shallower water it will move onshore. The model also suggests that in a sediment sample with a wide distribution of grain sizes, only one grain size can be at the null-point position: all grains coarser than this critical equilibrium grain size will have a stronger offshore component and will tend to shift offshore, while the finer grains will move onshore.

## The Hypothesis of Asymmetrical Thresholds under Waves

The hypothesis of asymmetrical thresholds under waves was first suggested by Cornish (1898), and has been reiterated many times since (Bagnold, 1940; Inman, 1949; King, 1972; Carter, 1988). This hypothesis is based on observations of the asymmetrical nature of the orbital motions under waves in shallow water: the onshore flow has a high velocity and a short duration, while the offshore flow has a lower velocity but a longer duration. The hypothesis suggests that the higher onshore nearbed velocities will produce a shear stress great enough to initiate motion for both large and small sediment particles, while the lower offshore velocities will only exceed the threshold shear stress for the smaller particles. This mechanism will act selectively to drive larger particles onshore, with a net offshore transport of finer sediment. Waves become progressively more asymmetrical as water depth decreases, and this increasing onshore asymmetry will be responsible for an onshore increase in sediment size. An assumption inherent in the hypothesis is that the grain size at a particular position on the beach profile represents an equilibrium response to maximum orbital velocities under the local waves.

Although the hypothesis of asymmetrical thresholds under waves is simpler than the null-point hypothesis, many researchers have suggested that the null-point model may have overestimated the importance of gravity on a sloping bed, because slopes are not sufficiently steep over much of the nearshore, and because it tends to be overwhelmed by other mechanisms that are more important in determining sediment size gradients and bottom slope (Cook & Gorsline, 1972; Swift, 1976). The importance of asymmetrical nearbed velocities was recognised by Wells (1967), who argued that wave-induced flow asymmetries alone, generated by the growth of second order effects, were sufficient to

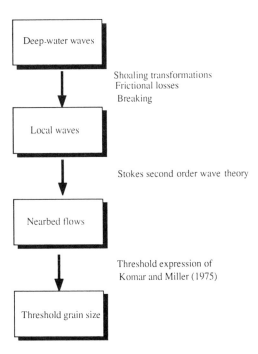

**Figure 18.1**   Flow chart of the asymmetrical threshold model

**Table 18.1** Equations used to calculate key variables in simulation program

| Variable | Equation | Source |
|---|---|---|
| 1 Local wavelength | $L = \dfrac{gT^2}{2\pi}\tanh\left(\dfrac{2\pi h}{L}\right)$ | Linear/Stokes second order theory |
| 2 Ratio of wave group to wave phase velocity | $n = \dfrac{1}{2}\left(1 + \dfrac{2kh}{\sinh(2kh)}\right)$ | Linear/Stokes second order theory |
| 3 Local wave celerity | $C = \dfrac{gT}{2\pi}\tanh\left(\dfrac{2\pi h}{L}\right)$ | Linear/Stokes second order theory |
| 4 Deep-water wave celerity | $C_o = \dfrac{gT}{2\pi}$ | Linear/Stokes second order theory |
| 5 Shoaling wave height | $\dfrac{H}{H_o} = \left(\dfrac{1}{2n}\dfrac{C_o}{C}\right)^{0.5}$ | Linear/Stokes second order theory |
| 6 Deep-water wavelength | $L_o = \dfrac{gT^2}{2\pi}$ | Linear/Stokes second order theory |
| 7 Deep-water depth | $h_o = \dfrac{L_o}{2}$ | Definition of wave base as $h/L = 0.5$ |
| 8 Neilsen's variable, $\psi$ | $\psi = \dfrac{f_e}{3\pi L_o\left(\dfrac{\delta h}{\delta x}\right)}$ | Neilsen (1982) |
| 9 Neilsen's variable, $I$ | $I = \dfrac{4L_o}{5h_o}\left(1 - \left(\dfrac{h_o}{h}\right)^{1.25}\right) + 2\pi\left(\left(\dfrac{h_o}{h}\right)^{0.25} - 1\right) + \dfrac{\pi^2 h_o}{2L_o}\left(1 - \left(\dfrac{h_o}{h}\right)^{-0.75}\right)$ | |
| 10 Wave height with frictional losses (after Neilsen, 1982) | $H_f = \dfrac{H_{oh}\left(\dfrac{h_o}{h}\right)^{0.25}\exp\left(\dfrac{\pi}{2L_o}(h - h_o)\right)}{1 + \psi H_{oh} I \exp\left(-\dfrac{\pi h_o}{2L_o}\right)}$ | |
| 11 Difference between shoaling wave height and wave height with frictional losses | $\Delta H = H - H_f$ | This paper |
| 12 Breaking wave | $\left(\dfrac{H_b}{h_b}\right) = 0.72\,(1 + 6.4\,\tan\beta)$ | Madsen (1976) |
| 13 Wave height shoreward of breakers | $H = \gamma h$ | Thornton and Guza (1982) |

(continues)

**Table 18.1**(continued)

| Variable | Equation | Source |
|---|---|---|
| 14 Local wave height | If $H_f > H_b$, then $H = H_b$ <br> If $H_f < H_b$, then $H = H_f$ | This paper |
| 15 Peak onshore velocity | $$u_{in} = \frac{\pi H}{T \sinh(kh)}$$ $$+ \frac{3}{4} \frac{(\pi H)^2}{LT \sinh^4(kh)}$$ | Stokes second order theory |
| 16 Peak offshore velocity | $$u_{ex} = \frac{\pi H}{T \sinh(kh)}$$ $$- \frac{3}{4} \frac{(\pi H)^2}{LT \sinh^4(kh)}$$ | Stokes second order theory |
| 17 Onshore $\theta_{cr}$ | $$\theta_{in} = \frac{\rho u_{in}}{(\rho_s - \rho) g T}$$ | Komar and Miller (1975) |
| 18 Offshore $\theta_{cr}$ | $$\theta_{ex} = \frac{\rho u_{ex}}{(\rho_s - \rho) g T}$$ | Komar and Miller (1975) |
| 19 Ratio of $(d_0/D)$ to $\theta_{in}$ | $$\left(\frac{d_0}{D}\right) = \left(\frac{0.463}{\theta_{in}}\right)^{1.333}$$ | Komar and Miller (1975) |
| 20 Ratio of $(d_0/D)$ to $\theta_{ex}$ | $$\left(\frac{d_0}{D}\right) = \left(\frac{0.463}{\theta_{ex}}\right)^{1.333}$$ | Komar and Miller (1975) |
| 21 Nearbed orbital diameter | $$d_0 = \frac{H}{\sinh(kh)}$$ | Linear theory |
| 22 Onshore threshold grain diameter in cm | $$D_{in} = \frac{d_0}{\left(\frac{d_0}{D}\right)}$$ | Komar and Miller (1975) |
| 23 Offshore threshold grain diameter in cm | $$D_{ex} = \frac{d_0}{\left(\frac{d_0}{D}\right)}$$ | Komar and Miller (1975) |
| 24 Onshore threshold grain diameter in mm | $D_{in} = 10(D_{in})$ | SI units |
| 25 Offshore threshold grain diameter in mm | $D_{ex} = 10(D_{ex})$ | SI units |
| 26 Onshore threshold grain diameter in phi units | $D_{in\,\phi} = -\log_2(D_{in\,mm})$ | Krumbein (1934) |
| 27 Offshore threshold grain diameter in phi units | $D_{ex\,\phi} = -\log_2(D_{ex\,mm})$ | Krumbein (1934) |

produce net sediment transport on a flat bed, without the necessity of including the effects of a sloping bed.

## A NUMERICAL MODEL FOR SHORE-NORMAL SEDIMENT SIZE VARIATION

### Description of the Model

The hypothesis of asymmetrical thresholds under waves has only been stated in a qualitative form, and the present author has attempted to put this hypothesis into a quantitative form as a simple numerical model, and to test the predictions of the hypothesis of asymmetrical thresholds against observed sediment size variations. The numerical model is described more fully in Horn (1992). Figure 18.1 shows a flow chart of the sequence of operations of the model. The equations used in the model are shown in Table 18.1. The model requires four initial input parameters: wave period, deep-water wave height, beach slope, and a wave friction energy dissipation coefficient (Jonsson, 1966). A number of quantities which are nearly constant are also supplied to the model at the outset: acceleration due to gravity, the density of water, and the density of quartz sand. SI units are used throughout. The model was designed to use easily available data as input parameters, without requiring detailed process measurements. As the most readily available wave data are measurements of wave period and wave height in deep water, the model predicts shoaling wave heights and nearbed velocities across the profile from deep water to the swash limit. However, observed shallow-water or breaking wave heights can also be used as input parameters if available.

The beach profile is represented by a series of regularly spaced points which are each assigned a corresponding depth value. The model varies the water depth over a beach profile, with depth decreasing from 20 metres to 1 metre. The model assumes that depth decreases monotonically onshore; however, a more realistic nearshore topography can be simulated by initialising the model with specific depths and offshore distances as input parameters. For each depth, the local wavelength and local wave celerity are calculated from linear wave theory. The threshold equations used in the model require a local wave height at fixed points on the nearshore profile. The local wave height, including wave height attenuation due to friction, is calculated according to the solution of Neilsen (1982). After calculating the local wave height, the water depth is decreased until the wave breaks. The breaking wave height is calculated using Madsen's (1976) equation (equation (12) in Table 18.1). After breaking, the saturation model of Thornton & Guza (1982) is applied. Their observations suggest that the waves shoreward of the breakpoint are saturated: the wave height at any point shoreward of the breakpoint is determined by the local depth. After breaking, therefore, the wave height is assumed to be a function of the depth. The local wave height required by the model is given by the wave height with frictional losses until breaking, and the saturated wave height after that. Once the local wave parameters have been calculated, the peak nearbed onshore and offshore velocities are calculated, using these local parameters, from Stokes' (1847, 1880) second order wave theory. The model identifies the depth at which the predictions of Stokes' second order theory cease to be realistic, based on an experimentally defined limiting criterion (Horn and Hardisty, 1990), and applies the attenuation factor to the predicted Stokes' flow velocities. The threshold grain size is calculated using the expression of

Komar and Miller (1975) for both the onshore and offshore peak flows. The threshold grain sizes calculated from the threshold expression of Komar & Miller (1975) are here called $D_{in}$ and $D_{ex}$, which represent the onshore threshold grain diameter and the offshore threshold grain diameter, respectively. They are expressed in both SI units and phi units. The sediment size on the equilibrium beach profile is assumed to be the grain size just at threshold for the flow velocity generated by the surface waves passing over the profile at that point. The model predicts a spatial variation of these threshold grain sizes across the beach profile.

The two threshold grain diameters predicted by the model can also be used to predict the limits of the grain size distribution at any particular water depth or location on the beach profile, and can be modified to allow for the changes in flow velocity that occur as water depth changes over a tidal cycle. As threshold grain size is proportional to nearbed flow velocity, the largest threshold grain sizes should be transported at the time at which maximum flows occur, and the lowest threshold grain sizes at the time at which minimum flows occur. When flows are varied at a point on the profile over a tidal cycle, the predicted sediment size distribution should vary between the largest onshore threshold grain size ($D_{in}$) and the smallest offshore threshold grain size ($D_{ex}$), based on the assumption that the maximum flows are onshore-directed and the minimum flows are offshore-directed. The tidal distribution version of the model predicts a sediment distribution at a landward position with a range of sediments between the largest and smallest threshold grain diameters ($D_{in}$ and $D_{ex}$), and a sediment distribution at a seaward position with sediments less than the smallest threshold grain diameter ($D_{ex}$). The seaward sediment distribution is better sorted than the landward distribution. Therefore, this simple conceptual model predicts the trends which have frequently been observed on the intertidal profile of a macrotidal beach, where the seaward sample (from the low-tide terrace and nearshore zone) are fine and well-sorted and the landward samples are coarser and poorly sorted (Wright et al., 1982; Bryant, 1984; Jago & Hardisty, 1984; Horn, 1991, 1993; Short, 1991).

## Predictions and Implications of the Model

Tests of the model's hydrodynamic and sedimentological predictions are described in Horn (1991, 1992). The sedimentological predictions of the model were tested against the data sets of Sato & Kishi (1954), Harrison & Alamo (1964), Perrett (1990), Foster (1991) and Horn (1991). The predictions of the asymmetrical threshold model described in this paper were also compared to the predictions of the null-point model, and these comparisons are reported fully in Horn (1992). Although the predictions of the asymmetrical threshold model seem to generate a theoretical shore-normal sediment size variation which is closer to measured grain sizes than the predictions of the null-point model, the actual agreement between measured and predicted sediment size is not particularly good. The model is more successful at predicting the general trend of onshore sediment size increase than at predicting actual sediment sizes. The predicted sediment sizes are significantly larger than the measured sediment sizes in shallow water, and the difference between measured and predicted sediment sizes can vary over several orders of magnitude. The gradient of the predicted sediment size increase in the onshore direction is considerably greater than the measured sediment size variation: the model predicts a grain size which is smaller than the measured sediment size in deep water, and significantly larger than the measured sediment size in shallow water.

Alterations could be made to the asymmetrical threshold model to improve its predictive capacity, and these have been discussed in detail in Horn (1991, 1992). The model in its present form represents a greatly simplified situation. Other factors are undoubtedly important in the field, such as the presence of a mixed-frequency spectrum, including long waves; wave set-up; mass transport; turbulence; wave–current interaction; suspended sediment transport; gravity-induced downslope transport; the presence of bed ripples; and the effects of a mixed sediment size distribution. However, the real problem does not lie in the model's treatment of the hydrodynamics of nearshore sediment transport, grossly simplified though that is. The model must be evaluated in terms of its ability both to predict and to explain. If the aim of modelling is to compare the model's predictions with the real world in order to evaluate the theory which gave rise to the model (in this case the theory of asymmetrical thresholds), then the numerical accuracy of the predictions is less important than the implications of the model's success or failure in prediction. No matter how much the predictive ability of the asymmetrical threshold model may be improved, it does not appear to explain sediment sorting on a real beach satisfactorily. Although the model has proved fairly successful at predicting the trend of increasing onshore grain size, it has not been possible to establish without doubt that this onshore size increase is due to an onshore increase in nearbed velocity and velocity asymmetry. The question, then, is whether or not the model in its present form has any utility other than in refuting hypotheses and suggesting directions for further research. Here, the problems with the model's predictions may prove useful to other researchers. For example, the problems that have been observed with the present model, which has attempted to go from known waves to known sediments, highlight the need for caution in attempting to reconstruct ancient sea conditions from sedimentary deposits. Most research of this nature has concentrated on the relationship between ripple length and nearbed orbital diameter, rather than on the relationship between grain size and maximum nearbed orbital velocity. It is in the interpretation of this relationship that the present model may prove most useful.

## APPLICATIONS OF THE THRESHOLD MODEL: ESTIMATION OF PALAEOWAVE CONDITIONS

Geologists have often attempted to use sedimentary structures to determine the environmental conditions under which these structures were deposited, and to reconstruct palaeohydraulic conditions (e.g. Newton, 1968; Harms, 1969; Tanner, 1971; Komar, 1974; Clifton, 1976; Miller and Komar, 1980; Allen, 1981a,b, 1984; Clifton & Dingler, 1984; Dupré, 1984; Diem, 1985). The most common approach is to describe an ancient deposit in terms of its ripple characteristics (spacing, steepness and asymmetry), and grain parameters (size, sorting, shape and density). Two flow parameters can be derived from these sediment parameters: the horizontal diameter of the water particle orbit at the bed and the maximum horizontal velocity at the bed. The primary relationships in which geologists are interested for the purposes of palaeohydraulic reconstruction are the relationship between the ripple wavelength or spacing and the horizontal diameter of the water particle orbit at the bed, and the relationship between the mean grain size of the sedimentary deposit and the nearbed orbital velocity. Once the nearbed flow parameters have been derived from ripple characteristics, these quantities can be related

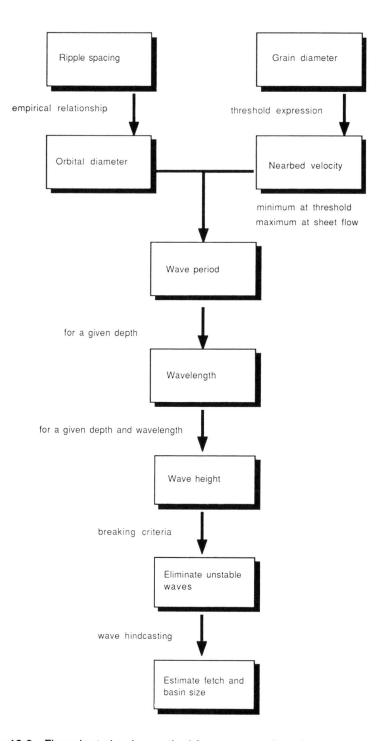

**Figure 18.2** Flow chart showing method for reconstruction of palaeowave conditions

to surface wave parameters by using one of several wave theories. From the derived wave parameters, geologists have attempted to infer the sea conditions which could have produced that orbital diameter and nearbed velocity. These ancient sea conditions are usually expressed in terms of combinations of wave height and water depth. From this, it may be possible to estimate the fetch and, therefore, the size of the basin in which the sediments were deposited. Figure 18.2 shows a flow chart of the method most commonly used by geologists for palaeohydraulic reconstruction.

The problems that have been observed in modelling the sediment size variation on modern beaches can provide guidelines which can help to identify some of the possible errors implicit in previous estimates of ancient flow and depositional systems, and to assess the extent to which individual parameters may influence the environmental interpretation. A number of errors concerning the palaeowave conditions can arise as a result of the initial assumptions. All of these effects are more pronounced as depth decreases.

## Calculation of Wave Height

The choice of wave theory is important in the estimation of wave height and water depth from inferred nearbed velocities. Incorrect estimates of wave height are most significant in the calculation of the possible fetch of the basin. An estimated wave height which is too small will result in an underestimation of the fetch as well. The general theory for waves of small amplitude was developed by Airy (1845). Airy theory applies to waves of small amplitude in deep water and treats waves as sinusoidal forms. It describes a wave that is symmetrical about the still-water level and has water particles that move in closed orbits. Most geologists use Airy wave theory to calculate the possible combinations of wave height and water depth from the sediment characteristics. However, Airy theory may not be appropriate for most palaeowave reconstruction. Tanner (1971) suggested that deep-water wave theory is not likely to be particularly useful in the reconstruction of the sea conditions under which wave-controlled sedimentary structures were formed, and argued that, for palaeogeographic purposes, it is more important to be able to make interpretations where water depths and wave heights were relatively small. In addition, Airy theory, which predicts symmetrical nearbed velocities, may not provide an accurate prediction of nearbed velocities in intermediate water depths. Field measurements show that a considerable flow asymmetry exists in the nearshore region, with the onshore flow beneath the wave crest having a greater magnitude but a shorter duration than the offshore flow beneath the wave trough (Huntley & Bowen, 1975; Greenwood & Sherman, 1984; Elgar et al., 1988). Stokes' second order wave theory describes a wave which has steeper crests and flatter troughs than predicted by Airy theory, and predicts nearbed orbital velocities which are closer to the characteristics of observed waves than Airy theory. Allen (1981b) recognised the conceptual importance of Stokes' theory, suggesting that it is reasonable to assume the operation of Stokes' waves in the formation of wave-ripple structures. The reconstruction of palaeowave conditions using Airy wave theory, rather than Stokes' theory, would either overestimate wave height or underestimate water depth. These effects are more pronounced for longer-period oceanic waves and shallow-water depths.

Each of the available wave theories is based upon certain assumptions which constrain the range of water depths over which the theory may be applied. The Stokes or Airy wave theories can adequately represent water motion in particular depth ranges, but limits exist beyond which their application is questionable. The velocity predictions of neither Stokes' second order nor Airy wave theory are realistic in very shallow water, and researchers should be aware of the ways in which these unrealistic predictions will influence palaeoenvironmental interpretation. Although Stokes' wave theory is preferable to Airy wave theory in intermediate depths, the use of either Stokes' or Airy theory in depths that are too shallow would lead to the overestimation of wave height or the underestimation of water depth.

Geological reconstructions generally attempt to predict wave heights and water depths from the horizontal diameter of the water particle orbit at the bed, assuming a constant wave height in all depths, and most palaeowave calculations produce only one wave height for the whole sedimentary basin, regardless of changing depth. In reality, shoaling waves are greatly affected by changed in water depth. As waves enter shallow water, the decrease in depth influences most of the characteristics of the waves: wave velocity and wavelength decrease, the wave steepens, and wave height increases until the waves become unstable and break. The wave profile becomes increasingly asymmetric. The shoaling wave height is smaller than the Airy wave height in deep water and larger than the Airy wave height in shallow water. However, since shoaling changes are generally not considered to be significant until relative depth ($h/L$) is less than 0.25, it is in shallow-water conditions that the calculation of shoaling wave heights will be most important. If the shoaling changes in wave height are not considered in the estimation of ancient sea conditions, wave height would be underestimated or water depth overestimated over most of the nearshore zone.

## Choice of Threshold Expression

Many equations have been proposed for the threshold of sediment motion under oscillatory flow. Although there is no one generally accepted relationship, the expression of Komar and Miller (1975) is often used in palaeohydraulic reconstruction. However, tests of the numerical model indicate that Komar and Miller's (1975) threshold expression predicts a gradient of sediment size variation which is too steep (Horn, 1991). The value of the empirical coefficient in Komar and Miller's (1975) threshold expression is inversely related to the predicted sediment size: as the coefficient decreases, the predicted grain size increases, and vice versa. If the coefficient in the threshold expression is too small, the predicted sediment diameters will be too large. For example, halving the coefficient in Komar & Miller's (1975) threshold expression from 0.463 to 0.2315 results in an increase of 252% in predicted maximum onshore threshold diameter. When Komar & Miller's (1975) threshold expression is rearranged to solve for grain size, the flow velocity is raised to the 4/3 power, which causes the grain size to respond more rapidly than flow velocity to changes in depth. Since the threshold expression is used in palaeoenvironmental reconstruction to calculate the minimum nearbed orbital velocity and, from that, the maximum wave period and the range of wave heights, the implication is that the gradient of predicted wave heights will also be too steep. In deep-water conditions, the palaeowave heights will be underestimated or the water depth overestimated; while in shallow-water conditions, the palaeowave height will be overestimated or the water depth underestimated.

## Assumption of a Plane Bed

The asymmetrical threshold model assumes a plane bed, following the precedent of most shore-normal transport models, and does not consider the effects of a rippled bed on the initiation of motion. Although this assumption is clearly not justified in the case of palaeoenvironmental interpretation of wave-formed ripples, the threshold expression of Komar and Miller (1975) also assumes plane bed conditions. This can lead to a miscalculation of the orbital velocity at threshold. The flow over a rippled bed is locally accelerated above a ripple crest and decelerated above a trough, which will alter the critical threshold velocity. There is evidence that the critical values for the initiation of motion are considerably smaller due to the generation of vortex-motions over ripples. For example, Carstens *et al.* (1969) reported a 50% reduction of the peak threshold velocity for a rippled bed. Dyer (1986) reasoned that the threshold of sand movement on a rippled bed is lower than the flat bed value because the bed shear stresses at the crests are enhanced above the average value. The effect of this on the numerical model's predictions is that the threshold velocity required to initiate motion for a particular grain size may have been overestimated, and thus the threshold grain diameter predicted by the model is too small. This means that, for a particular grain size, the threshold velocity may have been overestimated. If the threshold velocity is overestimated, the wave height inferred from this threshold velocity will also be too large. The implication of this for the interpretation of palaeoenvironmental conditions is that either wave height will be overestimated or water depth underestimated.

## Assumption of Uniform Sediment Size Distribution

The effects of a mixed grain size distribution on sediment entrainment, transport and deposition are not included in the asymmetrical threshold model. This topic has been considered to some extent under unidirectional flows, but has not yet been addressed for the case of shore-normal sediment transport under reversing flows. The sediment deposit is usually assumed to consist of spherical quartz grains of a uniform size (Clifton, 1976); however, the nature of the size distribution and the grain packing will influence the stability of the bed, which will affect sediment threshold. In particular, the effects of particle interaction and a situation analogous to sediment armouring in streams, where coarser particles are more exposed to the flow than the finer particles which are sheltered by the coarse partices, are not included. Most research on this topic has considered only unidirectional flow conditions. Van Rijn (1990) suggested that particle interaction can be corrected for by increasing the critical shear stress (and, by implication, the threshold velocity) of the finer particles and decreasing the critical shear stress for the coarser particles. Fenton & Abbott (1977) investigated the effect of the relative protrusion of sediment grains by pushing a grain gradually above other particles in a level bed and found that the value of $\theta_{cr}$ could be reduced by almost an order of magnitude, since the grain sheltered an increasingly large area. Komar and Li (1986) suggested that for large particles on a boundary of smaller particles, the threshold curve would be parallel to the Shields curve at a lower value of $\theta_{cr}$, while for fine particles $\theta_{cr}$ would be higher. This would give results similar to Fenton & Abbott, but caused by a different mechanism. If a mechanism similar to either of these occurs for oscillatory flow, the assumed threshold velocity would be too small, which would mean that the threshold grain diameter predicted by the model would also be too small. If the threshold velocity is overestimated for a particular grain size, the wave height inferred from this threshold velocity will also be overestimated. The implication of this for the interpretation of palaeoenvironmental conditions is that either wave height will be overestimated or water depth underestimated.

## Assumption of Maximum Energy Conditions

Finally, the sediment texture may not reflect the maximum energy level of the environment: the threshold velocity may be exceeded by a large percentage of the waves and flows acting on the beach profile, or the mean sediment diameter may not be in equilibrium with peak flow velocities. A prediction of palaeoenvironmental conditions based on the assumption that the observed grain size was in equilibrium with the maximum nearbed flow under waves would lead to the underestimation of wave height or period or the overestimation of water depth.

## Example of Palaeohydraulic Reconstruction

Komar (1974) gave an example of shallow-water conditions, in which the oscillatory ripples had an average spacing ($\lambda$) of 80 mm and a mean grain diameter of 0.125 mm. Komar's (1974) calculations are reworked here, and the palaeohydraulic values are shown in Table 18.2. The inferred value of $d_0$ is 0.123 m, the minimum orbital velocity is 0.115 m s$^{-1}$ and the maximum orbital velocity is 1.21 m s$^{-1}$. These values give a minimum wave period of 0.32 seconds and a maximum wave period of 3.35 seconds. The water depths were assumed to be between 5 m and 0.5 m. The wave heights calculated from the minimum wave period (which is itself unlikely) are unrealistically large, so only the maximum wave period is used for the wave height calculations. Using Airy theory, the predicted wave heights vary from 0.39 m (in 5 m depth) to 0.05 m (in 0.5 m depth). All of these waves are stable. With a maximum wave period of 3.35 seconds,

**Table 18.2**  Palaeowave reconstruction (ripple data from Komar, 1974)

|  | Without shoaling transformations | With shoaling transformations |
|---|---|---|
| Depth range (m) | 0.5–5.0 | 0.5–5.0 |
| $\lambda$ (mm) | 80 | 80 |
| $D$ (mm) | 0.125 | 0.125 |
| $d_0$ (m) | 0.123 | 0.123 |
| Minimum $u_m$ (m s$^{-1}$) | 0.115 | 0.115 |
| Maximum $u_m$ (m s$^{-1}$) | 1.21 | 1.21 |
| Minimum $T$ (s) | 0.32 | 0.32 |
| Maximum $T$ (s) | 3.35 | 3.35 |
| Minimum $H$ (m) | 0.050 | 0.670 |
| Maximum $H$ (m) | 0.392 | 0.764 |
| Wind velocity (m s$^{-1}$) | 10 | 10 |
| Maximum fetch (km) | 5.9 | 22.3 |
| Wind velocity (m s$^{-1}$) | 20 | 20 |
| Maximum fetch (km) | 1.5 | 5.6 |

water depths between 0.5 and 5.0 m, and assumed wind velocities of $10 \, m \, s^{-1}$ or less, the maximum estimated fetch of the basin in which these sediments were deposited would be approximately 5.9 km. With assumed wind velocities of $20 \, m \, s^{-1}$ (gale force), the maximum estimated fetch of the basin in which these sediments were deposited would be approximately 1.5 km.

If the shoaling transformations are calculated from linear theory, shoaling wave heights would vary between 0.67 m and 0.76 m, with the minimum wave height of 0.67 m occurring in 2 m depth. Calculations based on the shoaling wave height rather than the Airy wave height suggest that, with assumed wind velocities of $10 \, m \, s^{-1}$ or less, the maximum estimated fetch of the basin in which these sediments were deposited was approximately 22.3 km. For winds of $20 \, m \, s^{-1}$, the maximum fetch calculated on the basis of the shoaling wave heights was 5.6 km.

## POTENTIAL APPLICATIONS OF SIZE-DEPENDENT TRANSPORT MODELS

The results of the asymmetrical threshold model indicate a need to develop the threshold model further into a size-dependent transport model, and to consider the effect of a mixed sediment distribution on entrainment, transportation and deposition. If the threshold model can be developed into a shore-normal sediment transport model which includes a consideration of differential transport rates, it will have a number of useful applications.

### Beach Nourishment on Eroding Coasts

A potential use of the model lies in its predictive capability for coastal management decisions, particularly with respect to beach nourishment schemes. Beach nourishment is the process of placing sediment directly on an eroding beach to restore and maintain an adequate protective beach. Recent research indicates that a major disadvantage of beach nourishment schemes is the loss of sediment to the offshore zone, which cannot be predicted by a longshore transport model alone. Numerical modelling techniques can be used to predict the likely success of a beach nourishment scheme. A sediment transport model which can calculate the movement of individual sediment sizes would be used in a planning analysis of potential beach nourishment schemes to predict the movement and redistribution of beach fill sediment. The model could be presented with an input grain size distribution, corresponding either to the native sand sizes present across the profile of the beach in question or to the borrow sediment to be placed on the profile. The redistribution of the sediment on the beach could then be predicted as a response to the wave and tidal conditions acting upon the beach profile. Thus the changes in the shore-normal distribution of the borrow sediment could be predicted, particularly any loss of sediment to the offshore zone, which would provide useful information for the consideration of any potential beach nourishment scheme and would supplement the traditional type of planning analysis.

### Beach Response to Sea Level Rise

A size-dependent transport model could also be used in investigations of beach response to sea level change. In a recent review paper, Komar (1991) argued that there is a need for conceptual advances in predictive models of coastal response to changing sea level. The most commonly cited model is that of Bruun (1962, 1988), which relates shoreline retreat to an increase in local sea level. The Bruun model suggests that the equilibrium beach profile will move upward and landward as sea level rises. The conceptual model proposes that sediment will be eroded from the upper beach and deposited in the area immediately offshore. The model assumes that the amount of sediment eroded from the upper beach will be equal in volume to the sediment which is deposited offshore, and that this offshore deposition will keep pace with rising sea level, thus maintaining a constant water depth offshore and preserving the beach profile relative to rising sea level.

Although it is generally agreed that sea level rise will result in increased coastal erosion and a retreating shoreline, there is evidence to suggest that some coastlines are accreting despite sea level rise (Dean, 1990). Komar (1991) argued that the existence of shore-normal grain size variations makes beach response to sea level rise considerably more complex than proposed by the Bruun model. He suggested that, depending on the grain size distribution of the eroded sediment, onshore sediment transport may occur with rising sea level, with the beach deposit being maintained intact during a transgression. If a size-dependent transport model can be developed, it could be used to help establish the response of beaches to rising sea level, a problem which is becoming increasingly important.

## CONCLUSIONS

This paper describes a numerical model which was developed to predict and explain shore-normal sediment size variations on a macrotidal beach. Some problems which have been identified by the predictions of the model are discussed, and the implications of these findings for palaeoenvironmental reconstruction based on the sediment record are presented. A number of possible sources of error have been identified in the application of hydrodynamic principles to the interpretation of depositional environments. Some of these suggestions can be easily incorporated into a revised method for the reconstruction of palaeowave conditions. Other potential sources of error are not quantifiable at present; however, researchers should be aware of the effect that these assumptions will have on palaeoenvironmental interpretation.

## ACKNOWLEDGEMENTS

This work was undertaken when the author was a research student in the School of Geography at the University of Oxford. The research was supported by the British Geomorphological Research Group Research Fund; the Royal Geographical Society Henrietta Hutton Memorial Travel Fund; the Royal Society Dudley Stamp Memorial Fund; Sigma Xi, the Scientific Research Society; and Christ Church, Oxford.

## SYMBOLS

| | | |
|---|---|---|
| $C$ | wave celerity | ms$^{-1}$ |
| $C_0$ | deep-water wave celerity | ms$^{-1}$ |
| $d_0$ | wave orbital diameter at the bed | m |

| | | |
|---|---|---|
| $D$ | grain diameter | mm or $\phi$ |
| $D_{ex}$ | offshore threshold grain diameter | mm or $\phi$ |
| $D_{in}$ | onshore threshold grain diameter | mm or $\phi$ |
| $f_e$ | non-dimensional energy dissipation factor | — |
| $g$ | acceleration due to gravity | $ms^{-2}$ |
| $h$ | water depth | m |
| $h/L$ | relative depth | — |
| $h_o$ | depth at deep-water limit (taken as $L_o/2$) | m |
| $H$ | wave height | m |
| $H_b$ | breaking wave height | m |
| $H_f$ | local wave height with frictional losses | m |
| $H_o$ | wave height in deep water | m |
| $H_{oh}$ | wave height at deep-water depth ($h_o$) | m |
| $I$ | variable in Neilsen (1982) equation | — |
| $k$ | wave number ($2\pi/L$) | $m^{-1}$ |
| $L$ | wavelength | m |
| $L_o$ | deep-water wavelength | m |
| $n$ | ratio of wave group to wave phase velocity | — |
| $T$ | wave period | s |
| $u$ | horizontal component of water particle velocity | $ms^{-1}$ |
| $u_m$ | maximum nearbed orbital velocity | $ms^{-1}$ |
| $u_{ex}$ | offshore maximum nearbed orbital velocity | $ms^{-1}$ |
| $u_{in}$ | inshore maximum nearbed orbital velocity | $ms^{-1}$ |
| $x$ | distance offshore | — |
| $\tan\beta$ | beach slope | — |
| $\gamma$ | wave height divided by water depth ($H/h$) | — |
| $\delta h/\delta x$ | change in depth $h$ over distance $x$ (represents bedslope $\tan\beta$) | — |
| $\Delta H$ | difference between shoaling $H$ and $H$ with frictional losses | m |
| $\theta_{cr}$ | threshold Shields criterion of Komar and Miller (1975) | — |
| $\theta_{ex}$ | modified offshore Shields criterion of Komar and Miller (1975) | — |
| $\theta_{in}$ | modified onshore Shields criterion of Komar and Miller (1975) | — |
| $\lambda$ | ripple wavelength (spacing) | m |
| $\rho$ | density of water | $kg\,m^{-3}$ |
| $\rho_s$ | density of sediment | $kg\,m^{-3}$ |
| $\psi$ | variable in Neilsen (1982) equation | — |
| b | subscript denoting breaking wave conditions | — |
| o | subscript denoting deep-water conditions | — |
| 0 | subscript denoting conditions at the bed | — |

## REFERENCES

Airy, G. B. (1845). Tides and waves. *Encyclopaedia Metropolitana*, Volume 5, pp. 241–396.
Allen, P. A. (1981a). Some guidelines in reconstructing ancient sea conditions from wave ripplemarks. *Marine Geology*, **43**, M59–M67.

Allen, P. A. (1981b). Wave-generated structures in the Devonian lacustrine sediments of south-east Shetland and ancient wave conditions. *Sedimentology*, **28**, 369–379.

Allen, P. A. (1984). Reconstruction of ancient sea conditions with an example from the Swiss Molasse. *Marine Geology*, **60**, 455–473.

Bagnold, R. A. (1940). Beach formation by waves: some model-experiments in a wave tank. *Journal of the Institute of Civil Engineers*, Paper No. **5237**, 27–53.

Bruun, P. (1962). Sea level rise as a cause of shore erosion. *Journal of the Waterways and Harbors Division, American Society of Civil Engineers*, **88**, 117–130.

Bruun, P. (1988). The Bruun rule of erosion. *Journal of Coastal Research*, **4**, 627–648.

Bryant, E. (1984). Sediment characteristics of some eastern Australian beaches. *Australian Geographer*, **16**, 5–15.

Carstens, M. R., Neilson, F. M. and Altinbilek, H. D. (1969). Bed forms generated in the laboratory under an oscillatory flow: analytical and experimental study. *US Army Corps of Engineers, Coastal Engineering Research Center Technical Memo No. 28.*

Carter, R. W. G. (1988). *Coastal Environments. An Introduction to the Physical, Ecological and Cultural Systems of Coastlines*. Academic Press, London.

Clifton, H. E. (1976). Wave-generated structures – a conceptual model. In Davis, R. A. and Etherington, R. L. (eds), *Beach and Nearshore Processes*, Society of Economic and Paleontological Mineralogists Special Publication 24, pp. 126–148.

Clifton, H. E. and Dingler, J. R. (1984). Wave-formed structures and paleoenvironmental reconstruction. *Marine Geology*, **60**, 165–198.

Cook, D. O. and Gorsline, D. S. (1972). Field observations of sand transport by shoaling waves. *Marine Geology*, **13**, 31–55.

Cornaglia, P. (1889). On beaches. In Fisher, J. S. and Dolan, R. (eds) (1977). *Beach Processes and Coastal Hydrodynamics*, Dowden, Hutchinson and Ross, Inc., Stroudsburg, Pennsylvania, pp. 11–26. Translated by Felder, W. N. from Cornaglia, P. (1889). *Delle Spiaggie*, Accademia Nazionale dei Lincei Atti. Cl. Sciente Fisiche, Matematische e Naturali Mem. 5, ser. 4, pp. 284–304.

Cornish, V. (1898). On sea beaches and sand banks. *Geographical Journal*, **11**, 528–559, 628–647.

Dean, R. G. (1990). Beach response to sea level change. In Le Méhauté, B. and Hanes, D. M. (eds), *Ocean Engineering Science*, Wiley, New York, pp. 869–887.

Diem, B. (1985). Analytical method for estimating palaeowave climate and water depth from ripple marks. *Sedimentology*, **32**, 705–720.

Dupré, W. R. (1984). Reconstruction of paleo-wave conditions during the late Pleistocene from marine terrace deposits, Monterey Bay, California. *Marine Geology*, **60**, 435–454.

Dyer, K. R. (1986). *Coastal and Estuarine Sediment Dynamics*. Wiley, Chichester.

Eagleson, P. S. and Dean, R. G. (1959). Wave-induced motion of bottom sediment particles. *Proceedings of the American Society of Civil Engineers, Journal of the Hydraulics Division*, HY10, 53–79.

Eagleson, P. S. and Dean, R. G. (1961). Wave-induced motion of bottom sediment particles. *Transactions of the American Society of Civil Engineers*, **126** (1), 1162–1189.

Eagleson, P. S., Dean, R. G. and Peralta, L. A. (1958). The mechanics of the motion of discrete spherical bottom sediment particles due to shoaling waves. *US Army Corps of Engineers, Beach Erosion Board Technical Memo No. 104.*

Eagleson, P. S., Glenne, B. and Dracup, J. A. (1961). Equilibrium characteristics of sand beaches in the offshore zone. *US Army Corps of Engineers, Beach Erosion Board Technical Memo No. 126.*

Eagleson, P. S., Glenne, B. and Dracup, J. A. (1963). Equilibrium characteristics of sand beaches. *Proceedings of the American Society of Civil Engineers, Journal of the Hydraulics Division*, **89(HY1)**, Paper 3387, pp. 35–57.

Elgar, S., Guza, R. T. and Freilich, M. H. (1988). Eulerian measurements of horizontal accelerations in shoaling gravity waves. *Journal of Geophysical Research*, **93** (C8), 9261–9269.

Fenton, J. D. and Abbott, J. E. (1977). Initial movement of grains on a stream bed: The effect of relative protrusion. *Proceedings of the Royal Society of London*, **A352**, 523–527.

Foster, G. A. (1991). Beach nourishment from a nearshore dredge spoil dump at Mount Maunganui Beach, New Zealand. Unpublished M.Sc. Thesis, University of Waikato, 152 pp.

Fox, W. T., Ladd, J. W. and Martin, M. K. (1966). A profile of the four moment measures perpendicular to a shore line, South Haven, Michigan. *Journal of Sedimentary Petrology*, **36**, 1126–1130.

Greenwood, B. and Davidson-Arnott, R. G. D. (1972). Textural variations in sub-environments of the shallow-water wave zone, Kouchibouguac Bay, New Brunswick. *Canadian Journal of Earth Sciences*, **9**, 679–688.

Greenwood, B. and Sherman, D. J. (1984). Waves, currents, sediment flux and morphological response in a barred nearshore system. *Marine Geology*, **60**, 31–61.

Harms, J. C. (1969). Hydraulic significance of some sand ripples. *Bulletin of the Geological Society of America*, **80**, 363–396.

Harrison, W. and Alamo, R. M. (1964). Dynamic properties of immersed sand at Virginia Beach, Virginia. *US Army Corps of Engineers, Coastal Engineering Research Center Technical Memo Number 9*.

Horn, D. P. (1991). A numerical model for shore-normal sediment size variation. Unpublished D.Phil. Thesis, University of Oxford, 407 pp.

Horn, D. P. (1992). A numerical model for shore-normal sediment size variation on a macrotidal beach. *Earth Surface Processes and Landforms*, **17** (8), 755–773.

Horn, D. P. (1993). Sediment dynamics on a macrotidal beach. *Journal of Coastal Research*, **9** (1), 189–208.

Horn, D. P. and Hardisty, J. (1990). The applications of Stokes' wave theory under changing sea levels in the Irish Sea. *Marine Geology*, **94**, 341–351.

Huntley, D. A. and Bowen, A. J. (1975). Comparison of the hydrodynamics of steep and shallow beaches. In Hails, J. R. and Carr, A. P. (eds), *Nearshore Sediment Dynamics and Sedimentation*, Wiley, London, pp. 69–109.

Inman, D. L. (1949). Sorting of sediments in the light of fluid mechanics. *Journal of Sedimentary Petrology*, **19**, 51–70.

Inman, D. L. (1953). Areal and seasonal variations in beach and nearshore sediments at La Jolla, California. *US Army Corps of Engineers, Beach Erosion Board Technical Memo Number 39*.

Ippen, A. T. and Eagleson, P. S. (1955). A study of sediment sorting by waves shoaling on a plane beach. *US Army Corps of Engineers, Beach Erosion Board Technical Memo No. 63*.

Jago, C. F. and Hardisty, J. (1984). Sedimentology and morphodynamics of a macrotidal beach, Pendine Sands, SW Wales. *Marine Geology*, **60**, 123–154.

Jonsson, I. G. (1966). Wave boundary layers and friction factors. *Proceedings of the 10th Conference on Coastal Engineering*, pp. 127–148.

King, C. A. M. (1972). *Beaches and Coasts*, 2nd edition, Edward Arnold, London.

Komar, P. D. (1974). Oscillatory ripple marks and the evaluation of ancient wave conditions and environments. *Journal of Sedimentary Petrology*, **44**, 169–180.

Komar, P. D. (1991). The response of beaches to sea-level change: a review of predictive models. *Journal of Coastal Research*, **7**, 895–921.

Komar, P. D. and Li, Z. (1986). Pivoting analyses of the selective entrainment of sediments by shape and size with application to gravel threshold. *Sedimentology*, **33**, 425–436.

Komar, P. D. and Miller, M. C. (1975). On the comparison between the threshold of sediment motion under waves and unidirectional currents with a discussion of the practical evaluation of the threshold. *Journal of Sedimentary Petrology*, **45**, 362–367.

Krumbein, W. C. (1934). Size frequency distribution of sediments. *Journal of Sedimentary Petrology*, **4**, 65–77.

Madsen, O. S. (1976). Wave climate of the continental margin: elements of its mathematical description. In Swift, D. J. P. and Stanley, D. J. (eds), *Marine Sediment Transport and Environmental Management*, Wiley, New York, pp. 65–87.

Miller, M. C. and Komar, P. D. (1980). Oscillation sand ripples generated by laboratory apparatus. *Journal of Sedimentary Petrology*, **50**, 173–182.

Miller, R. L. and Zeigler, J. M. (1958). A model relating dynamics and sediment pattern in equilibrium in the region of shoaling waves, breaker zone, and foreshore. *Journal of Geology*, **66**, 417–441.

Miller, R. L. and Zeigler, J. M. (1964). A study of sediment distribution in the zone of shoaling waves over complicated bottom topography. In Miller, R. L. (ed.), *Papers in Marine Geology*, The Macmillan Company, New York, pp. 133–153.

Neilsen, P. (1982). Explicit formulae for practical wave calculations. *Coastal Engineering*, **6**, 389–398.

Niedoroda, A. W., Swift, D. J. P., Figueiredo, A. G. and Freeland, G. L. (1985). Barrier island evolution, Middle Atlantic Shelf, U.S.A. Part II: evidence from the shelf floor. *Marine Geology*, **63**, 363–396.

Newton, R. S. (1968). Internal structure of wave-formed ripple marks in the nearshore zone. *Sedimentology*, **11**, 275–292.

Perrett, T. L. (1990). Variations in sediment texture and biota off a wave dominated coast, Peka Peka beach, New Zealand. Unpublished M.Sc. Thesis, Victoria University of Wellington, New Zealand.

Sato, S. and Kishi, T. (1954). Shearing force on sea bed and movement of bed material due to wave motion. *Journal of Research of Public Works Research Institute*, **1**, 25–31.

Shipp, R. C. (1984). Bedforms and depositional sedimentary structures of a barred nearshore system, eastern Long Island, New York. *Marine Geology*, **30**, 235–259.

Short, A. D. (1991). Macro-meso tidal beach morphodynamics – an overview. *Journal of Coastal Research*, **7**, 417–436.

Stokes, G. G. (1847). On the theory of oscillatory waves. *Transactions of the Cambridge Philosophical Society*, **8**, 441–455.

Stokes, G. G. (1880). Supplement to a paper on the theory of oscillatory waves. In *Mathematical and Physical Papers 1*, Cambridge University Press, London, pp. 314–326.

Swift, D. J. P. (1976). Coastal sedimentation. In Swift, D. J. P. and Stanley, D. J. (eds), *Marine Sediment Transport and Environmental Management*. Wiley, New York, pp. 255–310.

Swift, D. J. P., Sanford, R. E., Dill, C. E. Jr and Avignone, N. F. (1971). Textural differentiation on the shoreface during erosional retreat of an unconsolidated coast, Cape Henry to Cape Hatteras, western North Atlantic shelf. *Sedimentology*, **16**, 221–250.

Tanner, W. F. (1958). The equilibrium beach. *Transactions of the American Geophysical Union*, **39**, 889–891.

Tanner, W. F. (1971). Numerical estimates of ancient waves, water depth and fetch. *Sedimentology*, **16**, 71–88.

Thornton, E. B. and Guza, R. T. (1982). Energy saturation and phase speeds measured on a natural beach. *Journal of Geophysical Research*, **87** (C12), 9499–9508.

van Rijn, L. C. (1990). *Handbook: Sediment Transport by Currents and Waves*, 2nd edition, Delft Hydraulics Report H 461, unpublished report.

Wells, D. R. (1967). Beach equilibrium and second-order wave theory. *Journal of Geophysical Research*, **72**, 497–504.

Wright, L. D., Neilsen, P., Short, A. D. and Green, M. O. (1982). Morphodynamics of a macrotidal beach. *Marine Geology*, **50**, 97–128.

Wright, L. D., Boon, J. D., Kim, S. C. and List, J. H. (1991). Models of cross-shore sediment transport on the shoreface of the Middle Atlantic Bight. *Marine Geology*, **96**, 19–51.

# Index

*Index compiled by Colin Will*